MUSCLE DEVELOPMENT AND GROWTH

This is Volume 18 in the

FISH PHYSIOLOGY series

Edited by William S. Hoar, David J. Randall, and Anthony P. Farrell

A complete list of books in this series appears at the end of the volume.

MUSCLE DEVELOPMENT AND GROWTH

Edited by

IAN A. JOHNSTON
School of Environmental and Evolutionary Biology
University of St. Andrews
St. Andrews, Fife
Scotland

ACADEMIC PRESS
A Harcourt Science and Technology Company

San Diego   San Francisco   New York   Boston   London   Sydney   Tokyo

Front cover photograph: The effect of human recombinant IGF1 on short term satellite cell proliferation. (For more details see Chapter 4, Figure 8.)

This book is printed on acid-free paper.

Copyright © 2001 by ACADEMIC PRESS

All Rights Reserved.
No part of this publication may be reproduced or transmitted in any form or by any means, electronic or mechanical, including photocopy, recording, or any information storage and retrieval system, without permission in writing from the publisher.

Requests for permission to make copies of any part of the work should be mailed to: Permissions Department, Harcourt Inc., 6277 Sea Harbor Drive, Orlando, Florida 32887-6777

Explicit permission from Academic Press is not required to reproduce a maximum of two figures or tables from an Academic Press chapter in another scientific or research publication provided that the material has not been credited to another source and that full credit to the Academic Press chapter is given.

Academic Press
A Harcourt Science and Technology Company
525 B Street, Suite 1900, San Diego, California 92101-4495, USA
http://www.academicpress.com

Academic Press
Harcourt Place, 32 Jamestown Road, London NW1 7BY, UK
http://www.academicpress.com

Library of Congress Catalog Card Number: 00-103963

International Standard Book Number: 0-12-350442-2

PRINTED IN THE UNITED STATES OF AMERICA
00 01 02 03 04 05   QW 9 8 7 6 5 4 3 2 1

CONTENTS

CONTRIBUTORS ix
PREFACE xi

1. Induction and Patterning of Embryonic Skeletal Muscle Cells in The Zebrafish
   Peter D. Currie and P. W. Ingham

   I. Introduction 1
   II. The Process of Myogenesis in Zebrafish 2
   III. Myotomal Patterning Mutants and the Molecular Mechanisms Controlling Slow Muscle Cell Specification 6
   IV. Mutations Affecting Muscle Differentiation 8
   V. Muscles of Paired Fins 9
   VI. Head Muscles 11
   VII. Conclusions 14
   References 14

2. Myogenic Regulatory Factors
   Shugo Watabe

   I. Abstract 19
   II. Introduction 20
   III. Fish Myogenic Regulatory and Related Factors 23
   IV. Expression of Fish MRFs During Ontogeny 27
   V. Expression of Fish MRFs During Temperature Acclimation 32
   VI. Conclusion 35
   References 36

3. Myosin Expression During Ontogeny, Post-Hatching Growth, and Adaptation
Geoffrey Goldspink, David Wilkes, and Steven Ennion

I.	Abstract	43
II.	Introduction	44
III.	The Myocin hc Genes	46
IV.	Expression of Myosin Isogenes During Early Development and Muscle Regeneration	48
V.	A Myosin Gene that is Associated with Adult Muscle Hyperplasia in Carp	52
VI.	Molecular Signals Involved in Muscle Gene Expression During Pre- and Post-Hatching Growth	55
VII.	Expression of Myosin Isogenes in Response to Temperature	57
VIII.	Regions Determining Functional Diversity in Myosin hc Isoforms	60
IX.	Conclusions	64
	References	66

4. Muscle Satellite Cells in Fish
Benoit Fauconneau and Gilles Paboeuf

I.	Abstract	73
II.	Introduction	74
III.	Characteristics of Satellite Cells *In Vivo*	75
IV.	Characteristics of Satellite Cells *In Vitro*	78
V.	Origin of Satellite Cells and Related Questions	87
VI.	Contribution of Satellite Cells to Muscle Growth	89
VII.	Effect of Intrinsic Factors	89
VIII.	Effect of Extrinsic Factors	95
IX.	Conclusions	97
	References	98

5. Cellular Mechanisms of Post-Embryonic Muscle Growth in Aquaculture Species
A. Rowlerson and A. Veggetti

I.	Abstract	103
II.	Introduction	104
III.	Muscle Growth Mechanisms	109
IV.	Factors Affecting Muscle Growth	120
V.	Trends for Future Research on Muscle Growth in Fish, Especially Aquaculture Species	127
	References	128

6. Genetic and Environmental Determinants of Muscle Growth Patterns
Ian A. Johnston

I.	Introduction	141
II.	Myogenesis in Salmonids	145
III.	Methods for Quantifying Growth Patterns in Muscle	153
IV.	Genetic Variation in Muscle Growth Patterns	155
V.	Environmental Influences on Muscle Fiber Growth Patterns	161
VI.	Exercise as a Stimulus for Muscle Growth	173
VII.	Mechanisms Underlying Differences in Muscle Growth Patterns	174
VIII.	Implications for Flesh Quality in Farmed Fish	176
	References	179

7. Muscle Fiber Diversity and Plasticity
Alexandra M. Sänger and W. Stoiber

I.	Introduction	187
II.	Muscle Fiber Diversity	192
III.	Plasticity of Muscle Phenotype	222
IV.	Summary	236
	References	237

8. Hormonal Regulation of Muscle Growth
Thomas P. Mommsen and Thomas W. Moon

I.	Introduction	251
II.	Growth in Fish	252
III.	Non-Hormonal Regulation of Growth	260
IV.	Hormonal Effects	265
V.	Future Directions	291
	References	294

INDEX 309

OTHER VOLUMES IN THE FISH PHYSIOLOGY SERIES 317

CONTRIBUTORS

Numbers in parentheses indicate the pages on which the authors' contributions begin.

PETER D. CURRIE *(1), Comparative and Developmental Genetics Section, MRC Human Genetics Unit, Western General Hospital, Edinburgh EH4 2XU, United Kingdom*

STEVEN ENNION *(43), Anatomy and Developmental Biology, Royal Free and University College Medical School, London University, London NW3 2PF, United Kingdom*

BENOIT FAUCONNEAU *(73), Department of Hydrobiology and Wildlife, INRA, Laboratoire e Physiollogie des Poisons, Campus de Beaulieu 35 042, Rennes, France*

GEOFFREY GOLDSPINK *(43), Anatomy and Developmental Biology, Royal Free and University College Medical School, London University, London NW3 2PF, United Kingdom*

P. W. INGHAM *(1), Department of Biomedical Sciences, MRC Intercellular Signalling Group, Developmental Genetics Programme, University of Sheffield, Sheffield S10 2TN, United Kingdom*

IAN A. JOHNSTON *(141), School of Environmental and Evolutionary Biology, Gatty Marine Laboratory, University of St. Andrews, St. Andrews, Fife KY16 8LB, Scotland*

THOMAS P. MOMMSEN *(251), Department of Biochemistry and Microbiology, University of Victoria, Victoria, British Columbia V8W 3P6, Canada*

THOMAS W. MOON *(251), Department of Biology, University of Ottawa, Ottawa, Ontario K1N 6N5, Canada*

GILLES PABOEUF *(73), Department of Hydrobiology and Wildlife, INRA, Laboratoire e Physiollogie des Poisons, Campus de Beaulieu 35 042, Rennes, France*

ANTHEA ROWLERSON *(103), Applied Clinical Anatomy, GKT School of Biomedical Sciences, King's College, London SE1 1UL, United Kingdom*

ALEXANDRA M. SÄNGER *(187), Department of Experimental Zoology, Institute of Zoology, University of Salzburg, A-5020 Salzburg, Austria*

WALTER STOIBER *(187), Department of Experimental Zoology, Institute of Zoology, University of Salzburg, A-5020 Salzburg, Austria*

ALBA VEGGETTI *(103), Departmento di Morfofisiologia Veterinaria e Produzioni Animali, Università di Bologna, 1-40064, Ozzano-Emilia, Italy*

SHUGO WATABE *(19), Laboratory of Aquatic Molecular Biology and Biotechnology, Graduate School of Agricultural and Life Sciences, The University of Tokyo, Bunkyo, Tokyo 113, Japan*

DAVID WILKES *(43), Anatomy and Developmental Biology, Royal Free and University College Medical School, London University, London NW3 2PF, United Kingdom*

PREFACE

Studies of skeletal muscle have an important role in understanding mechanisms specifying cell fates in vertebrate embryos and for exploring the control of cell proliferation and differentiation. Recently, transcriptional regulators of the myogenic pathway have been identified that are beginning to reveal new insights into the molecular mechanisms responsible for the establishment of muscle-specific gene expression and myogenic lineages. The zebrafish *(Danio rerio)* has become one of the standard organisms for the study of vertebrate development and considerable effort is being invested in sequencing its genome. Large-scale chemical mutagenesis screens in zebrafish have led to the isolation of a large number of mutants of enormous value in the study of embryonic development in vertebrates. The anatomical separation of the different muscle fiber types in fish facilitates many kinds of study that are difficult or impracticable in other vertebrates. Muscle growth is dependent on the division of a population of stem cells, although the origin and nature of these cells are as yet poorly understood. Postembryonic muscle growth involves the hypertrophy of the embryonic fibers, and new fiber production. Thus unlike in birds and mammals, fiber number continues to increase throughout much of the life in fish. The relative importance of fiber hypertrophy and recruitment to growth varies between different species, between different populations of the same species, with forced exercise, and with a wide range of environmental factors. The phenotypic plasticity of muscle fiber types to environmental change also varies considerably during ontogeny and between species according to their habitat and evolutionary history. Interestingly, the temperature during early development can have both short-term and lasting impacts on muscle growth characteristics.

The skeletal muscle constitutes the edible part of the fish. Studies of muscle growth are therefore important for the development of fish farming. Selective breeding programs and advances in the understanding of nutritional requirements have succeeded in dramatically increasing muscle growth rates in salmonid fish. However, in some cases, rapid growth has been associated with problems of flesh quality and this is attributable, at least in part, to changes in muscle structure and cellularity.

The present volume brings together contributions from some of the leading workers in the field of muscle development and growth in fish. The chapters explore muscle growth from the molecular to the whole animal level covering fundamental, environmental, and applied aspects. I wish to thank the authors for their cooperation and dedication to this volume and also to express my gratitude to the numerous reviewers who have helped improve the quality of the final presentations.

IAN A. JOHNSTON

1

INDUCTION AND PATTERNING OF EMBRYONIC SKELETAL MUSCLE CELLS IN THE ZEBRAFISH

PETER D. CURRIE
P. W. INGHAM

I. Introduction
II. The Process of Myogenesis in Zebrafish
   A. Generating Somitic Muscle Precursors
   B. Myotome Formation and Muscle Differentiation
III. Myotomal Patterning Mutants and the Molecular Mechanisms Controlling Slow Muscle Cell Specification
IV. Mutations Affecting Muscle Differentiation
V. Muscles of Paired Fins
VI. Head Muscles
VII. Conclusions
   References

I. INTRODUCTION

This chapter describes the embryonic origins and molecular events which control the formation of different populations of fish muscle cells. We focus primarily on myogenesis in the zebrafish and recent studies which have taken advantage of its simple embryology and genetic tractability to dissect the complex series of inductive cues and cell movements that underlie muscle cell specification. The optical clarity of the zebrafish embryo makes it especially well suited to the application of sophisticated cell labeling techniques, approaches that facilitate the direct visualization of muscle cell ontogeny. The identification of the genes that control this process is well underway following two large-scale mutant screens of the zebrafish genome; these have uncovered numerous mutants that disrupt muscle cell formation and differentiation in various ways (Driever *et al.,* 1996; Haffter *et al.,* 1996; van Eden *et al.,* 1996; Granato *et al.,* 1996). The combination of these two powerful approaches, together with strategies for analyzing the *in vivo* activities of cloned genes, is yielding an increasingly detailed understanding of the embryology of teleost muscle formation. Although the majority of the studies to date have focused on the mechanisms deployed in generating the axial musculature, we also briefly discuss the current state of understanding of the ways in which the other skeletal muscle populations, such as those present in the paired fins or attached to the embryonic cartilage of the segmented head skeleton, are generated. We chart the complex series of cell movements and behaviors that give rise to different populations of muscle cells and discuss what is known about the inductive cues that define these different populations.

II. THE PROCESS OF MYOGENESIS IN ZEBRAFISH

A. Generating Somitic Muscle Precursors

In fish, as in other vertebrates, skeletal muscle of the trunk and tail derives from a specific embryological compartment, the myotome, which in amniotes has been shown to be induced within the segmented mesoderm of the somite. Somites condense from mesoderm immediately adjacent to the central body axis, the so-called paraxial mesoderm, and segment in a stereotypic rostral to caudal progression. Painstaking cell labeling and fate studies in zebrafish have shown that paraxial mesoderm derives from a specific region of the embryo identifiable just prior to gastrulation (Kimmel *et al.,* 1990; Fig. 1). These cells undergo a complex series of cell movements before they initiate the myogenic program which is signaled by the onset of expression of the myogenic family of transcription factors.

The process in which cells, scattered around the hemisphere of the gastrulating zebrafish embryo, come to lie in axial and paraxial positions involves a set of cell movements termed convergence and extension. This process is central to the generation of the boundary between axial and paraxial populations and hence to the appropriate specification of muscle cells. Mutational analysis in the zebrafish has identified two genes that are critically required for the correct allocation of cells to the paraxial and axial mesoderm: the *spadetail* (*spt*) gene, which encodes a T-box transcription factor (Griffin *et al.,* 1998), and the *floating head* (*flh*) gene, which encodes a homeodomain-containing transcription factor (Talbot *et al.,* 1995). Activity of SPT is specifically required in the cells of the paraxial mesoderm, from which the axial musculature will derive. In the absence of SPT function, cells of the paraxial primordia migrate aberrantly during gastrulation and become associated with the tail, creating the characteristic "spadetail" after which the mutation was named (Ho and Kane, 1990; Warga *et al.,* 1998). As a result, homozygous mutant *spt* embryos fail to form trunk somites and exhibit a severe lack of axial muscle (Kimmel *et al.,* 1989). The myogenic transcription factor *myoD* (see below) also fails to be expressed during gastrulation and is very much

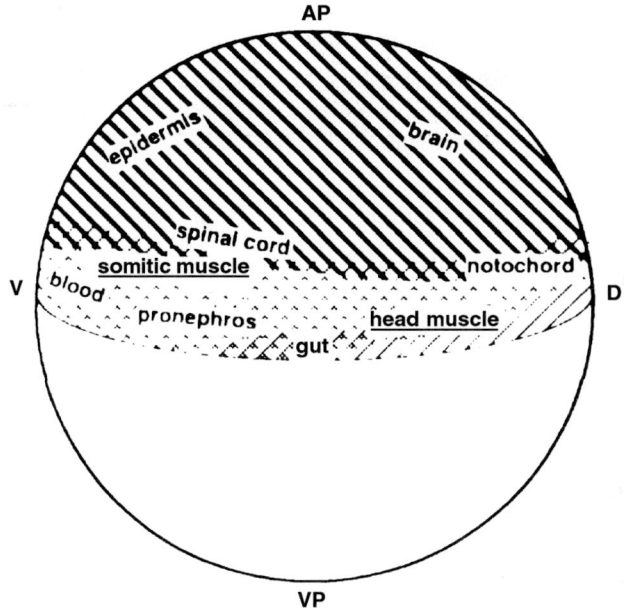

Fig. 1. Origin of zebrafish skeletal muscle cells in the early embryo. Cell labeling studies have revealed that the cells that are fated to give rise to zebrafish skeletal muscle (underlined) arise from a specific location within the zebrafish blastula. Reproduced from Kimmel et al. (1990), with the kind permission of the Company of Biologists.

reduced in expression during somitogenesis. Recent evidence has suggested that SPT may control the morphogenesis of paraxial mesoderm by regulating the differential expression of the cell adhesion molecule paraxial protocadherin (PAPC); (Yamamoto et al., 1998). The fact that PAPC function has been placed genetically downstream of SPT activity and the similarities in expression between the two genes suggest that SPT itself may directly regulate *papc* transcription. PAPC activity may in turn regulate the cell affinity differences between paraxial mesoderm precursors required for their correct migration.

The role of *flh* in mesoderm development is complementary to that of *spt*, being required for correct formation of the notochord, the most axial mesodermal fate. Embryos mutant for *flh* entirely lack notochord and instead form blocks of somitic muscle which fuse underneath the neural tube (Talbot et al., 1995). Lineage analysis has demonstrated that cells normally fated to become notochord differentiate instead into muscle in the absence of FLH activity; these cells initially express markers of both notochord and muscle differentiation (Halpern et al., 1995), but subsequently lose expression of the former. Consistent with this loss of function phenotype, overexpression of the *Xenopus* homolog of *flh*,

XNot-2, in frog embryos results in excessive notochord formation (Gont *et al.*, 1996). Taken together these results suggest that FLH acts to promote notochord development and repress muscle formation. Thus, the spatially regulated expression of *spt* and *flh* together creates the boundary between axial and paraxial mesoderm that is required for somite morphogenesis (Amacher *et al.*, 1998).

B. Myotome Formation and Muscle Differentiation

Once cells have been allocated to either axial or paraxial mesoderm, the next phase of axial muscle specification begins. Cells become committed to the myogenic fate remarkably early in fish embryos, with the onset of expression of the myogenic transcription factor *MyoD* heralding restriction to the muscle lineages at the end of gastrulation. This contrasts greatly with the timing of initiation of myogenesis in amniote embryos, which does not occur until somitogenesis is well established. There is an obvious advantage for fish embryos, which develop from external fertilization, to generate the ability to move as quickly as possible after hatching in order to evade predation. By comparison amniotes have little need for the skeletal musculature until after birth.

In zebrafish, *MyoD* expression initiates in two triangular-shaped fields on either side of the forming axial midline (Weinberg *et al.*, 1996). It is believed, but not proven, that the cells within this field converge to the midline during axis extension to generate a row of cells, termed the "adaxial cells," which flank the forming notochord. This location adjacent to the notochord, as well their large cuboidal morphology, singles out these cells from the rest of the paraxial mesoderm (Figs. 2A and 2B, see color plate). The first elongating and striating cells of the teleost myotome, termed the muscle pioneers, arise from the adaxial cells (Waterman, 1969; Felsenfeld *et al.*, 1991); these are clearly distinguishable from other muscle cells by their expression of the zebrafish *engrailed* (*eng*) *1* and *2* genes (Ekker *et al.*, 1992; Fig. 2C). These cells differentiate in an anteroposterior wave mirroring somite differentiation, forming adjacent to the notochord, at the dorsoventral midline of the zebrafish myotome. At this level, a specialized structure of the myotome—the horizontal myoseptum—forms (Fig. 2D). Horizontal myosepta are present in all gnathostome fishes and divide the differentiating myotome into dorsal, nominally epaxial and ventral, nominally hypaxial muscle masses, but they are not found in cyclostomes (jawless fish; Bone, 1989). Structurally similar to vertical myosepta that separate adjacent mature myomeres in all vertebrates, horizontal myosepta are composed of connective tissue sheets that extend medially to become attached to the axial column, notochord, or vertebral column depending on the stage of development. The close association between muscle pioneer cell differentiation and horizontal myoseptum formation has led to speculation that these cells may play a role in controlling horizontal myoseptum formation (Halpern *et al.*, 1993). The striation and elongation of the muscle pio-

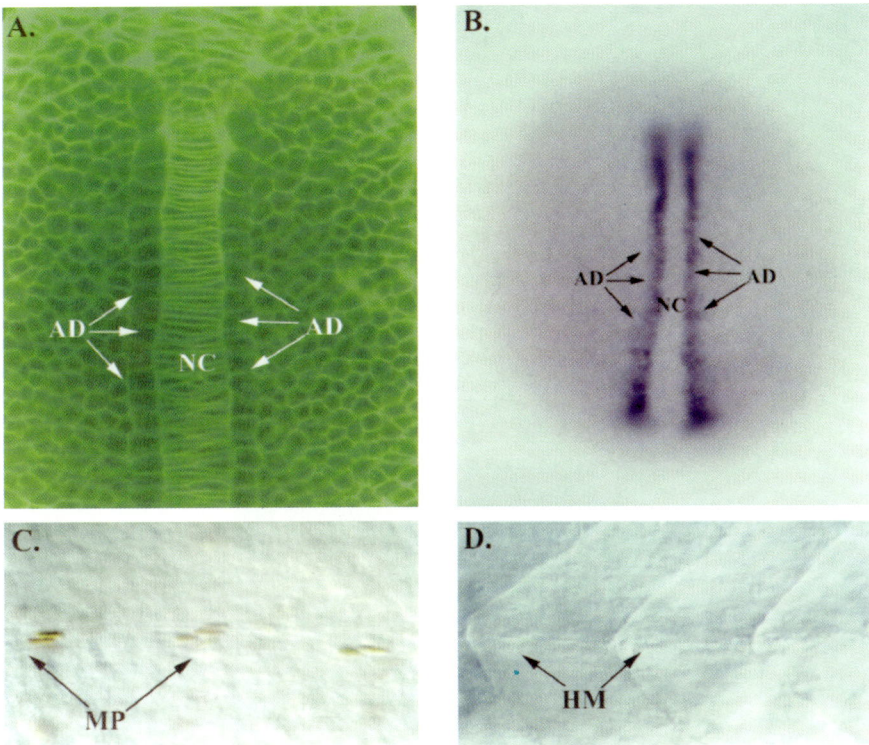

Figure 2. Formation of different muscle cell types in the zebrafish myotome. (A) Whole embryo bodipy ceramide membrane staining of a segmentation stage zebrafish embryo. Adaxial cells (AD) form a single cell row of cells flanking the notochord (NC) and take on a large cuboidal morphology which distinguishes them from the remainder of the paraxial mesoderm. Dorsal view at the level of the notochord, anterior to the top of the page. (B) *In situ* hybridization against the zebrafish *MyoD* gene in a presegmentation stage embryo. Adaxial cells are also distinguished by their precocious expression of the myogenic transcription factor *MyoD*, which occurs even prior to segmentation. (C) Lateral view of a 26-somite-stage embryo reacted with an antibody against the Engrailed protein (yellow), which is expressed within the nuclei of muscle pioneer cells (MP) that form from the adaxial cells at the level of the horizontal myoseptum. 4-Somite width view at the level of the trunk somites anterior to the left. (D) Lateral view of the zebrafish somite within a living 26-somite-stage embryo. The characteristic chevron form of the embryonic myotome can be seen with the horizontal myoseptum (HM) forming at the dorsoventral midline of the somite. 4-Somite-width view at the level fo the trunk somites anterior to the left.

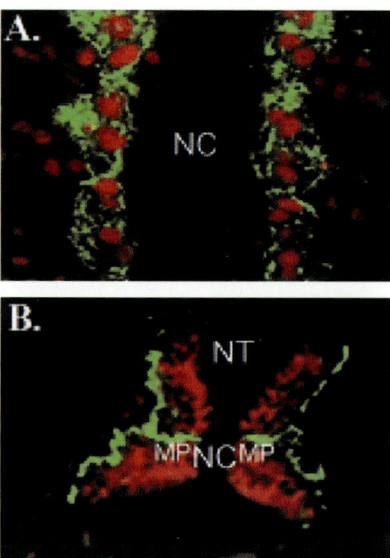

Figure 3. (A) Adaxial cells differentiate exclusively as slow myosin heavy chain (MyHC) expressing muscle cells. A confocal microscope view at the level of the notochord (NC) 15-somite-stage embryo reacted with a slow myosin-specific antibody (green) and an antibody against myogenic transcription factor positive nuclei (red). Dorsal view, anterior to the top of the page. (B) Double antibody staining reveals that slow muscle (green) migrates to the lateral extent of the myotome (except for the muscle pioneer cells, MP) and fast muscle (red) differentiates behind this wave of migration to form the remainder of the muscle cells of the myotome. Cross-section at the level of the trunk somites, dorsal to the top of the page. (NT, neural tube).

neer cells are also coincident with a change in myotomal shape from an early block-like form to a characteristic chevron shape. How the muscle pioneer cells might control these different morphogenetic events is poorly understood, but it is proposed that generation of this chevron form is critical for allowing relatively short muscle fibers to collectively extend over several skeletal segments and generate the powerful alternate lateral muscle undulations, which are the major propulsive force of the fish (Hardistry, 1979).

Differentiated muscles can be further subdivided into two distinct classes on the basis of the type of myosin expressed within each fiber. All the derivatives of the adaxial cells, including the muscle pioneer cells, express a slow isoform of myosin heavy chain (MyHC) and differentiate into slow twitch muscle (Fig. 3A, see color plate). Midway through segmentation most adaxial cells begin a striking migration from their site of origin next to the notochord, traversing the entire extent of the myotome and coming to lie at its most lateral surface (Fig. 3B; Devoto *et al.*, 1996). Here, these slow MyHC expressing myoblasts form a superficial subcutaneous layer of slow twitch muscle which will form the aerobic red muscle of the adult fish (Bone, 1978; Devoto *et al.*, 1996; Blagden *et al.*, 1997). By contrast, the muscle pioneer cells, which also express slow MyHC, remain in their medial location up to 48 h after differentiation, reinforcing their distinct role in somite morphogenesis (Waterman, 1969; Devoto *et al.*, 1996; Blagden *et al.*, 1997). The rest of the lateral aspect of the myotome differentiates as fast twitch muscle and this differentiation begins after the migration of adaxially derived slow muscle cells which have already begun to express slow forms of myosin heavy chain prior to their migration (Devoto *et al.*, 1996; Blagden *et al.*, 1997). Differentiation of fast twitch muscle does not require slow muscle differentiation, as mutants which completely lack slow muscle are still able to form fast muscle at its stereotypic stage of differentiation (Blagden *et al.*, 1997).

The simultaneous development of both muscle layers in the free swimming larvae has been suggested to endow the organism with the abilities to search for food sources at low swimming velocity and to avoid predators by escape at high velocity (Koumans and Akster, 1995). For most rapidly developing fish such as cyprinids (of which zebrafish is a member), coregonids, and herrings, the transition to free feeding occurs early in development due to a relatively small contribution of yolk that the spawner makes to the egg. The timing of this transition correlates with the differentiation of slow twitch muscle before hatching, and in zebrafish this occurs remarkably early, 24 h before hatching. Fish species that possess large amounts of yolk, such as salmonids, often do not develop the superficial slow twitch red muscle layer until after hatching (Nag and Nursall, 1972; Proctor *et al.*, 1980).

This arrangement of fast and slow twitch muscle cells differs markedly from the musculature of amniote embryos. Here, fast and slow twitch muscle cells do not form in topographically separable populations, but instead arise in a salt and pepper array throughout the formed myotome. The overall architecture of the forming myotome also differs in a number of important ways. In amniotes, the maturing somite is arranged such that the neural tube is directly opposed to the medial component of the dermomyotome while ventrally, the sclerotome (the precursors of the axial skeleton) forms adjacent to the notochord. This relationship is altered in teleosts and other fishes. Here, the notochord and neural tube both form opposed to the medial myotome, with the large vacuolated notochord aligned at the dorsoventral midline. As in amniotes, sclerotome forms ventrally within the zebrafish somite; however, the altered positional relationships in teleost somites mean that sclerotome arises at some distance from the notochord (Morinkensicki and Eisen, 1997) and represents a very much reduced portion of the somitic derivatives. Muscle consequently constitutes the bulk of the derivatives of the somitic lineage. The explanation for this almost certainly lies in the fact that unlike tetrapods, which utilize the appendicular muscles of the limb to move, the major propulsive force in teleosts is supplied by the axial muscles, with few muscles migrating to populate the fin rudiment (see below).

III. MYOTOMAL PATTERNING MUTANTS AND THE MOLECULAR MECHANISMS CONTROLLING SLOW MUSCLE CELL SPECIFICATION

Systematic screens for mutants have identified many of the genes required for development of the zebrafish embryo (Haffter *et al.,* 1996; Driever *et al.,* 1996). A number of mutants, in addition to *flh* described above, that affect formation of the notochord have been isolated (Halpern *et al.,* 1993; Talbot *et al.,* 1995; Odenthal *et al.,* 1996; Stemple *et al.,* 1996; Table I). Invariably, mutant embryos that lack a differentiated notochord also show defects in their paraxial mesoderm. In particular, such embryos fail to form the horizontal myoseptum, lack muscle pioneer cells, and display a disrupted U-shaped somite morphology, suggesting that the notochord in some way controls the formation of these cell types (see below). In the case of two such mutants, *no tail* (*ntl*) and *doc,* transplantation of wild-type notochord precursors into these mutant embryos has been found to rescue the ability of mutant cells to form horizontal myosepta and muscle pioneer cells, but only within the somite directly adjacent to transplanted notochord cells (Halpern *et al.,* 1993; Odenthal *et al.,* 1996). These observations strongly suggest that notochord-derived signal(s) acts at short range to control specification within the teleost myotome.

A number of proteins with inductive properties are known to be expressed within the zebrafish notochord. Notable among these are members of the Hedgehog (HH) family of secreted glycoproteins, known to be required for patterning a variety of tissues in developing vertebrate embryos (reviewed by Ingham, 1995; Hammerschmidt *et al.,* 1997). In the zebrafish, two different hedgehog proteins,

Table I.
Phenotypic Classes of Zebrafish Muscle Mutants

Mutation class	Muscle specification	Muscle differentiation and development[a]	Muscle maintenance
Complementation groups	Notochord defective mutants. *no tail, doc, bozozok, floating head.* *you*-type mutants. *you-too; chameleon; you; sonic you; u-boot; choker; iguana*	Fibers reduced or absent— *sloth, frozen, fibrils unbundled;* fibers disorganized— *turtle, buzz-off, faulpelz, slow motion, schnecke, hermes, duesentrieb, mach two, slop, jam,* and *slinky*	*sapje, softy, schwammerl,* and *runzel*
Muscle phenotype	Majority of mutations affect muscle pioneer cells and slow muscle formation. Horizontal myosepta absent.	Muscle fibers reduced or absent completely immotile, birefringence of muscle reduced or absent; fibers disorganized, movement reduced, reduced muscle birefringence, some with heart defects	Muscle fibers form normally and then degenerate; somitic muscle specific

[a] A number of mutations in these classes remain to be resolved into complementation groups.

Sonic Hedgehog (SHH) and Echidna Hedgehog (EHH), are expressed in and presumably secreted by the notochord (Krauss *et al.,* 1993; Currie and Ingham, 1996). Analysis of the expression of the genes encoding these proteins in *ntl* and *flh* embryos, two of the notochordless mutants that lack muscle pioneer cells, has shown that while *shh* is robustly expressed early in embryogenesis, *ehh* expression is absent, a finding that suggests that EHH might be required for muscle pioneer cell formation. More direct evidence for a role of Hedgehog proteins in muscle specification comes from analysis of embryos homozygous for a mutation in the *shh* gene itself, *sonic you (syu)*. Such embryos also lack muscle pioneers; in addition, they also exhibit a reduction, though not complete elimination, of the superficial slow muscle fibers that derive from the adaxial cells (Schauerte *et al.,* 1998; P.C., unpublished observations). Conversely, the ubiquitous expression of *shh* during early embryogenesis (achieved by injecting *shh* sense mRNA into newly fertilized eggs) results in the activation of *myoD* throughout the presomitic paraxial mesoderm (Weinberg *et al.,* 1996; Concordet *et al.,* 1996) and the subsequent differentiation of the entire myotome into slow twitch muscle fibers (Blagden *et al.,* 1997; Du *et al.,* 1997). Taken together, these data indicate that the specification of all slow muscle cells depends upon inductive signals from the notochord mediated by hedgehog proteins. Strong support for this interpretation comes from the finding that the *you-too (yot)* mutant, which inactivates the GLI-2 protein (Karlstrom *et al.,* 1999), a transcription factor required for transduction of HH signals, also causes a similar loss of muscle pioneers and horizontal myosepta. Interestingly in this case, the effect on superficial slow twitch muscle is much more severe, suggesting that the combined activity of other HH proteins expressed within the midline contributes to their specification. Several other mutants phenotypically similar to *syu* and *yot* have also been isolated and collectively constitute a specific phenotypic class know as the u-type mutants (Table I). The mutants within this class are named for their altered somite morphology in which the somite takes on a u-shaped appearance rather than the characteristic chevron of wild-type embryos. There is good evidence to suggest that most of these mutants identify genes required for the generation or transduction of the Hedgehog signals, and collectively exhibit defects in muscle pioneer cell formation and fiber type specification (van Eden *et al.*, 1996; Schauerte *et al.*, 1998; Lewis *et al.*, 1999). One exception is the *choker* (*cho*) mutant, which although it lacks the horizontal myoseptum (and therefore has a u-shaped somite phenotype) differentiates muscle pioneer cells, suggesting that the wild-type function of the *cho* gene may be to regulate or interpret the signal from the muscle pioneer cells that controls horizontal myosepta formation. Another member of this class of mutants that similarly seems to act downstream of the HH pathway is the mutant *u-boot*. In embryos homozygous for this mutation, adaxial cell induction seems to proceed normally, but differentiation of muscle pioneers is subsequently blocked. Collectively, mutants that disrupt formation of zebrafish myotome reveal a stepwise series of inductive events originating in the notochord that act to pattern the myotome.

In contrast, notochord-derived HH signals are not required for the formation of fast muscle cells which differentiate normally in mutations that lack HH gene expression (Blagden *et al.*, 1997). A separate, yet to be defined mechanism therefore controls the specification of these cells.

IV. MUTATIONS AFFECTING MUSCLE DIFFERENTIATION

Several mutations which affect the formation and differentiation of mature myofibrils have been isolated (Felsenfeld *et al.*, 1991; Granato *et al.*, 1996). The *fibrils unbundled* (*fub*) mutation, originally described nearly a decade ago, was the first mutation that disrupts the formation of myofibrils in the zebrafish to be isolated. In homozygous mutant embryos, muscle pioneer cells are unaffected, reinforcing the different morphogenesis that these cells undergo in comparison to the remainder of the myotomally derived muscle cells (Felsenfeld, 1991). Subsequent to this analysis, the large-scale screens of the zebrafish genome have isolated 63 mutants, identifying 18 different genes, that disrupt the differentiation of the forming musculature. Embryos homozygous for these mutations exhibit reduced mobility and fail to hatch from their chorions. The birefringence of normal muscle, which is due to the parallel threadlike myofibrils within muscle fibers, is disrupted or absent in mutant homozygotes of this class, suggesting the absence of striated muscle tissue. These mutants have been subdivided into four different classes. The first of these is composed of *sloth* (*slo*) and *frozen* (*fro*), together with newly isolated alleles of *fub*. Mutants of all three loci have reduced or absent muscle striation caused by a block at the initial stages of myoblast differentiation. The initial expression of early markers of myoblast specification is unaffected, indicating that all three genes act after the commitment of paraxial mesodermal cells into the myogenic lineage. The nuclear morphology of muscle cells in *slo* and *fro* homozygotes is typical of younger, less differentiated wild-type myocytes, suggesting that muscle development in these mutants is blocked at an earlier stage than in *fub* mutants, which exhibit the more flattened nuclear morphology typical of wild-type muscle cells of a similar stage.

Two classes of mutants disrupt the progression of differentiated myocytes into myotubes and fusion into muscle fibers. These classes are distinguished by whether or not heart defects are also evident. In particular, three mutations, *slop* (*slp*), *jam,* and *slinky* (*sky*), show a reduction in the overall number of fibers as well as a reduced heart beat, although it remains unresolved how these mutations affect cardiac muscle formation. The *duesentriech* (*dus*) mutants have a much greater reduction in fiber number but no heart defect. Another member of this class, *buzz-off* (*buf*), has little reduction in fiber number, but sarcomeric segmentation is irregular compared to wild-type embryos. The final class of mutations affecting muscle differentiation includes those which have reduced muscle striation as a consequence of muscle degeneration. Muscle formation in these mutants is initially indistinguishable from wild type; however, 4 days after fertilization degenerative lesions appear within muscle cells of the myotome. In a number of mutants of this class, lesions are restricted to a small region of a few somitic segments, while in others they spread to every myotomal segment. The lesions appear to be restricted to somitically derived muscle, as heart and jaw muscles are not obviously affected. The late onset of the muscle degeneration has obvious parallels to dystrophic muscle-wasting diseases in humans.

In summary, muscle cell formation in the zebrafish can be dissected genetically at a number of discrete steps in the ontogeny of the mature muscle fiber. Mutants affecting specification, differentiation, and maintenance of the different muscle populations have been isolated and their characterization will tell us much about how teleosts generate their axial musculature.

V. MUSCLES OF PAIRED FINS

While the majority of the mechanical force required for locomotion is generated by the contraction of the axial musculature, a small amount of muscle is deployed within the paired fins (pectoral and pelvic) to control their movement. Unlike the axial musculature, the embryonic origins and ontogeny of the muscle of fins are poorly characterized. In fact, the most detailed understanding of the origins of the muscle-forming cells of the fins comes from studies performed at the turn of the century on the pectoral fins of Selacians (Dohrn, 1884; Braus, 1899). Here, a detailed characterization of embryos of different stages suggests that the muscle of the pectoral fins is derived from direct extensions of cells of the myotome into the developing fin bud. These somite extensions, originally termed "muskelknopsen," or muscle buds, have also been described in teleost species (Corning, 1842; Harrison, 1895), the most recent study with trout utilizing the resolving power of the electron microscope to chart the origin of these cells (Grimm, 1973). This study revealed that short extensions of epithelially organized cells, derived from the ventrolateral margins of the mature myotomes directly opposed to the fin bud, grow toward the proliferating fin bud mesenchyme. These extensions are composed of undifferentiated cells characterized by free ribosomes present within the cytoplasm. The process of colonization by the muskelknopsen is temporally regulated, with extensions present on the cranial side of the fin bud disappearing first while the caudal extensions remain to populate the medial base of the fin bud. These descriptive studies lack any confirmation via fate mapping studies within rostral myotomes of zebrafish; however, they do suggest that different myotomes contribute cells which form distinct muscle populations of the teleost fin. Such a mechanism of colonization of the fin by myotomal extensions differs markedly from the way in which amniote embryos generate limb musculature. No such myotomal extensions can be detected in amniote embryos; rather, the muscles of the limb are generated by the recruitment of a small number of myogenic precursor cells, which are released from the lateral hypaxial myotome to migrate the comparatively lengthy distance to the limb bud. This mechanism of recruiting a small number of actively dividing muscle precursors to the limb bud has been argued to be necessary to generate the larger amount and diversity of the appendicular muscle of the limbs of amniotes (Amthor *et al.,* 1998). Cues from the local environment of the limb then control proliferation and specification of different muscles of the limb.

A number of recent observations in zebrafish, however, cast doubt on the notion that the generation of the musculature of the fin proceeds via a mechanism wholly distinct from that of amniotes. Analysis of the expression of the myogenic transcription factor *myoD* in zebrafish reveals that it is initiated within the fin mesenchyme in a phase temporally distinct from its myotomal expression. Cells expressing *myoD* are no longer proliferative; in zebrafish, expression is contained within the fin mesenchyme, suggesting that if muscle precursor cells do migrate into the zebrafish fin bud they do so as proliferating precursors, akin to those in amniotes.

By comparison, there is virtually no information on the origin and derivation of the muscles of the pelvic fins. Pelvic fin development in zebrafish begins during the third week postfertilization (6.5 mm, 18 days postfertilization) with muscle differentiation evident when fry are at 9.0 mm in length (Grandel and Schulte-Merker, 1998). The late onset of pelvic fin development means that morphogenesis leads directly to the formation of an adult appendage without the larval intermediate form evident in pectoral fin development. This large temporal displacement in pelvic fin development also is at odds with amniote hindlimb formation which, although delayed in comparison to forelimb development, exhibits a relatively short developmental lag and seemingly employs similar molecular cues to drive muscle formation in the hindlimb bud. Whether these morphological differences translate into different molecular cues for muscle cell formation in pelvic and pectoral fins remains to be elucidated.

VI. HEAD MUSCLES

Muscles that attach to the head skeleton are generated by a series of morphogenetic events distinct from those that control formation of the trunk and fin musculature. Muscles of the head are responsible for, among other things, generating eye movements, controlling opening and closing of the jaw, and regulating movement of the operculum, which controls the movement of water over the gills. Head muscles and their associated cartilages differentiate within the reiterated segments of the pharyngeal arches. Pharyngeal muscle differentiation occurs during a period of development temporally distinct from that required for trunk muscle formation, beginning at 58 h postfertilization (hpf), reinforcing the suggestion that separate mechanisms control muscle specification in the head (Schilling and Kimmel, 1998).

There are seven pharyngeal arches in the zebrafish embryo consisting of both dorsal and ventral cartilage elements and their corresponding muscles (Fig. 4). While the precise embryonic origin of the cells that give rise to each of the individual head muscles of the zebrafish embryo has yet to be elucidated by lineage analysis, the examination of myoblast formation during jaw elongation suggests that specific muscles arise from a small, finite number of myoblast precursors within the arch environment. Separate muscles within an individual arch seem to develop either by the subdivision of this early precursor pool or by the formation of individual pools for each muscle. Evidence for the progressive subdivision of an initial myoblast pool comes from the examination of muscles derived from the first or mandibular arch mesoderm. As early as 28 hpf, expression of the Engrailed (EN) protein begins in a small group of dorsal mesodermal cells and expression of EN continues in two dorsal muscles of the mandibular arch, the levator arcus palatini and the dilator operculi, after these muscle differentiate (Hatta *et al.*, 1990). Surprisingly, unlike the expression of EN within muscle pioneer cells, EN expression in the first arch mesoderm occurs before the onset of transcription of *MyoD* within these cells, suggesting that EN may be controlling primary

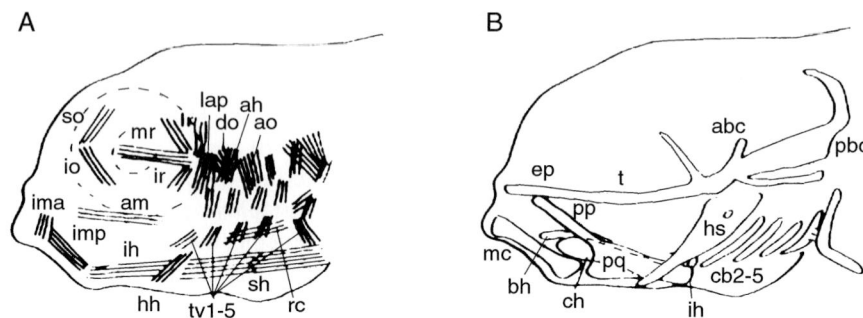

Fig. 4. Embryonic origins of zebrafish head muscles and their associated cartilage attachments. Camera lucida drawings of a lateral view of a 120-h-postfertilization stage zebrafish embryo stained for muscle and cartilage elements. Abbreviations: (A) Muscles: ah, adductor hyoideus; am, adductor mandibulae; ao, adductor operculi; do, dilator operculi; hh, hyohydoideus; ih, interhyoideus; ima, intermandibularis anterior; imp, intermandibularis posterior; io, inferior oblique; ir, inferior rectus; lap, levator arcus palatini; lr, inferior rectus; mr, medial rectus; rc, rectus communis; sh, sternohyoideus; so, superior oblique; tv, transversus ventralis. (B) Cartilage elements: abc, anterior basicranial commissure; bh, basihyal; cb, ceratobranchial; ch, ceratohyal; ep, ethmoid plate; hs, hyosymplectic; ih, interhyal; mc, Meckel's cartilage; pbc, posterior basicranial commisure; pq, palatoquadrate; pp, pterygoid process and t, trabeculae. Reproduced from Schilling and Kimmel (1998), with the kind permission of the Company of Biologists.

specification events for these cells to become muscle. *MyoD* expression in head myoblasts begins at around 50 hpf and precedes the division of the dorsal muscles of the mandibular arch, and a similar partitioning seems to occur for the non-EN positive ventral muscles of this arch which also seem to stem from a single *MyoD* pool (Schilling and Kimmel, 1998). Confirmation of the model of division of a myoblast precursor pool in forming the musculature of the first arch, however, awaits definitive cell labeling experiments.

Muscles of other arches appear to arise from the *de novo* differentiation of individual myoblast pools. Hyoid and branchial muscles arise from small groups of precursors located in positions that will give rise to specific differentiated muscle. The analysis of expression pattern of genes such as *MyoD* and *tropomyosin*, the expression of which marks individual myoblast precursor pools, supports the notion that there is no splitting of precursors and no consequent contribution of these myoblast pools to separate muscles (Schilling and Kimmel, 1998). Analyses in other teleosts have revealed a third mechanism for the formation of a subset of the hypobranchial muscles, the sternohyals. Studies in salmon have suggested that these muscles form by the fusion of three different sets of founder cells derived from the first few somitically derived myotomes. These founders are thought to migrate anteriorly beneath the branchial arches and attach to the hyoid arch (Winterbottom, 1974).

The signals that control the specification of individual myoblast populations within the head remain ill defined. Myoblasts have an intimate association with cartilage precursors from their inception. Indeed, within the arch environment the neural crest-derived cartilage precursors surround a mesodermal core, which consequently gives rise to muscle. Differentiation of the two cell types occurs in close synchrony, suggesting the mechanisms which control the differentiation of these different cells may be linked.

Mutations in the zebrafish gene *chinless* (*chn*) disrupt patterning of structures derived from the pharyngeal arches (Schilling *et al.*, 1996). Cell transplantation studies have revealed that the gene is required cell autonomously for cartilage formation and nonautonomously for the formation of the cranial muscles, suggesting that cartilage formation is required for muscle formation and that neural crest-derived cartilage precursors may provide instructive signaling which determines the type of muscle that may form within an individual arch. Cell labeling experiments have revealed that neural crest from particular anteroposterior levels within the hindbrain migrates into a specific branchial arch (Schilling *et al.*, 1996). Hence the identity of the arch primordium may be controlled by regional identity signals within the hindbrain which control specification of neural crest. Anteroposterior patterning in the developing neural tube is controlled by the activity of the homeotic or *HOX* genes (for review see Lumsden and Krumlauf, 1996). This code may be imparted to migrating neural crest cells derived from a particular anteroposterior region of the hindbrain. In support of this notion, over-expression of *Hoxa1* anterior to its normal limit in developing zebrafish embryos can lead to duplication and fusion of first arch structures, suggesting that in zebrafish neural crest cells are specified by the HOX code present within the hindbrain (Alexandre *et al.*, 1996).

While anteroposterior identity of the branchial muscles may rely on signaling outside the arch environment, dorsoventral identity may be supplied by signals within the arch, similar to those which provide identity in the trunk axial musculature. *Shh* is expressed in the ventral endoderm of all pharyngeal arches, and its receptor *Patched* (*Ptc*) is found expressed within the ventral arch mesoderm (Krauss *et al.*, 1993; Concordet *et al.*, 1996). While there is little functional evidence for SHH activity in controlling patterning of ventral arch elements, its restricted expression suggests that such a role may be possible. BMP4, another signaling peptide that has been implicated in the patterning of muscle in amniotes, is expressed within the dorsal region of the arch. It is tempting to speculate that these two secreted signaling peptides may work in tandem to polarize the arch environment to control dorsoventral cell type specification within the arch. That different signals may control dorsoventral identities within the arch is borne out by the analysis of mutations that have defects in arch development. In the large-scale screens of the zebrafish genome, 158 mutations affecting cranial development were isolated, representing at least 30 separate loci (Schilling *et al.*, 1996; Neuhauss *et al.*, 1996; Piotrowski *et al.*, 1996). In mutants of a particular class, dorsal elements remain intact while ventral muscles and cartilage are absent (Schilling, 1997). It is hoped that the analyses of mutations within these separate classes will lead to an understanding of the molecular nature of the signals which generate the complex musculoskeletal patterning of the head.

VII. CONCLUSIONS

Recent advances have significantly expanded our knowledge of the mechanisms which control the origins of distinct muscle populations within fish embryos. Some of these studies have highlighted the morphogenetic differences between amniote and fish species in the formation of distinct muscle populations. The formation of topologically separate regions of fast and slow muscle cell differentiation within the mature fish myotome is a clear architectural peculiarity of teleost embryogenesis. Also, the emphasis on axial muscle populations in generating locomotor force in most teleost species differs greatly from the reliance on muscles attached to the appendicular skeleton to control movement in tetrapod species. Despite these overt morphological differences, however, dissection of the molecular events that control formation of different fish muscle cell populations in the zebrafish embryo has illustrated that similar molecular cues trigger the formation of related muscle populations in tetrapod and fish species. Such a realization has important implications not only for understanding how these different populations are derived, but also for the mechanisms that are involved in the evolution of these different muscle populations.

REFERENCES

Alexandre, D., Clarke, J. D., Oxtoby, E., Yan, Y. L., Jowett, T., and Holder, N. (1996). Ectopic expression of Hoxa-1 in the zebrafish alters the fate of the mandibular arch neural crest and phenocopies a retinoic acid-induced phenotype. *Development* **122,** 735–746.

Amacher, S. L., and Kimmel, C. B. (1998). Promoting notochord fate and repressing muscle development in zebrafish axial mesoderm. *Development* **125,** 1397–1406.

Amthor, H., Christ, B., Weil, M., and Patel, K. (1998). The importance of timing differentiation during limb muscle development. *Curr. Biol.* **21,** 642–652.

Blagden, C. S., Currie, P. D., Ingham, P. W., and Hughes, S. M. (1997). Notochord induction of zebrafish slow muscle is mediated by Sonic Hedgehog. *Genes Dev.* **11,** 2163–2175.

Bone, Q. (1978). "Fish Physiology" (Hoar, W. S., and Randall, D. J., eds.), Vol. 7, pp. 361–424. Academic Press, New York.
Bone, Q. (1989). Evolutionary patterns of axial muscle systems in some invertebrates and fish. *Am. Zool.* **29**, 5–18.
Braus, H. (1899). Beitrage zur entwicklung der musculatur unddes peripheren nervensystems der selachier. *Morphol. J. B.* **27**, 501–629.
Concordet, J. P., Lewis, K. E., Moore, J. W., Goodrich, L. V., Johnson, R. L., Scott, M. P., and Ingham, P. W. (1996). Spatial regulation of a zebrafish patched homolog reflects the roles of sonic-hedgehog and protein-kinase-a in neural-tube and somite patterning. *Development* **122**, 2835–2846.
Corning, H. K. (1842). Uber die ventralen urwirbelknopsen in der brustflosse der teleostier. *Morphol. J. B.* **22**, 79–98.
Currie, P. D., and Ingham, P. W. (1996). Induction of a specific muscle cell type by a hedgehog-like protein in zebrafish. *Nature* **382**, 452–455.
Devoto, S. H., Melancon, E., Eisen, J. S., and Westerfield, M. (1996). Identification of separate slow and fast muscle precursor cells *in vivo,* prior to somite formation. *Development* **122**, 3371–3380.
Dohrn, A. (1884). Die paarigen und unpaaren Flossen der Selachier. *Mitt. Zool. Stat. Neapel.* **5**, 161–195.
Driever, W., Solnica-Krezel, L., Schier, A. F., Neuhauss, S.C. F., Malicki, J., Stemple, D. L., Stainier, D. Y. R., Zwartkruis, F., Abdelilah, S., Rangini, Z., Belak, J., and Boggs, C. (1996). A genetic screen for mutations affecting embryogenesis in zebrafish. *Development* **123**, 37–46.
Du, S. J., Devoto, S. H., Westerfield, M., and Moon, R. T. (1997). Positive and negative regulation of muscle cell identity by members of the hedgehog and TGF-beta gene families. *J. Cell Biol.* **139**, 145–156.
Ekker, M., Wegner, J., Akimenko, M. A., and Westerfield, M. (1992). Coordinate embryonic expression of three zebrafish engrailed genes. *Development* **116(4)**, 1001–1010.
Fan, C.-M., and Tessier-Lavigne, M. (1994). Patterning of mammalian somites by surface ectoderm and notochord: Evidence for sclerotome induction by a hedgehog homolog. *Cell* **79**, 1175–1186.
Felsenfeld, A. L., Curry, M., and Kimmel, C. B. (1991). The fub-1 mutation blocks intial myofibril formation in zebrafish muscle pioneer cells. *Dev. Biol.* **148**, 23–30.
Gont, L. K., Fainsod, A., Kim, S. H., and De Robertis, E. M. (1996). Overexpression of the homeobox gene Xnot-2 leads to notochord formation in *Xenopus. Dev. Biol.* **174**, 174–178.
Granato, M., van Eden, F. J. M., Schach, U., Trowe, T., Brand, M., Furutaniseiki, M., Haffter, P., Hammerschmidt, M., Heisenberg, C. P., Jiang, Y. J., Kane, D. A., Kelsh, R. N., Mullins, M. C., Odenthal, J., and Nusslein-Volhard, C. (1996). Genes controlling and mediating locomotion behavior of the zebrafish embryo and larva. *Development* **123**, 399–413.
Grandel, H., and Schulte-Merker, S. (1998). The development of the paired fins in the zebrafish. *Mech. Dev.* **79**, 99–120.
Griffin, K. J., Amacher, S. L., Kimmel, C. B., and Kimmelman, D. (1998). Molecular identification of spadetail: Regulation of zebrafish trunk and tail mesoderm formation by T-box genes. *Development* **125**, 3379–3388.
Grimm, M. (1973) Origin of the muscle blastemas in the developing pectoral fin of the rainbow trout (*Salmo gairdneri*). *Folia Morphol.* **21**, 197–199.
Haffter, P., Granato, M., Brand, M., Mullins, M. C., Hammerschmidt, M., Kane, D. A., Odenthal, J., van Eeden, F. J. M., Jiang, Y.-J., Heisenberg, C.-P., Kelsh, R. N., Furutani-Seiki, M., Vogelsang, E., Beuchle, D., Schach, U., Fabian, C., and Nusslein-Volhard, C. (1996). The identification of genes with unique and essential functions in the development of the zebrafish, *Danio rerio. Development* **123**, 1–36.
Halpern, M. E., Ho, R. K., Walker, C., and Kimmel, C. B. (1993). The induction of muscle pioneers and floor plate is distinguished by the zebrafish mutation no tail. *Cell* **75**, 99–111.

Halpern, M. E., Thisse, C., Ho, R. K., Thisse, B., Riggleman, B., Trevarrow, B., Weinberg, E. S., Postlethwait, J. H., and Kimmel, C. B. (1995). Cell autonomous shift from axial to paraxial mesodermal development in zebrafish floating head mutants. *Development* **121,** 4257–4264.

Hammerschmidt, M., Brook, A., and McMahon, A. P. (1997). The world according to hedgehog. *Trends Genet.* **13,** 14–21.

Hardistry, M. W. (1979). "Biology of the Cyclostomes." Chapman & Hall, London.

Harrison, R. G. (1895). Die entwicklung der unpaaren und paarigen flossen der teleostier. *Arch. mikr. Anat.* **46,** 500–578.

Hatta, K., Schilling, T. F., BreMiller, R. A., and Kimmel, C. B. (1990). Specification of jaw muscle identity in zebrafish: Correlation with engrailed-homeoprotein expression. *Science* **250,** 802–805.

Ho, R. K., and Kane, D. A. (1990). Cell-autonomous action of zebrafish spt-1 mutation in specific mesodermal precursors. *Nature* **348,** 728–730.

Ingham, P. W. (1995). Signalling by hedgehog family proteins in *Drosophila* and vertebrate development. *Curr. Opin. Genet. Dev.* **5,** 492–498.

Jensen, A. J., and Wallace, V. A. (1997). Expression of Sonic hedgehog and its putative role as a precursor cell mitogen in the developing mouse retina. *Development* **124,** 363–371.

Johnson, R. L, Laufer, E., Riddle, R. D., and Tabin, C. (1994). Ectopic expression of sonic hedgehog alters dorsal-ventral patterning of somites. *Cell* **79,** 1165–1173.

Karlstrom, R. O., Talbot, W. S., and Schier, A. F. (1999). Comparative synteny cloning of zebrafish you-too: Mutations in the hedgehog target gli2 affect ventral forebrain patterning. *Genes Dev.* **15,** 388–393.

Kimmel, C. B., Kane, D. A., Walker, C., Warga, R. M., and Rothman, M. B. (1989). A mutation that changes cell movement and cell fate in the zebrafish embryo. *Nature* **337,** 358–362.

Kimmel, C. B and Warga, R. M. (1987). Cell lineages generating axial muscle in the zebrafish embryo. *Nature* **327,** 234–237.

Kimmel, C. B., Warga, R. M. and Schilling, T. F. (1990). Origin and organization of the zebrafish fate map. *Development* **108,** 581–594.

Koumans, J. T. M., and Akster, H. A. (1995). Myogenic cells in development and growth of fish. *Comp. Biochem. Physiol. A* **110,** 3–20.

Krauss, S., Concordet, J.-P., and Ingham, P. W. (1993). A functionally conserved homolog of the *Drosophila* segment polarity gene hh is expressed in tissues with polarizing activity in zebrafish embryos. *Cell.* **75,** 1431–1444.

Lewis, K. E., Currie, P. D., Roy, S., Schauerte, H., Haffter, P., and Ingham, P. W. (1999). Control of muscle cell-type specification in the zebrafish embryo by Hedgehog signalling. *Dev. Biol.* **15;** **216(2),** 469–480.

Lumsden, A., and Krumlauf, R. (1996). Patterning the vertebrate neuraxis. *Science* **15,** 1109–1115.

Morinkensicki, E. M., and Eisen, J. S. (1997). Sclerotome development and peripheral nervous system segmentation in embryoic zebrafish. *Development* **124,** 159–167.

Nag, A. C., and Nursall, J. R. (1972). Histogenesis of white and red muscle fibres of trunk muscles of a fish *Salmo gairdneri. Cytobios* **6,** 227–246.

Neuhauss, S. C., Solnica-Krezel, L., Schier, A. F., Zwartkruis, F., Stemple, D. L., Malicki, J., Abdelilah, S., Stainier, D. Y., Driever, W. (1996). Mutations affecting craniofacial development in zebrafish. *Development* **123,** 357–367.

Odenthal, J., Haffter, P., Vogelsang, E., Brand, M., van Eeden, F. J. M., Furutani-Seiki, M., Granato, M., Hammershmidt, M., Heisenberg, C.-P., Jiang, Y.-J., Kane, D. A., Kelsh, R. N., Mullins, M. C., Warga, R. M., Allende, M. L., Weinberg, E. S., and Nusslein-Volhard, C. (1996). Mutations affecting the formation of the notochord in the zebrafish, *Danio rerio. Development* **123,** 103–115.

Piotrowski, T., Schilling, T. F., Brand, M., Jiang, Y. J., Heisenberg, C. P., Beuchle, D., Grandel, H., van Eeden, F. J., Furutani-Seiki, M., Granato, M., Haffter, P., Hammerschmidt, M., Kane, D. A., Kelsh, R. N., Mullins, M. C., Odenthal, J., Warga, R. M., and Nusslein-Volhard, C. (1996). Jaw and branchial arch mutants in zebrafish. II. Anterior arches and cartilage differentiation. *Development* **123**, 345–356.

Proctor, C., Mosse, P. R. L., and Hudson, R. C. L. (1980). A histochemical and ultrasturctural study of the development of the propulsive musculature of the brown trout, *Salmo trutta* L., in relation to its swimming behaviour. *J. Fish Biol.* **16**, 309–329.

Stemple, D. L., Solnica-Krezel, L., Zwartkruis, F., Neuhauss, S. C. F., Schier, A. F., Malicki, J., Stainier, D. Y. R., Abdelilah, S., Rangini, Z., Mountcastle-Shah, E., and Driever, W. (1996). Mutations affecting development of the notochord in zebrafish. *Development* **123**, 117–128.

Schilling, T. F. (1997). Genetic analysis of craniofacial development in the vertebrate embryo. *Bioessays* **19**, 459–468.

Schilling, T. F., and Kimmel, C. B. (1998). Musculoskeletal patterning in the pharyngeal segments of the zebrafish embryo. *Development* **124**, 2945–2960.

Schilling, T. F., Piotrowski, T., Grandel, H., Brand, M., Heisenberg, C. P., Jiang, Y. J., Beuchle, D., Hammerschmidt, M., Kane, D. A., Mullins, M. C., van Eeden, F. J., Kelsh, R. N., Furutani-Seiki, M., Granato, M., Haffter, P., Odenthal, J., Warga, R. M., Trowe, T., and Nusslein-Volhard, C. (1996). Jaw and branchial arch mutants in zebrafish. I. Branchial arches. *Development* **123**, 329–344.

Schilling, T. F., Walker, C., and Kimmel, C. B. (1996). The chinless mutation and neural crest cell interactions in zebrafish jaw development. *Development* **122**, 1417–1426.

Schauerte, H. E., van Eeden, F. J., Fricke, C., Odenthal, J., Strahle, U., and Haffter, P. (1998). Sonic hedgehog is not required for the induction of medial floor plate cells in the zebrafish. *Development* **125**, 2983–2993.

Talbot, W. S., Trevarrow, B., Halpern, M. E., Melby, A. E., Farr, G., Postlethwait, J. H., and Jowett, T., Kimmel, C. B., and Kimelman, D. (1995). A homeobox gene essential for zebrafish notochord development. *Nature* **378**, 150–157.

Thisse, C., Thisse, B., Schilling, T. F., and Postlethwait, J. H. (1993). Structure of the zebrafish snail1 gene and its expression in wild-type, spadetail and no tail mutant embryos. *Development* **119**, 1203–1215.

van Eeden, F. J. M., Granato, M., Schach, U., Brand M., Furutani-Seiki, M., Haffter, P., Hammerschmidt, M., Heisenberg, C. -P., Jiang, Y.-J., Kane, D. A., Kelsh, R. N., Mullins, M. C., Odenthal, J., Warga, R.M., Allende, M. L., Weinberg, E. S., and Nusslein-Volhard, C. (1996). Mutations affecting somite formation and patterning in the zebrafish, *Danio rerio*. *Development* **123**, 153–164.

Warga, R. M., and Nusslein-Volhard, C. (1998). Spadetail-dependent cell compaction of the dorsal zebrafish blastula. *Dev. Biol.* **203**, 116–121.

Waterman, R. E. (1969). Development of the lateral musculature in the teleost, *Brachydanio rerio*: A fine structure study. *Am. J. Anat.* **125**, 457–493.

Weinberg, E. S., Allende, M. L., Kelly, C. S., Abdelhamid, A., Murakami, T., Andermann, P., Doerre, O. G., Grunwald, D. J., and Riggleman, B. (1996). Developmental regulation of zebrafish MyoD in wild-type, no tail and spadetail embryos. *Development* **122**, 271–280.

Winterbottom, R. (1974). A descriptive synomy of the striated muscle of the Teleostei. *Proc. Acad. Nat. Sci. Philadelphia* **125**, 225–317.

Yamamoto, A., Amacher, S. L., Kim, S. H., Geissert, D., Kimmel, C. B., and De Robertis, E. M. (1998). Zebrafish paraxial protocadherin is a downstream target of spadetail involved in morphogenesis of gastrula mesoderm. *Development* **125**, 3389–3397.

2

MYOGENIC REGULATORY FACTORS

SHUGO WATABE

I. Abstract
II. Introduction
III. Fish Myogenic Regulatory and Related Factors
    A. Myogenin
    B. MyoD
    C. Myf-5
    D. Comparison of MyoD Family Members
    E. MEF2 Family
    F. E Protein and Other Transcription Factors
IV. Expression of Fish MRFs During Ontogeny
    A. MyoD and MEF2 Families
    B. E Protein
V. Expression of Fish MRFs During Temperature Acclimation
VI. Conclusion
    Acknowledgments
    References

I. ABSTRACT

Members of the MyoD family of basic helix-loop-helix (bHLH) transcription factors have a central role in the determination and differentiation of vertebrate skeletal muscle. Collectively referred to as myogenic regulatory factors (MRFs), their mammalian counterparts include the unlinked genes MyoD and myogenin, and the closely linked genes myf-5 and MRF4 (myf-6). MRFs form heterodimers with E protein products such as E12 and E47 of the ubiquitously expressed E2A gene family and bind to the E-box DNA sequences (CANNTG) present in the regulatory regions of many muscle-specific genes. Another important regulator of skeletal muscle differentiation is the myocyte-specific enhancer factor 2 (MEF2) family of transcription factors which form homodimers and bind to an A/T-rich sequence present in many muscle-specific promotors and enhancers. MEF2 isoforms regulate myogenic bHLH genes and cooperate with MRFs in activating skeletal muscle-specific transcription. We have recently cloned cDNAs encoding MRFs of myogenin, MyoD, and myf-5, together with E12, from embryos and larvae of the common carp. MEF2 cDNAs were also isolated from an adult carp cDNA library. During the period of somite formation in carp, myf-5 was first expressed followed by MEF2C and MyoD, then myogenin and MEF2A, finally skeletal myosin heavy chain (MyHC) and α-actin. E12 was detected from embryos at various developmental stages as well as in juveniles. We also examined by Northern blot analysis the changes in the accumulated mRNA levels of MyHC isoforms in carp fast skeletal muscle during water temperature acclimation from 20 to 30°C in relation to those of MyoD family and MEF2 family members. There was a dramatic decrease in the transcripts of the MyHC isoform predominantly expressed in cold-acclimated carp and a significant increase in the transcripts of the dominant MyHC isoform of warm-acclimated carp. Over such an acclimation period, the transcription levels of myogenin, MEF2A, and MEF2C were changed significantly, whereas MyoD transcripts were rather constant. MRFs and related transcription factors probably play important roles not only during development, but also during temperature acclimation of adult fish.

II. INTRODUCTION

Members of the MyoD family of transcription factors with a bHLH structure have been extensively studied with skeletal muscles from higher vertebrates and shown to have a central role in the determination and differentiation of these muscles (see Weintraub, 1993; Fig. 1). Collectively referred to as MRFs they include the unlinked genes MyoD and myogenin, and the closely linked genes

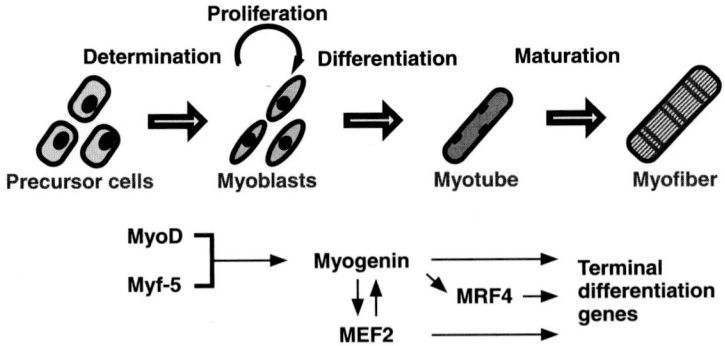

Fig. 1. Myogenic lineage determination and differentiation in relation to expression of myoD family and MEF2 family transcription factors together with muscle-specific genes. (Modified from Watabe, 1999.)

myf-5 and MRF4 (myf-6) which have only 8.5 kb separating their coding sequences on human chromosome 12 (Braun et al., 1990).

In mammals, forced expression of each of the MRFs can convert non-muscle cells to myoblasts, suggesting they may be functionally interchangeable (Weintraub et al., 1991), although "knock-out" and expression studies with mice indicate that each has distinct but overlapping functional roles (Rudnicki et al., 1992; Smith et al., 1994). Although double mutant knock-out mice deficient in both MyoD ($-/-$) and myf-5 ($-/-$) are lethal and lack all myogenic progenitor (precursor) cells, those deficient in either one of the two genes apparently grow well (Rudnicki et al., 1992) (Fig. 1). Since the MyoD ($-/-$) mutant mice are severely deficient in muscle regenerative ability (Megeney et al., 1998), this transcription factor plays an important role at least in the activation of satellite cells and *de novo* formation of muscle fibers. On the other hand, myogenin ($-/-$) mutant mice maintained certain, though decreased, levels of many muscle-specific transcripts (Hasty et al., 1993). There is evidence that *cis*-acting interactions are important in regulating the expression of myf-5 and MRF4 (Yoon et al., 1997).

MRFs form heterodimers with E protein products and regulate many muscle-specific genes (Fig. 2) (see Olson, 1992). Transcription factor E proteins, E12 and E47, are generated from the E2A gene by alternative splicing in the exon coding for the DNA binding domain (Murre et al., 1989; Sun and Baltimore, 1991) and are ubiquitously expressed in various tissues (Roberts et al., 1993; Wulbeck et al., 1994). They have the bHLH domain which is essential to form homodimers, or heterodimers with the other tissue-specific bHLH proteins (Fig. 2). Homo- and heterodimers bind to a common DNA sequence (CANNTG), termed

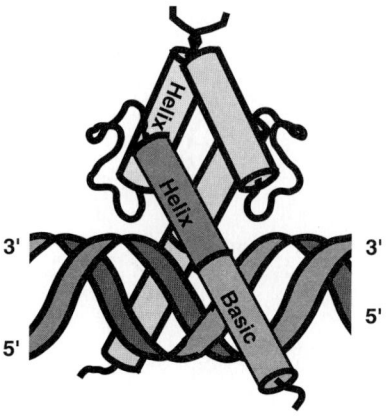

Fig. 2. Formation of heterodimer between MRFs and E protein and its binding to DNA. (Modified from Ma et al., 1994.) The loops are located between two helix regions, forming the conserved bHLH domain together with a basic region.

E-box, which is present in tissue-specific gene enhancer, and activate transcription (Murre et al., 1989, 1994; Lassar et al., 1991). E-box was first identified as the element in the immunoglobulin heavy chain gene enhancer (Church et al., 1985). Subsequently, it has been revealed that one or more E-boxes are present in muscle-specific gene enhancers, including those of myosin light chains 1 and 3 (Wentworth et al., 1991), troponin (Lin et al., 1991), cardiac α-actin (Sartorelli et al., 1990), α subunit of the acetylcholine receptor (Piette et al., 1990), and creatine kinase (Lassar et al., 1989; Braun et al., 1990). The muscle-specific gene enhancers are specifically recognized by MRFs which form heterodimers with E12 or E47 (Murre et al., 1989; Benezra et al., 1990; Sun and Baltimore, 1991).

MRF activity is controlled through both positive and negative regulatory pathways involving, among other factors, LIM protein (MLP) (Kong et al., 1997) and Id proteins (Benezra et al., 1990), respectively.

Another important regulator of skeletal muscle differentiation is the MEF2 family of transcription factors (Rhodes and Konieczny, 1989), which form homodimers and bind to an A/T-rich sequence present in many muscle-specific promotors and enhancers (Fig. 3) (Gossett et al., 1989). Multiple isoforms of MEF2 have been identified in vertebrates, all of which contain the DNA-binding sequence characteristic of the MADS gene family and an adjacent highly conserved MEF2-specific domain. MEF2 isoforms regulate myogenic bHLH genes and cooperate with MRFs in activating skeletal muscle-specific transcription (see Olson 1992 for a review).

Andres and Walsh (1996) claimed that myogenin-positive C2C12 myoblasts remained capable of replicating DNA. In contrast, subsequent expression of the cell cycle inhibitor p21 in differentiating myoblasts correlated with the establishment of the post-mitotic state. Later during myogenesis, post-mitotic mononucleated myoblasts activated the expression of the muscle structural proteins

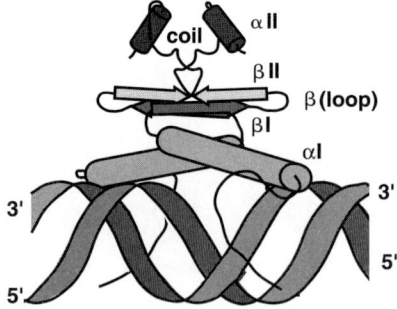

Fig. 3. Formation of homodimer formation of MEF2 family members and its binding to DNA. (Modified from Pellegrini et al., 1995.) α-Helices and β-pleated sheet structures are represented by rods and arrows, respectively.

such as MyHC and α-actin, and then fused to form a multinucleated myotube. It seems from these results that muscle-specific genes are upregulated by myogenin. While forced expression of myogenin or MyoD in 10T1/2 fibroblasts induced MEF2 DNA binding activity, it has been shown that MEF2 proteins lack myogenic activity on their own, but that they are able to act synergistically with MRFs during myogenic conversion of 10T1/2 fibroblasts in culture. It is thought that the synergy is mediated by direct protein-protein interactions between MEF2 factors and heterodimers formed between MRFs and E proteins (Ludolph and Konieczny, 1995; Molkentin and Olson, 1996). However, the overall cascade for expression of MRFs and induction of muscle-specific genes during myogenesis still remains unclear.

III. FISH MYOGENIC REGULATORY AND RELATED FACTORS

Most members of MyoD and MEF2 families identified in skeletal muscles of higher vertebrates have been recognized in zebrafish. However, some of the members still have remained not determined and associated works have been largely focused on histological aspects. Rescan and co-workers (Rescan et al., 1994, 1995; Rescan and Gauvry, 1996) cloned MRFs from rainbow trout, but unfortunately no information is available on MEF2 family members for this fish. Recently, cDNA clones encoding MyoD and MEF2 families have been systematically isolated from the common carp (Kobiyama et al., 1998). The following sections are mainly highlighted with structural properties elucidated and residue numbers discussed are those from the N terminus of carp transcription.

A. Myogenin

The bHLH structure is essential to DNA binding of MRFs and heterodimerization with E proteins in higher vertebrates (Lasser et al., 1989; Davis et al., 1990). The corresponding region in carp myogenin was highly conserved (Kobiyama et al., 1998) and showed almost the same sequence (93–98% identity) as those of other vertebrates including rainbow trout (*Oncorhynchus mykiss;* Rescan et al., 1995). While myogenin was also well conserved at residues 81–94 and 156–164, the entire sequence of carp myogenin showed 69, 55, and 51% sequence identity with those of rainbow trout (Rescan et al., 1995), chicken (Fujisawa-Sehara et al., 1990), and mouse (Edmondson et al., 1989), respectively (Table I). In addition, carp myogenin had additional sequences at residues 52–56 and 68–79 which were not observed in chicken and mouse.

Table I.
Homology in Amino Acid Sequences of Carp Myogenin, MyoD, and Myf-5
to Those of Other Vertebrates (% identity)

Carp	Zebrafish	Rainbow trout	Xenopus	Chicken	Mouse
myogenin	—[a]	68	—	55	51
MyoD	93	81	73	71	60
myf-5	—	—	56	57	56

Note: Sequences cited were for carp myogenin, MyoD, and myf-5 (Kobiyama *et al.*, 1998); zebrafish MyoD (Weinberg *et al.*, 1996; rainbow trout myogenin (Rescan *et al.*, 1995) and MyoD (Rescan *et al.*, 1994); *Xenopus* MyoD (Hopwood *et al.*, 1989) and myf-5 (Hopwood *et al.*, 1991); chicken myogenin (Fujisawa-Sehara *et al.*, 1990), MyoD (Lin *et al.*, 1989), and myf-5 (Saitoh *et al.*, 1993); mouse myogenin (Edmondson *et al.*, 1989), MyoD (GenBank accession #M84918), and myf-5 (Buonanno *et al.*, 1992).

[a] Not available.

B. MyoD

The bHLH region of carp MyoD was highly conserved with a sequence identity to other vertebrates, including zebrafish, in the range from 93–98% (Kobiyama *et al.*, 1998). In contrast to myogenin, however, MyoD from carp contained several other conserved regions at 27–53, 77–101, 189–211, and 246–261 amino acid residues from the N terminus, resulting in higher homology of the whole molecule in carp. MyoD in carp showed 93, 81, 73, and 71% identities with MyoD from zebrafish (Weinberg *et al.*, 1996), rainbow trout (Rescan *et al.*, 1994), *Xenopus* (Hopwood *et al.*, 1989), and chicken (Lin *et al.*, 1989), respectively (Table I). Carp MyoD was relatively similar to other fish MyoDs sequenced containing deletions at 9–21, 66–76, 171–175, and 306–312 amino acids from the N terminus that are not observed in higher vertebrates.

C. Myf-5

The cloning of cDNAs encoding the full-length fish myf-5 was only accomplished for carp (Kobiyama *et al.*, 1998). The bHLH region of carp myf-5 was highly conserved, showing an identity with other vertebrate myf-5s of 82–84%. Carp myf-5 contained several other conserved regions at 64–77, 139–151, and 205–216 amino acid residues and the whole molecule showed 56, 57, and 56% identities with myf-5 from *Xenopus* (Hopwood *et al.*, 1991), chicken (Saitoh *et al.*, 1993), and mouse (Buonanno *et al.*, 1992), respectively (Table I). Carp myf-5 contained deletions at 4–8, 28–30, and 37–45 amino acids relative to mouse and chicken.

D. Comparison of MyoD Family Members

Comparison of the deduced amino acid sequences of carp MRFs with those from other vertebrates revealed that MyoD was more highly conserved than myogenin and myf-5, including the bHLH DNA-binding region. It is thought that myogenic activity is dependent on two amino acid residues (alanine and threonine) in the center of the MyoD basic domain and one amino acid (lysine) in the junction region of the first helix of MyoD (Davis and Weintraub, 1992). Black *et al.* (1998) found that the active site residues, alanine and threonine, were required for MyoD to synergistically activate transcription with MEF2, but they were not required for interaction with MEF2. Marked differences in the amino acid sequence were observed between carp myogenin, MyoD, and myf-5, showing only 73–87% identity in the bHLH region and 31–47% identity in the whole coding region. Similar sequence divergence has also been reported between myogenin, MyoD, and myf-5 in higher vertebrates (Lin *et al.*, 1989; Fujisawa-Sehara *et al.*, 1990; Edmondson *et al.*, 1989; Pinney *et al.*, 1988; Saitoh *et al.*, 1993; Buonanno *et al.*, 1992).

E. MEF2 Family

The cloning of the full-length cDNAs encoding the carp MEF2 family was accomplished through screening of a cDNA library constructed from adult carp muscle acclimated to 10°C (Kobiyama *et al.*, 1998). A comparison of the predicted amino acid sequences of MEF2A and MEF2C from carp showed 93% identity in the MADS box and MEF2 domain and 58% identity in the whole coding region. The region of these MADS boxes and MEF2 domains from carp MEF2A and MEF2C showed 94–100% sequence identity with other vertebrate MEF2 isoforms. Carp MEF2A contained deletions at 211–229 and 424–437 amino acid residues from the N terminus, the entire sequence showing 91, 72, and 70% sequence identity with those of zebrafish (Ticho *et al.*, 1996), mouse (GenBank database accession #U30823), and human (Yu *et al.*, 1992), respectively (Table II). Carp MEF2C contained a deletion at 345–348 amino acids. The entire amino acid sequence of zebrafish (Ticho *et al.*, 1996), mouse (Martin *et al.*, 1993), and human (McDermott *et al.*, 1993) MEF2C are 85, 80, and 80% identical to the carp MEF2C sequence, respectively (Table II).

F. E Protein and Other Transcription Factors

Nihei *et al.* (1999) isolated a cDNA clone encoding a part of E12 by RT-PCR from larvae at hatching of carp. The deduced amino acid sequence of carp E12 showed 81, 51, and 55% identities with corresponding sequences of zebrafish (Wulbeck *et al.*, 1994), *Xenopus* (Rashbass *et al.*, 1992), and human (Nourse

Table II.
Homology in Amino Acid Sequences of Carp MEF2A and
MEF2C to Those of Zebrafish, Mouse, and Human (% identity)

Carp	Zebrafish	Mouse	Human
MEF2A	91	72	70
MEF2C	85	80	80

Note: Sequences cited were for carp MEF2A and MEF2C (Kobiyama *et al.*, 1998); zebrafish MEF2A and MEF2C (Ticho *et al.*, 1996); mouse MEF2A (GenBank database accession #I30823) and MEF2C (Martin *et al.*, 1993); human MEF2A (Yu *et al.*, 1992) and MEF2C (McDermott *et al.*, 1993).

et al., 1990), respectively (Table III). The bHLH domain of carp E12 was highly conserved, showing approximately 90% identity with those of E12 from other vertebrate. The two helix regions in the bHLH domain of carp E12 were rich in hydrophobic amino acid residues which are supposed to form hydrophobic clusters, each on opposite sites of the two helices. Such structural features are important for E proteins to form heterodimers with MRFs through hydrophobic amino acid residues (Fig. 2; Ellenberger *et al.*, 1994; Ma and Rould, 1994). The C terminal part next to the bHLH domain of carp E12 contained 23 amino acid residues which were identical among species compared (Nihei *et al.*, 1999).

HLH proteins related to the inhibitor of DNA binding/differentiation (Id) serve as general antagonists of cell differentiation. They lack a basic amino acid domain necessary in binding DNA and are thought to function in a dominant negative manner by sequestering bHLH transcription factors such as MRFs, thereby blocking the binding of dimerized bHLH proteins to DNA. Rescan (1997) cloned two rainbow trout Ids, one of which encoded a trout putative Id1 with 63% identity to the human Id1 protein over the entire length and 78% identity in the HLH region. The other one encoding a trout putative Id2 protein showed 82% identity to the human Id2 protein and was conservative over the HLH region with only one amino acid substitution. In juveniles and adults, trout Id1 and Id2 tran-

Table III.
Homology in Partial Amino Acid Sequences of Carp E12
to Those of Zebrafish, *Xenopus,* and Human (% identity)

	Zebrafish	*Xenopus*	Human
Carp	81	51	55

Note: Sequences cited were for carp (Nihei *et al.*, 1999), zebrafish (Wulbeck *et al.*, 1994), *Xenopus* (Rashbass *et al.*, 1992), and human (Nourse *et al.*, 1990).

scripts were abundant in the slow oxidative fibers, but were absent in the fast glycolytic fibers. These expression patterns suggest that rainbow trout Id genes play a role in the regulation of muscle fiber phenotype in addition to controlling early myogenesis.

Members of the cysteine-rich protein (CRP) family are featured by the presence of two tandemly arrayed LIM proteins linked to a short glycine-rich motif. The LIM domains thus organized are proposed to be involved in controlling gene expression and cell differentiation. These structures work as specific adapter elements and promote protein-protein interactions with identical or different LIM domains as well as with other protein motifs (Dawid et al., 1998). Delalande and Rescan (1998) cloned a rainbow trout homolog that displayed 86, 76, and 67% identity with chicken CRP2, CRP1, and MLP/CRP3 proteins, respectively (Weiskirchen et al., 1995). The whole-mount in situ hybridization showed that trout CRP transcripts were first detected just before somitogenesis in the paraxial mesoderm, but not in the axial structures.

IV. EXPRESSION OF FISH MRFs DURING ONTOGENY

A. MyoD and MEF2 Families

In all known skeletal muscle lineage of higher vertebrates including mammals and avians, myf-5/MyoD expression is followed by upregulation of myogenin and MEF2 family factors, the latters enhancing expression of differentiation genes (Yun and Wold, 1996) (see Fig. 1). The last myogenic regulatory factor to be activated in most muscle type is MRF4 which is expressed until adult. MEF2 isoforms regulate myogenic bHLH genes and cooperate with MRFs in activating skeletal muscle-specific transcription.

In zebrafish, comparable works on skeletal muscle lineage have been carried out. Although zebrafish is an excellent system for studying early muscle development and differentiation, it does not provide a general model for post-larval muscle growth of fish because of its modest ultimate body size (3–5 cm). Muscle growth in small fish species primarily involves the hypertrophy of the fibers formed in the embryo and early during the larval stages (Weatherley et al., 1988). Such phenotype of muscle growth in these fish is typically observed in high vertebrates, where hyperplasia or muscle fiber recruitment is terminated at the neonatal stage (Rowe and Goldspink, 1969) (Fig. 4). In contrast, for fish species with indeterminate growth, including carp, new muscle fibers are continuously recruited during juvenile and adult stages from a stem cell population present under the basal lamina of muscle fibers (Stickland, 1983; Alami-Durante et al., 1997; Koumans et al., 1993; Johnston et al., 1998) (Fig. 4). In addition, carp fast

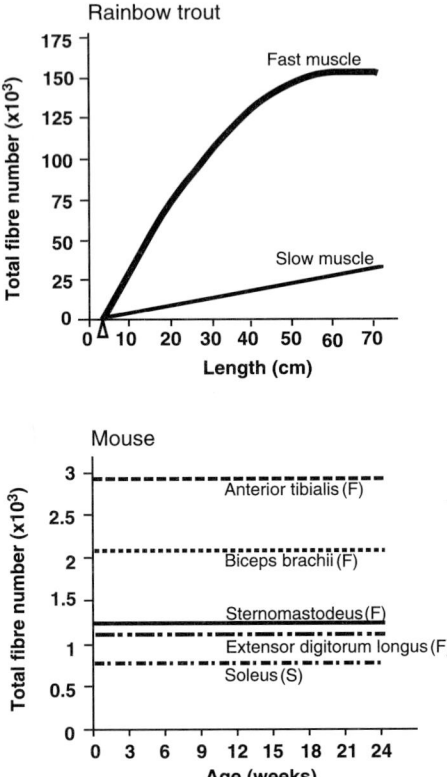

Fig. 4. Fish muscle growth during development and their post-larval muscle growth involving hypertrophy and hyperplasia in comparison with mammalian muscle. (Modified from Stickland, 1983; Rowe and Goldspink, 1969.) An arrowhead indicates an approximate hatching period for rainbow trout. In the panel for mouse, F and S represent fast- and slow-fiber rich muscular tissues, respectively.

muscle fibers change MyHC isoforms following temperature acclimation (Guo et al., 1994; Watabe et al., 1995; Imai et al., 1997; Hirayama and Watabe, 1997), suggesting the participation of MRFs in muscle fiber recruitment even at the adult. However, almost nothing has been documented about the expression patterns and role of MRFs and MEF2 isoforms during post-larval growth of fish.

Recently, total RNAs from carp at various developmental stages including embryos, larvae, and juveniles, have been analyzed by Northern blot for myogenin, MyoD, myf-5, MEF2A, and MEF2C mRNAs as well as for muscle-specific genes encoding skeletal MyHC and α-actin (Fig. 5) (Kobiyama et al., 1998).

2. MYOGENIC REGULATORY FACTORS

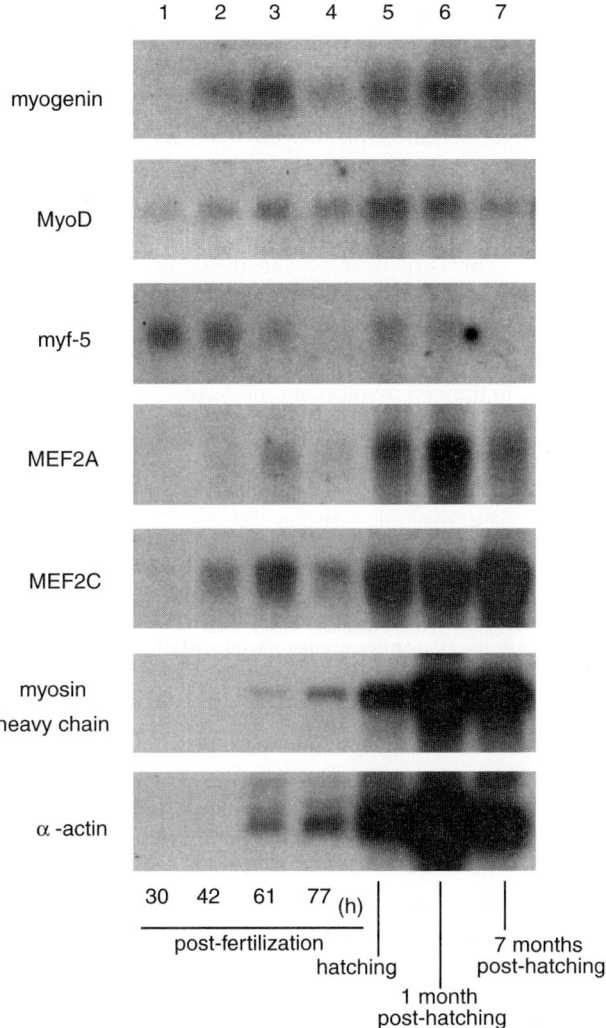

Fig. 5. Northern blot analyses of mRNAs encoding carp MyoD, myogenin, myf-5, MEF2A, and MEF2C in comparison with those for carp skeletal MyHC and α-actin (Kobiyama *et al.,* 1998). Total RNAs of 20 μg for MyoD, myf-5, MEF2A, MEF2C, skeletal MyHC, and α-actin, or 30 μg for myogenin were electrophoresed in 1.4% agarose gels and transferred onto nylon membranes. Then, the membranes were hybridized with ^{32}P-labeled PCR products of MyoD, myogenin, myf-5, MEF2A, MEF2C, skeletal MyHC, and α-actin. Lanes 1–7 contain total RNAs from embryos at 30, 42, 61, and 77 hr PFT, larvae at hatching, and juveniles aged 1 and 7 months post-hatching, respectively.

A single mRNA band encoding carp myogenin was first detected in embryos 42 hr post-fertilization (PFT) at the 15-somite stage. Strong transcription signals were detected in carp embryos 61 hr PFT at around the time when the heartbeat was first observed as well as in 1-month-old juveniles. In contrast, weaker signals were obtained for myogenin transcripts in eyed-stage embryos (77 hr PFT), larvae at hatching, and in 7-month-old juveniles.

Carp MyoD mRNA was detected in embryos 30 hr PFT at the 3-somite stage, appearing earlier than myogenin mRNA. The developmental stages which showed the maximal expression levels were different between myogenin and MyoD. While the accumulated levels of myogenin transcripts showed two peaks, those of MyoD transcripts had nearly one. The abundance of the MyoD transcripts increased during development until hatching and thereafter declined, although they were still present in the fast myotomal muscle of 7-month-old juveniles.

Carp myf-5 was already present in 3-somite stage embryos (30 hr PFT), which were the first samples collected in these experiments. While the transcripts were still weakly detectable in 7-month-old juveniles, the strongest signal for myf-5 mRNA appeared in early embryonic samples, 30 and 42 hr PFT.

Transcripts of MEF2A were first detected in 15-somite stage embryos (42 PFT) in concert with the appearance of myogenin mRNA. The expression pattern of MEF2A during development was also similar to myogenin with the maximal level detected in 1-month-old juveniles. MEF2C was first detected in 3-somite stage embryos, and gradually increased during development, although the signal obtained from eyed-stage embryos 77 hr PFT was weak. Carp MEF2A and MEF2C transcripts both increased after hatching. In contrast to MRFs, a strong signal for MEF2 family transcription factors was detected in the fast muscles of 7-month-old juveniles. The signals of the two muscle-specific mRNAs encoding MyHC and α-actin were first detected in carp embryos 61 hr PFT at around the time the heartbeat was first observed.

A comparison with other expression data reveals differences in the transcriptional regulation of the MyoD and MEF2 families between species (Fig. 6). For example, the first MRF to be expressed in the precursor cells for myotomal muscle is myf-5 in mouse (Ott et al., 1991), but it is MyoD in the quail (Pownall and Emerson, 1992). MyoD expression appears earlier than that of myogenin in quail (Pownall and Emerson, 1992). In contrast, MyoD transcripts are first detected around 2 days after myogenin transcripts begin to accumulate in the precursor cells for the myotomal muscle of the mouse (Sasoon et al., 1989). MEF2C transcripts appear first followed by MEF2A and MEF2D in the mouse (Edmondson et al., 1994). Xenopus first expresses myf-5 and MyoD and then myogenin and MRF4 (Hopwood et al., 1991; Takahashi et al., 1998).

In zebrafish, MyoD (Weinberg et al., 1996) and MEF2D (Ticho et al., 1996) transcripts are first detected at mid-gastrulation and are present in the adaxial cells

2. MYOGENIC REGULATORY FACTORS

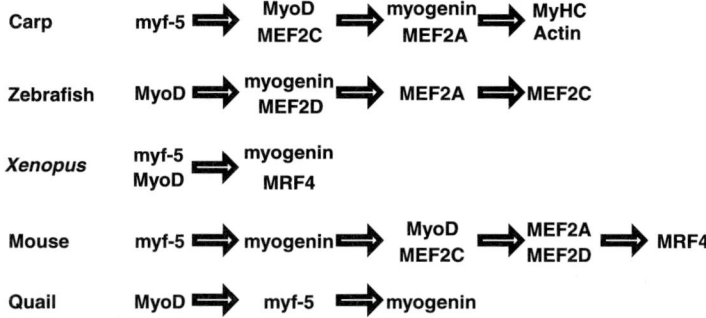

Fig. 6. Species-specific differences in expression patterns of MyoD family and MEF2 family members during development; MyHC—the myosin heavy chain.

adjacent to the notochord in the pre-somitic mesoderm. During somitogenesis, MEF2D-expressing cells of zebrafish are present on either side of the transverse myoseptum and expression occurs in a rostral to caudal sequence (Ticho *et al.*, 1996). MEF2A and MEF2C transcripts appear later and follow the same expression pattern within the somites as MEF2D (Ticho *et al.*, 1996). Myogenin appears in the somites at around the same time as MEF2D does (Ticho *et al.*, 1996; Weinberg *et al.*, 1996). The same sequential activation of myogenin functioning downstream of MyoD has been also reported for rainbow trout (Rescan *et al.*, 1994, 1995).

In carp, the pattern in the accumulation of mRNA of MyoD family is therefore similar to that described in quail, rainbow tout, and zebrafish, whereas expression patterns of the MEF2 family are similar to those in the mouse.

Transcripts of MEF2 family were expressed in both somatic and cardiac cells in zebrafish (Ticho *et al.*, 1996). It was noted that myogenin transcripts concentrated within slow muscle fibers in adult rainbow trout, but only low concentration in fast fibers (Rescan *et al.*, 1995). Such selective accumulation of myogenin transcripts contrasted with the distribution of MyoD counterparts which were found equally in both muscle fiber types (Rescan *et al.*, 1994).

B. E Protein

In situ hybridization on rat has shown that E2A transcripts are found in most embryonic and adult tissues except for the heart and liver (Roberts *et al.*, 1993). However, it is clear that the muscle-specific gene enhancers are specifically recognized by MRFs which form heterodimers with E12 or E47 (Benezra *et al.*, 1990; Murre *et al.*, 1989; Sun and Baltimore, 1991), as described before (see

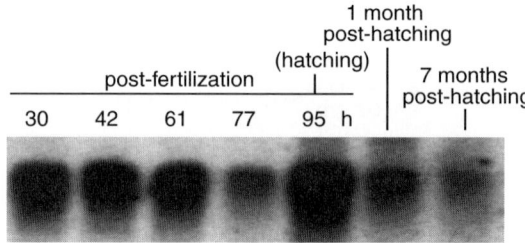

Fig. 7. Northern blot analyses of mRNAs encoding carp E12 during development (Nihei *et al.*, 1999). Refer to the legend of Fig. 5 for the developmental stages of carp embryos.

Fig. 2). Therefore, it is considered that E12 is also involved in muscle cell differentiation of fish as E proteins from higher vertebrates.

Carp E12 mRNAs were detected from embryos 30 hr PFT to 7-month-old juveniles (Nihei *et al.*, 1999) (Fig. 7). In zebrafish embryos, E12 mRNAs were found in all tissues except for the notochord (Wulbeck *et al.*, 1994). In the case of carp, intense signals of E12 transcripts were detected from embryos at very early developmental stages (Fig. 7), suggesting that carp E12 may also play important roles in cell muscle proliferation and differentiation. The whole embryos and larvae at hatching of carp were subjected to the extraction of total RNA, whereas myotomal muscle samples were used for juveniles. However, the expression patterns of MyoD and MEF2 family have been shown for carp at the same developmental stages (Kobiyama *et al.*, 1998) (see Fig. 5). Therefore, it seems that carp E12 expression occurs also in somites and myofibers. The transcripts for E12 were found in the fast myotomal muscle from 1- and 7-month-old juveniles as in the case of carp MyoD family members (Fig. 7). On the other hand, zebrafish E12 transcripts were no longer detectable in the trunk and tail region when somitegenesis was completed, although the transcripts were particularly abundant in somites in early stages of embryos (Wulbeck *et al.*, 1994). Such differences in expression pattern during development between carp and zebrafish probably reflect different muscle growth phenotypes for these two fish species (see Fig. 4).

V. EXPRESSION OF FISH MRFs DURING TEMPERATURE ACCLIMATION

Skeletal muscle phenotype can be modified by temperature acclimation in cyprinid fish, including carp. Cold acclimation results in an increase in myofibrillar ATPase activity in fast muscle fibers, whereas warm acclimation induces myofibrils with low ATPase in the same muscle (Johnston *et al.*, 1975; Sidell, 1980; Heap *et al.*, 1985, 1986). The maximum initial velocity (V_{max}) of actin-activated Mg^{2+}-ATPase activity of myosin from carp accclimated to 10°C was 1.6-fold higher than that of carp acclimated to 30°C at the same assay temperature of 20°C (Hwang *et al.*, 1990, 1991; Watabe *et al.*, 1992, 1994, 1998; Guo *et al.*, 1994). These results suggest that different myofibrillar ATPase activities from cyprinid fish acclimated to altered water temperatures are due to changes in myosin isoforms. A minimum of three isoforms of fast skeletal muscle MyHC have been identified in common carp, which are expressed in an acclimation-temperature-dependent fashion each with different Mg^{2+}-ATPase activities and thermal stabilities (Watabe *et al.*, 1992, 1994; Guo *et al.*, 1994; Nakaya *et al.*, 1995, 1997; Kakinuma *et al.*, 1998). We named 10- and 30°C-type isoforms which were expressed predominantly in fast skeletal muscles of carp acclimated to 10 and 30°C, respectively (Watabe *et al.*, 1995; Imai *et al.*, 1997; Hirayama and Watabe, 1997). The other isoform, called the intermediate type, had an intermediate structure between the 10 and 30°C types. The mRNA levels of three MyHC isoforms were dramatically changed following temperature acclimation irrespective of probes degenerated from the regions encoding myosin subfragment-1 and light meromyosin (Imai *et al.*, 1997; Hirayama and Watabe, 1997). These results suggest dynamic reorganization of muscle fibers for adult carp in association with environmental temperature fluctuation.

Carp contains an MyHC gene of approximately 12 kbp which is half the size of the corresponding mammalian and avian MyHC genes (Ennion *et al.*, 1995). The flanking region of carp MyHC gene has been further analyzed and found to contain a consensus sequence of E-box and MEF2-binding site (Gauvry *et al.*, 1996). Therefore, MyoD family and MEF2 family transcription factors of carp can bind to E-boxes and MEF2-binding sites through heterodimers with proteins encoded by the E2A genes, thus regulating expression of MyHC genes. Thus, we examined expression patterns of carp MyoD and MEF2 families during temperature acclimation by Northern blot analyses and compared them with those of fast skeletal MyHC isoforms (Kobiyama *et al.*, 2000).

When rearing temperature for carp was raised from 20 to 30°C, mRNAs encoding the 10°C-type MyHC was rapidly decreased (Fig. 8; Kobiyama *et al.*, 2000). On the other hand, the 30°C type was increased concomitantly. While myofibrillar ATPase activity was decreased gradually and reached a steady state after 4–5 weeks, α-actin transcripts were not markedly changed during temperature acclimation.

The mRNAs encoding carp myogenin maintained high levels within 5 days, then decreased to very low levels (Fig. 8). Such changes of myogenin transcripts were, however, not statistically significant. MyoD transcripts were also increased within one day after temperature raise. On the other hand, myf-5 transcripts could not be detected during these experimental periods. Besides MRFs, carp MEF2A and MEF2C (Kobiyama *et al.*, 2000) as well as E12 (Nihei, personal communication) decreased gradually during temperature acclimation. Therefore, continuously expressed MyoD transcripts at certain levels in carp seem to be related to

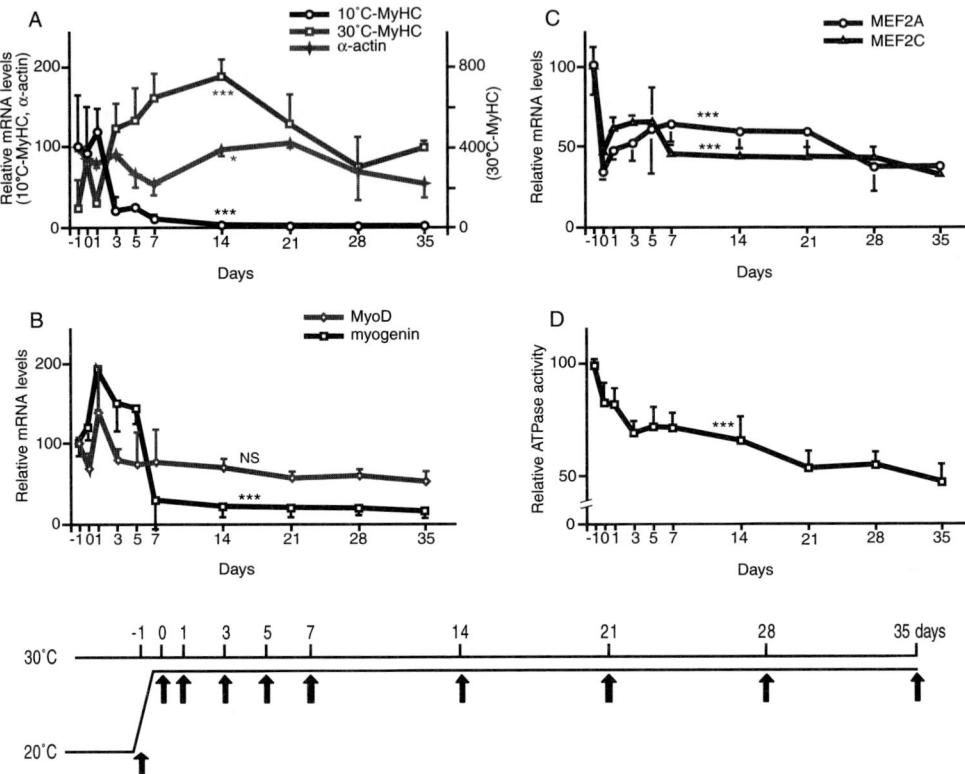

Fig. 8. Changes in transcription levels of the 10- and 30°C-type MyHC isoforms and α-actin together with those of transcription factors of MyoD and MEF2 families during warm acclimation of carp (Kobiyama et al., 2000). Northern blots were performed using probes specific to the 10- and 30°C-type MyHC isoforms (A), α-actin (A), myogenin (B), MyoD (B), MEF2A (C), and MEF2C (C). Changes in myofibrillar ATPase activity of carp fast muscle are shown in panel D. The time-course of raising water temperature from 20–30°C and sampling schedule indicated with arrows are given at the bottom. Fish had been previously acclimated to 20°C for 5 weeks and temperature was changed from day 1 to day 0. Three samples were analyzed per group. Significant levels at $p < 0.05$ (∗) and $p < 0.005$ (∗∗∗) with ANOVA are indicated, whereas NS denotes the levels not significant at $p > 0.05$.

recruitment of an associated type of fast skeletal MyHC isoform during temperature acclimation. However, it is still ambiguous whether a new MyHC type is recruited within the same muscle fiber previously present before temperature raise, or in a new muscle fiber probably formed from a stem cell population present under the basal lamina of muscle fibers (Fig. 9).

2. MYOGENIC REGULATORY FACTORS

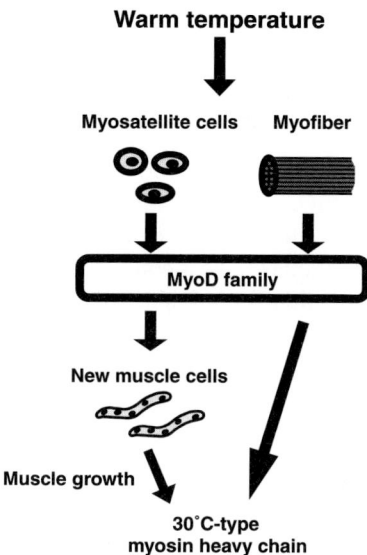

Fig. 9. Two presumed mechanisms underlying recruitment of carp fast skeletal MyHC isoforms during temperature acclimation.

VI. CONCLUSION

In mammals, the number of muscle fibers is fixed at birth and post-natal muscle growth involves the hypertrophy of muscle fibers (Rowe and Goldspink, 1969) (see Fig. 4). The expanding muscle fibers require a source of additional nuclei which are provided by a pool of undifferentiated myogenic stem cells located beneath the basal lamina (Moss and Leblond, 1971). These myogenic stem cells (satellite cells) are also activated following muscle injury. Following such activation the division products of the myogenic stem cell population express myogenic bHLH proteins (Yablonka-Reuveni and Rivera, 1994; Megeney *et al.*, 1996). Muscle growth in carp and other fish involves new muscle fiber production in addition to fiber hypertrophy (see Fig. 4; Stickland, 1983; Koumans *et al.*, 1993; Johnston *et al.*, 1998). The continued expression of members of the MyoD family and MEF2 isoforms in the juveniles of carp (see Fig. 5) may reflect activated myogenic cells. Such cells constitute a declining proportion of the total number of muscle nuclei during the larval and juvenile stages as growth proceeds (Koumans *et al.*, 1993; Johnston *et al.*, 1998).

Cornelison and Wold (1997) used a multiplex single-cell RT-PCR assay to monitor the expression of MRFs in mice following injury and found that activated satellite cells began to express either MyoD or myf-5 first, followed by myogenin and MRF4. MRF4 is one of MRFs in mammals which claims to play important roles in muscle maturation and maintain the adult muscle phenotype (Hinterberger *et al.,* 1991). However, MRF4 has not yet been cloned from any fish species.

Muscle is a post-mitotic tissue and post-embryonic growth involves a population(s) of undifferentiated myoblasts (Johnston, 1999). As fibers expand, they absorb nuclei in order to maintain a relatively constant nuclear to cytoplasmic ratio (Koumans *et al.,* 1994). New fibers form on the suface of existing fibers by fusing myoblasts into multinucleated myotubes. The decision of a myoblast to hold in the cell cycle or to differentiate is determined by antagonistic signals for proliferation and differentiation that are incompletely understood in mammals and hardly studied at all in fish. Members of the MyoD gene family are considered to activate the muscle differentiation program and inhibit proliferation of the myoblast producer cells. However, these transcription factors are components of a highly redundant and poorly understood regulatory system for modulating muscle growth. It is urgent to reveal the role of these factors in fish myogenesis. Differences in regulation of myogenesis such as myosin isoform expression in the earliest stages and during temperature acclimation in the adult stages are to be seen.

ACKNOWLEDGMENTS

The present study on carp was supported by Grant-in-Aids for Scientific Research from the Ministry of Education, Science, Sports, and Culture of Japan, and from the Asahi Glass Foundation, Japan. The author is indebted to Dr. Ian A. Johnston for invaluable discussion. Thanks are also due to A. Kobiyama and Y. Nihei for their excellent assistance in preparing the manuscript.

REFERENCES

Alami-Durante, H., Fauconneau, B., Rouel, M., Escaffre, A. M., and Bergot, P. (1997). Growth and multiplication of white skeletal muscle fibres in carp larvae in relation to somatic growth rate. *J. Fish Biol.* **50,** 1285–1320.

Andres, V., and Walsh, K. (1996). Myogenin expression, cell cycle withdrawal, and phenotypic differentiation are temporally separable events that precede cell fusion upon myogenesis. *J. Cell Biol.* **132,** 657–666.

Benezra, R., Davis, R. L., Lockshon, D., Turner, D. L., and Weintraub, H. (1990). The protein Id: a negative regulator of helix-loop-helix DNA binding proteins. *Cell* **61,** 49–59.

Black, B. L., Molkentin, J. D., and Olson, E. N. (1998). Multiple roles for the MyoD basic region in transmission of transcriptional activation signals and interaction with MEF2. *Mol. Cell Biol.* **18,** 69–77.

Braun, T., Bober, E., Winter, B., Rosenyhal, N., and Arnold, H. H. (1990). Myf-6, a new member of the human gene family of myogenic determination factors: evidence for a gene cluster on chromosome 12. *EMBO J.* **9,** 821–831.

Buonanno, A., Apone, L., Morasso, M. I., Beers, R., Brenner, H. R., and Eftimie, R. (1992). The

MyoD family of myogenic factors is regulated by electrical activity: isolation and characterization of a mouse Myf-5 cDNA. *Nucleic Acids Res.* **20,** 539–544.
Church, G. M., Ephrussi, A., Gilbert, W., and Tonegawa, S. (1985). Cell-type-specific contacts to immunoglobulin enhancers in nuclei. *Nature* **313,** 798–801.
Cornelison, D. D. W., and Wold, B. J. (1997). Single-cell analysis of regulatory gene expression in quiescent and activated mouse skeletal muscle satellite cells. *Dev. Biol.* **191,** 270–283.
Davis, R. L., Cheng, P.-F., Lassar, A. B., and Weintraub, H. (1990). The MyoD DNA binding domain contains a recognition code for muscle-specific gene activation. *Cell* **60,** 733–746.
Davis, R. L., and Weintraub, H. (1992). Acquisition of myogenic specificity by replacement of three amino acid residues from MyoD into E12. *Science* **256,** 1027–1030.
Dawid, I. B., Breen, J. J., and Toyama, R. (1998). LIM domains: multiple roles as adapers and functional modifiers in protein interactions. *Trends Genet.* **14,** 156–161.
Delalande, J. M., and Rescan, P. Y. (1998) Expression of a cysteine-rich protein (CRP) encoding gene during early development of the trout. *Mech. Dev.* **76,** 179–183.
Edmondson, D. G., and Olson, E. N. (1989). A gene with homology to the *myc* similarity region of MyoD1 is expressed during myogenesis and is sufficient to activate the muscle differentiation program. *Genes Dev.* **3,** 628–640.
Edmondson, D. G., Lyons, G. E., Martin, J. F., and Olson, E. N. (1994). *Mef2* gene expression marks the cardiac and skeletal muscle lineages during mouse embryogenesis. *Development* **120,** 1251–1263.
Ellenberger, T., Fass, D., Arnaud, M., and Harrison, S. C. (1994). Crystal structure of transcription factor E47: E-box recognition by a basic region helix-loop-helix dimer. *Genes Dev.* **8,** 970–980.
Ennion, S., Gauvry, L., Butterworth, P., and Goldspink, G. (1995). Small-diameter white myotomal muscle fibres associated with growth hyperplasia in the carp (*Cyprinus carpio*) express a distinct myosin heavy chain gene. *J. Exp. Biol.* **198,** 1603–1611.
Fujisawa-Sehara, A., Nabeshima, Y., Hosoda, Y., Obinata, T., and Nabeshima, Y. (1990). Myogenin contains two domains conserved among myogenic factors. *J. Biol. Chem.* **265,** 15219–15223.
Gauvry, L., Ennion, S., Hansen, E., Butterworth, P., and Goldspink, G. (1996). The characterisation of the 5′ regulatory region of a temperature-induced myosin-heavy–chain gene associated with myotomal muscle growth in the carp. *Eur. J. Biochem.* **236,** 887–894.
Gossett, L. A., Kelvin, D. J., Sternberg, E. A., and Olson, E. N. (1989). A new myocyte-specific enhancer-binding factor that recognizes a conserved element associated with multiple muscle-specific genes. *Mol. Cell Biol.* **9,** 5022–5033.
Guo, X.-G., Nakaya, M., and Watabe, S. (1994). Myosin subfragment-1 isoforms having different heavy chain structures from fast skeletal muscle of thermally acclimated carp. *J. Biochem.* **116,** 728–735.
Hasty, P., Bradley, A., Morris, J. H., Edmondson, D. G., Venuti, J. M., Olson, E. N., and Klein, W. H. (1993). Muscle deficiency and neonatal death in mice with a targeted mutation in *myogenin* gene. *Nature* **364,** 501–506.
Heap, S. P., Watt, P. W., and Goldspink, G. (1985) Consequences of thermal change on the myofibrillar ATPase of five freshwater teleosts. *J. Fish Biol.* **26,** 733–738.
Heap, S. P., Watt, P. W., and Goldspink, G. (1986). Alterations to the swimming performance of carp, *Cyprinus carpio,* as a result of temperature acclimation. *J. Exp. Biol.* **123,** 373–382.
Hinterberger, T. J., Sasoon, D. A., Rhodes, S. J., and Konieczny, S. F. (1991). Expression of the mouse regulatory factor MRF4 during somite and skeletal myofiber development. *Dev. Biol.* **147,** 144–156.
Hirayama, Y., and Watabe, S. (1997). Structural differences in the crossbridge head of temperature-associated myosin subfragment-1 isoforms from carp fast skeletal muscle. *Eur. J. Biochem.* **246,** 380–387.
Hopwood, N. D., Pluck, A., and Gurdon, J. B. (1989). MyoD expression in the forming somites is an early response to mesoderm induction in *Xenopus* embryos. *EMBO J.* **8,** 3409–3417.

Hopwood, N. D., Pluck, A., and Gurdon, J. B. (1991). *Xenopus* Myf-5 marks early muscle cells and can activate muscle genes ectopically in early embryos. *Development* **111,** 551–560.
Hwang, G.-C., Watabe, S., and Hashimoto, K. (1990). Changes in carp myosin ATPase activity induced by temperature acclimation. *J. Comp. Physiol. B* **160,** 233–239.
Hwang, G.-C., Ochiai, Y., Watabe, S., and Hashimoto, K. (1991). Changes of carp myosin subfragment-1 induced by temperature acclimation. *J. Comp. Physiol. B* **161,** 141–146.
Imai, J., Hirayama, Y., Kikuchi, K., Kakinuma, M., and Watabe, S. (1997). cDNA cloning of myosin heavy chain isoforms from carp fast skeletal muscle and their gene expression associated with temperature acclimation. *J. Exp. Biol.* **200,** 27–34.
Johnston, I. A. (1999) Muscle development and growth: potential implications for flesh quality in fish. *Aquaculture* **177,** 99–115.
Johnston, I. A., Davison, W., and Goldspink, G. (1975). Adaptations in Mg^{2+}-activated myofibrillar ATPase activity induced by temperature acclimation. *FEBS Lett.* **50,** 293–295.
Johnston, I. A., Cole, N. J., Abercomby, M., and Vieira, V. L. A. (1998). Embryonic temperature modulates muscle growth characteristics in larval and juvenile herring. *J. Exp. Biol.* **201,** 623–646.
Kakinuma, M., Nakaya, M., Hatanaka, A., Hirayama, Y., Watabe, S., Maeda, K., Ooi, T., and Suzuki, S. (1998). Thermal unfolding of three acclimation temperature-associated isoforms of carp light meromyosin expressed by recombinant DNAs. *Biochemistry* **37,** 6606–6613.
Kobiyama, A., Nihei, Y., Hirayama, Y., Kikuchi, Y., Suetake, H., Johnston, I. A., and Watabe, S. (1998). Molecular cloning and developmental expression patterns of the MyoD and MEF2 families of muscle transcription factors in the carp. *J. Exp. Biol.* **201,** 2801–2813.
Kobiyama, A., Nihei, Y., Hirayama, Y., Kikuchi, K., and Watabe, S. (2000). Expression changes of transcripts for fast skeletal myosin heavy chain isoforms in relation to those of myoD and MEF2 families during warm temperature acclimation of carp. *Fish. Sci.* **66,** (in press).
Kong, Y., Flick, M. J., Judla, A. J., and Konieczny, S. F. (1997). Muscle LIM protein promotes myogenesis by enhancing the activity of MoyD. *Mol. Cell. Biol.* **17,** 4750-4760.
Koumans, J. T. M., Akster, H. A., Booms, G. H. R., and Osse, J. W. M. (1993). Growth of carp (*Cyprinus carpio*) white axial muscle; hyperplasia and hypertrophy in relation to the myonucleus/sarcoplasm ratio and the occurrence of different subclasses of myogenic cells. *J. Fish Biol.* **43,** 69–80.
Koumans, J. T. M., Akster, H. A., Witkam, A., and Osse, J. W. M. (1994) Numbers of muscle nuclei and myosatellite cell nuclei in red and white axial muscle during growth of the carp (*Cyprinus carpio*). *J. Fish Biol.* **44,** 391–408.
Lassar, A. B., Buskin, J. N., Lockshon, D., Davis, R. L., Apone, S., Hauschka, S. D., and Weintraub, H. (1989). MyoD is a sequence-specific DNA binding protein requiring a region of *myc* homology to bind to the muscle creatine kinase enhancer. *Cell* **58,** 823–831.
Lassar, A. B., Davis, R. L., Wright, W. E., Kadesch, T., Murre, C., Voronova, A., Baltimore, D., and Weintraub, H. (1991). Functional activity of myogenic HLH proteins requires heterooligomerization with E12/E47-like proteins *in vivo*. *Cell* **66,** 305–315.
Lin, H., Yutzey, K. E., and Konieczny, S. F. (1991). Muscle-specific expression of the troponin I gene requires interactions between helix-loop-helix muscle regulatory factors and ubiquitous transcription factors. *Mol. Cell. Biol.* **11,** 267–280.
Lin, Z. Y., Dechesne, C. A., Eldridge, J., and Paterson, B. M. (1989). An avian muscle factor related to MyoD1 activates muscle-specific promoters in nonmuscle cells of different germ-layer origin and in BrdU-treated myoblasts. *Genes Dev.* **3,** 986–996.
Ludolph, D. C., and Konieczny, S. F. (1995). Transcription factor families: muscling on the myogenic program. *FASEB J.* **9,** 1595–1604.
Ma, P.-C., Rould, M. A., Weintraub, H., and Pabo, C. O. (1994). Crystal structure of MyoD bHLH

domain-DNA complex: perspectives on DNA recognition and implications for transcriptional activation. *Cell* **77,** 451–459.

Martin, J. F., Schwarz, J. J., and Olson, E. N. (1993). Myocyte enhancer factor (MEF) 2C: tissue-restricted member of the MEF-2 family of transcription factors. *Proc. Natl. Acad. Sci. U.S.A.* **90,** 5282–5286.

McDermott, J. C., Cardoso, M. C., Yu, Y.-T., Anderes, V., Leifer, D., Krainc, D., Lipton, S. A., and Nadal-Ginard, B. (1993). hMEF2C gene encodes skeletal muscle- and brain-specific transcription factors. *Mol. Cell. Biol.* **13,** 2564–2577.

Megeney, L. A., Kablar, B., Garrett, K., Anderson, J. E., and Rudnicki, M. A. (1996). MyoD is required for myogenic stem cell function in adult skeletal muscle. *Genes Dev.* **10,** 1173–1183.

Molkentin, J. D., and Olson, E. C. (1996). Combinational control of muscle development by basic helix-loop-helix and MADS-box transcription factors. *Proc. Natl. Acad. Sci. U.S.A.* **93,** 9366–9373.

Moss, F. P., and Leblond, C. P. (1971). Satellite cells as the source of nuclei in muscles of growing rats. *Anat. Rec.* **170,** 421–436.

Murre, C., McCau, P. S., and Baltimore, D. (1989). A new DNA binding and dimerization motif in immunoglobulin enhancer binding, daughterless, MyoD, and myc proteins. *Cell* **56,** 777–783.

Murre, C., Bain, G., Van Dijk, M. A., Engel, I., Furnari, B. A., Massari, M. E., Matthews, J. R., Quong, M. W., Rivera, R. R., and Stuiver, M. H. (1994). Structure and function of helix-loop-helix proteins. *Biochim. Biophys. Acta* **1218,** 129–135.

Nakaya, M., Watabe, S., and Ooi, T. (1995). Differences in the thermal stability of acclimation temperature-associated types of carp myosin and its rod on differential scanning calorimetry. *Biochemistry* **34,** 3114–3120.

Nakaya, M., Kakinuma, M., Watabe, S., and Ooi, T. (1997). Differential scanning calorimetry and CD spectrometry of acclimation-temperature-associated types of carp light meromyosin. *Biochemistry* **36,** 9179–9184.

Nihei, Y., Kobiyama, A., Hirayama, Y., Kikuchi, K., and Watabe, S. (1999). mRNA expression patterns of a transcription factor E12 in various developmental stages and adult tissues of carp. *Fish. Sci.* **64,** 600–605.

Nourse, J., Mellentin, J. D., Galili, N., Wilkinson, J., Stanbridge, E., Smith, S. D., and Cleary, M. L. (1990) Chromosomal translocation t (1;19) results in synthesis of a homeobox fusion mRNA that codes for a potential chimeric transcription factor. *Cell* **60,** 535–545.

Olson, E. N. (1992). Interplay between proliferation and differentiation within the myogenic lineage. *Dev. Biol.* **154,** 261–272.

Olson, E. N., and Klein, W. H. (1994). BHLH factors in muscle development: dead lines and commitments, what to leave in and what to leave out. *Genes Dev.* **8,** 1–8.

Ott, M.-O., Bober, E., Lyons, G., Arnold, H., and Buckingham, M. (1991). Early expression of the myogenic regulatory gene, *myf5,* in precursor cells of skeletal muscle in the mouse embryo. *Development* **111,** 1097–1107.

Pellegrini, L., Tan, S., and Richmond, T. J. (1995). Structure of serum response factor core bound to DNA. *Nature* **376,** 490–498.

Piette, J. Bessereau, J.-L., Huchet, M., and Changeux, J.-P. (1990). Two adjacent MyoD1-binding sites regulate expression of the acetylcholine receptor α-subunit. *Nature* **345,** 353–355.

Pinney, D. F., Pearson-White, S. H., Konieczny, S. F., Latham, K. E., and Emerson, C. P. (1988). Myogenic lineage determination and differentiation: evidence for a regulatory gene pathway. *Cell* **53,** 781–793.

Pownall, M. E., and Emerson, C. P. (1992). Sequential activation of three myogenic regulatory genes during somite morphorgenesis in quail embryos. *Dev. Biol.* **151,** 67–79.

Rashbass, J., Taylor, M. V., and Gurdon J. B. (1992). The DNA-binding protein E12 co-operates with

XMyoD in the activation of muscle-specific gene expression in *Xenopus* embryos. *EMBO J.* **11**, 2981–2990.

Rescan, P.-Y. (1997). Identification in a fish species of two Id (inhibitor of DNA binding/differentiation)-related helix-loop-helix factors expressed in the slow oxidative muscle fibers. *Eur. J. Biochem.* **247**, 870–876.

Rescan, P.-Y., and Gauvry, L. (1996). Genome of the rainbow trout (*Oncorhynchus mykiss*) encodes two distinct muscle regulatory factors with homology to MyoD. *Comp. Biochem. Physiol.* **113B**, 711–715.

Rescan, P.-Y., Gauvry, L., Paboeuf, G., and Fauconneau, B. (1994). Identification of a muscle factor related to MyoD in fish species. *Biochim. Biophys. Acta* **1218**, 202–204.

Rescan, P.-Y., Gauvry, L., and Paboeuf, G. (1995). A gene of homology to myogenin is expressed in developing myotomal musculature of the rainbow trout and *in vitro* during the conversion of myosatellite cells to myotubes. *FEBS Lett.* **362**, 89–92.

Rhodes, S. J., and Konieczny, S. F. (1989). Identification of MRF4: a new member of the muscle regulatory factor gene family. *Genes Dev.* **3**, 2050–2061.

Roberts, V. J., Steenbergen, R., and Murre, C. (1993) Localization of E2A mRNA expression in developing and adult rat tissues. *Proc. Natl. Acad. Sci. U.S.A.* **90**, 7583–7587.

Rowe, R. W. D., and Goldspink, G. (1969). Muscle fibre growth in five different muscles in both sexes of mice. *J. Anat.* **104**, 519–530.

Rudnicki, M. A., Braun, T., Hinuma, S., and Jaenisch, R. (1992). Inactivation of *MyoD* in mice leads to upregulation of the myogenic HLH gene *Myf-5* and results in apparently normal muscle development. *Cell* **71**, 383–390.

Saitoh, O., Fujisawa-Sehara, A., Nabeshima, Y., and Periasamy, M. (1993). Expression of myogenic factors in denervated chicken breast muscle: isolation of the chicken myf5 gene. *Nucleic Acids Res.* **21**, 2503–2509.

Sartorelli, V., Webster, K. A., and Kedes, L. (1990). Muscle-specific expression of the cardiac α-actin gene requires MyoD1, CArG-box binding factor, and Sp-1. *Genes Dev.* **4**, 1811–1822.

Sassoon, D., Lyons, G., Woodring, G. L., Wright, W. E., Lin, V., Lassar, A., Weintraub, H., and Buckingham, M. (1989). Expression of two myogenic regulatory factors myogenin and MyoD1 during mouse embryogenesis. *Nature* **341**, 303–307.

Sidell, B. D. (1980). Response of goldfish (*Carassius auratus* L.) to temperature acclimation: alterations in biochemistry and proportions of different fibre types. *Physiol. Zool.* **53**, 98–107.

Smith, T. H., Kachinsky, A. M., and Miller, J. B. (1994). Somite subdomains, muscle cell origins, and the four muscle regulatory factor proteins. *J. Cell Biol.* **127**, 95–105.

Stickland, N. C. (1983). Growth and development of muscle fibres in the rainbow trout (*Salmo gairdneri*). *J. Anat.* **137**, 323–333.

Sun, X.-H., and Baltimore, D. (1991). An inhibitory domain of E12 transcription factor prevents DNA binding in E12 homodimers but not in E12 heterodimes. *Cell* **64**, 459–470.

Takahashi, S., Esumi, E., Nabeshima, Y., and Asashima, M. (1998) Regulation of the *Xmyf-5* and *XmyoD* expression pattern during early *Xenopus* development. *Zool. Sci.* **15**, 231–238.

Ticho, B. S., Stainier, D. Y. R., Fishman, M. C., and Breitbart, R. E. (1996). Three zebrafish MEF2 genes delineate somitic and cardiac muscle development in wild-type and mutant embryos. *Mech. Dev.* **59**, 205–218.

Watabe, S. (1999). Myogenic regulatory factors and muscle differentiation during ontogeny in fish. *J. Fish Biol.* **55 (Suppl. A)**, 1–18.

Watabe, S., Hwang, G.-C., Nakaya, M., Guo, X.-F., and Okamoto, Y. (1992) Fast skeletal myosin isoforms in thermally acclimated carp. *J. Biochem.* **111**, 113–122.

Watabe, S., Guo, X.-F., and Hwang, G.-C. (1994). Carp express specific isoforms of the myosin cross-bridge head, subfragment-1, in association with cold and warm temperature acclimation. *J. Therm. Biol.* **19**, 261–268.

Watabe, S., Imai, J., Nakaya, M., Hirayama, Y., Okamoto, Y., Masaki, H., Uozumi, T., Hirono, I., and Aoki, T. (1995). Temperature acclimation induces light meromyosin isoforms with different primary structures in carp fast skeletal muscle. *Biochem. Biophys. Res. Commun.* **208,** 118–125.

Watabe, S., Hirayama, Y., Nakaya, M., Kakinuma, M., Kikuchi, K., Guo, X.-F., Kanoh, S., Chaen, S., and Ooi, T. (1998). Carp expresses fast skeletal myosin isoforms with altered motor functions and structural stabilities to compensate for changes in environmental temperature. *J. Therm. Biol.* **22,** 375–390.

Weatherley, A. H., Gill, H. S., and Lobo, A. F. (1988). Recruitment and maximal diameter of axial muscle fibres in teleosts and their relationship to somatic growth and ultimate size. *J. Fish Biol.* **33,** 851–859.

Weinberg, E. S., Allende, M. L., Kelly, C. S., Abdelhamid, A., Murakami, T., Andermann, P., Doerre, O. G., Grunwald, D. J., and Riggleman, B. (1996). Developmental regulation of zebrafish *MyoD* in wild-type, *no tail* and *spadetail* embryos. *Development* **122,** 271–280.

Weintraub, H. (1993). The MyoD family and myogenesis: redundancy, networks, and thresholds. *Cell* **75,** 1241–1244.

Weintraub, H., Davis, R., Tapscott, S., Thayer, M., Krause, M., Benezra, R., Blackwell, T. K., Turner, D., Rupp, R., Hollenberg, S., Zhuang, Y., and Lasser, A. (1991). The myoD gene family: nodal point during specification of the muscle cell lineage. *Science* **251,** 761–766.

Weiskirchen, R., Pino, J. D., Macalma, T., Bister, K., and Beckerle, M. C. (1995). The cysteine-rich protein family of highly related LIM domain proteins. *J. Biol. Chem.* **270,** 28946–28950.

Wentworth, B. M., Donoghue, M., Engert, J. C., Berglund, E. B., and Rosenthal, N. (1991). Paired MyoD-binding sites regulate myosin light chain gene expression. *Proc. Natl. Acad. Sci. U.S.A.* **88,** 1242–1246.

Wulbeck, C., Fromental-Ramain, C., and Campos-Orgtega, J. A. (1994) The HLH domain of a zebrafish HE12 homologue can partially substitute for functions of the HLH domain of *Drosophila* DAUGHTERLESS. *Mech. Dev.* **46,** 73–85.

Yablonka-Reuveni, Z., and Rivera, A. J. (1994). Temporal expression of regulatory and structural muscle proteins during myogenesis of satellite cells on isolated adult rat fibres. *Dev. Biol.* **164,** 588–603.

Yoon, J. K., Olson, E. N., Arnold, H.-H., and Wold, B. J. (1997). Different *MRF4* knockout alleles differentially distrupt Myf-5 expression: *cis*-regulatory interactions at the *MRF4/Myf-5* locus. *Dev. Biol.* **188,** 349–362.

Yu, Y.- T., Breitbart, R. E., Smoot, L. B., Lee, Y., Mahdavi, V., and Nadal-Ginard, B. (1992). Human myocyte-specific enhancer factor 2 comprises a group of tissue-restricted MADS box transcription factors. *Genes Dev.* **6,** 1783–1798.

Yun, K., and Wold, B. (1996). Skeletal muscle determination and differentiation: story of a core regulatory network and its context. *Curr. Opin. Cell Biol.* **8,** 877–889.

3

MYOSIN EXPRESSION DURING ONTOGENY, POST-HATCHING GROWTH, AND ADAPTATION

GEOFFREY GOLDSPINK
DAVID WILKES
STEVEN ENNION

I. Abstract
II. Introduction
III. The Myosin hc Genes
IV. Expression of Myosin Isogenes During Early Development and Muscle Regeneration
V. A Myosin Gene that is Associated with Adult Muscle Hyperplasia in Carp
VI. Molecular Signals Involved in Muscle Gene Expression During Pre- and Post-Hatching Growth
VII. Expression of Myosin Isogenes in Response to Temperature
VIII. Regions Determining Functional Diversity in Myosin hc Isoforms
    A. Regulatory Elements of the Myosin hc Genes
IX. Conclusions
    Acknowledgments
    References

I. ABSTRACT

Muscle is the most abundant tissue and myosin is the most abundant protein in this tissue. It is also the protein that generates the force for contraction and exists in different isoforms. The myosin heavy chains (hc) represent a range of different molecular motors which are encoded in a family of different genes. These endow different contractile properties on the swimming and other muscles and because fish show such a wide diversity in modes of life and habitats, the adaptation of the molecular motors is particularly interesting. Velocity, efficiency, and economy of contraction requirements alter during ontogeny and growth as well as for survival at different temperatures. As in mammals there is sequential gene expression with two different genes expressed *in ovo*. The structure of the myosin genes is reviewed with regard to the regulatory regions as well as the coding regions that differ between isogenes. The present day knowledge of the signals involved in gene switching is also discussed. Fish muscle development is contrasted with that in mammals in which the origins of the fast and slow fibers is somewhat different. Also during piscine muscle development hyperplasia continues throughout life and a myosin that is expressed only in the putative newly formed fibers has been cloned. Temperature has a marked effect on the rate and type of muscle that is produced. However, some fish species also have the ability to rebuild their myofibrils for low temperature swimming by expressing different myosins that enable them to produce more power at these low temperatures.

II. INTRODUCTION

In order to fully understanding the regulation of growth at the molecular level, it is necessary to understand the control processes for the expression of genes coding for structural proteins. Myosin is an obvious choice for study since it is the most abundant protein of muscle, which is the most abundant tissue in the organism as well as being the tissue of commercial importance in terms of food production. The myosins are not only structural proteins, but they are also the molecular motors which produce the force for muscular contraction. Locomotion and the maintenance of vital processes such as feeding and respiration are dependent on the appropriate design of the muscle including the type of myosin molecular motor expressed.

The superfamily of myosin motor proteins is ubiquitous in all eukaryotic cell types and is involved in a diverse range of cell and intercellular movements including cytokinesis, phagocytosis, cell division, vesicle transport, and muscle contraction. Twelve distinct classes of myosin have been described to date (for review see Sellers *et al.,* 1996) of which only class II are involved in smooth, cardiac, and skeletal muscle contraction. The latter will be the subject of this review.

Myosin II is a hexamer consisting of two "heavy" protein chains (approximately 220 kDa each) and four "light" chains (approximately 17–20 kDa each; Weeds and Lowey, 1971). The two heavy chains are intertwined at their carboxyl regions forming an α-helical coiled-coil, or "rod region" which anchors the molecule to the thick filament of the sarcomere. The N terminal region of each heavy chain consists of a globular head where force is generated by the "crossbridge" to pull the thin filaments in over the thick filaments. This globular head contains an actin-binding domain, the sites where the myosin light chains interact and is the region where the ATPase activity of the molecule is also located.

In vertebrate muscle, the myosin hc genes which encode the molecular motors exist as a family of individual genes (Weiss and Leinwand, 1996). These encode different isoforms which in mammals include embryonic, neonatal, extraocular, cardiac (α and β), and adult fast (2A, 2B, and 2X) isoforms. The slow type I muscle fibers express a gene which is also expressed in cardiac muscle (the β cardiac). Mammalian skeletal muscle fibers are distinguished mainly by the type of myosin hc they express. Therefore the myosin hcs are the main marker when studying the phenotypic determination of fibers and whole muscles during growth and adaptation.

Different fiber types also exist in fish muscle, but the arrangement is somewhat different to mammalian muscle where the different fibers tend to be intermingled to form a mosaic. The bulk of the myotomal muscle in most fish consists of fast contracting white glycolytic fibers which are utilized mainly for burst speed swimming; for example, to escape predation or to capture prey. For slow cruise swimming the fish utilizes slow contracting oxidative (red) muscle fibers which are located in a thin V-shaped wedge running along the lateral line just underneath the skin. Bordering these main fiber types there is usually a layer of intermediate "pink" fibers.

EMG studies (Johnston *et al.,* 1997; Rome *et al.,* 1984; 1988) have shown that the red muscle is activated at slow swimming speed, the pink muscle becomes recruited at the faster cruising speed, and the white musculature is activated only at the fastest swimming speeds. Thus there is a sort of three-geared system which enables the fish to swim efficiently at different speeds, although the high burst speed cannot be maintained for long periods at a time as ATP cannot be supplied at the rate at which the fast white muscle type molecular motors use the energy.

Fish that have evolved to live at different temperatures have contractile systems adapted to function effectively within that temperature range. For this reason we have studied the way in which the myosin hc gene, particularly the region that codes for the ATPase site of the molecular motor, differs between Antarctic and tropical fish. Earlier work (Johnston *et al.,* 1973; Johnston *et al.,* 1975) also showed that some fish can adapt to different environmental temperatures on a seasonal basis by rebuilding their myofibrillar system for warm temperature or cold temperature swimming. Using isolated myofibrils from species that live in habitats ranging from the Antarctic at around 0°C to equatorial hot springs at 40°C, it was found that at a given measurement temperature the specific ATPase activity was much higher in myofibrils from the Antarctic fish than temperate or tropical fish. The thermostability of the ATPase was, however, much lower in the Antarctic fish. Thus during evolution there seems to have been a "trade off" between the higher ATPase activity which enables the Antarctic fish to move more efficiently at the lower temperatures and the thermostability of the ATPase site.

The question then arose as to what happens in pond fish such as the carp which may experience temperatures of near 0°C in the winter to about 30°C in the summer. Interestingly, we (Johnston *et al.,* 1975) found that the specific ATPase activity of the myofibrils from carp muscle changes during acclimation to warm and cold temperatures. In carp where the water temperature is gradually lowered from 18 to 8°C, the myofibrillar ATPase more than doubles within a month. The thermostability of the system also significantly decreased. The opposite was true for carp acclimated to higher environmental temperatures. These changes in the contractile system might have been explained by post-translational modification of the myosin protein. However, more recent research by our group and Professor Shugo Watabe's group in Japan have shown that this type of seasonal adaptation is due to the expression of different myosin isoform genes; in other words the molecular motors are changed. Several other species of cyprinid fish have also been shown to have acquired the same method of reducing the effects of lower environmental temperature (Heap et al., 1985), which in the carp is reflected in the increased power output from the red muscle at 8°C (Rome et al., 1984).

Our studies have shown that there is even greater diversity in the myosin gene in fish than in mammals. In addition to studying the way the different isogenes are regulated by environmental temperature, we have also been interested in their expression during growth. As the zebrafish has become one of the main models for studying molecular genetics of development, other information is emerging such as the role of myogenic and growth factors in determining muscle development in fish. We are now beginning to understand something about the signal molecules that bind to specific membrane and nuclear receptors to mediate hormonal influences and the way anoxia, pollutants, and noxious chemicals interfere with the normal processes can be best understood using molecular biology methods. Therefore the study can then be extended in several different ways once the basis mechanisms of the regulation of genes that encode major structural proteins, such as the myosin hcs, have been elucidated.

III. THE MYOSIN hc GENES

In the vertebrates the myosin hc isoforms which form different molecular motors are encoded by separate genes that form a family (Weiss and Leinwand, 1996; Schiaffino and Reggiani, 1996). In man, the embryonic, neonatal, and fast type II genes are on chromosome 17 and the slow type I β cardiac gene is in tandem with the α cardiac myosin gene on chromosome 14. (Leinwand et al., 1983a,b). There are also two smooth muscle myosin hc genes on chromosome 16 (Deng et al., 1993). In all there are at least 10 different class II myosin hc genes expressed in mammalian muscle. In tetraploid fish species the number of myosin hc genes seems to be greater than in mammalian species. In carp (Cyprinus carpio) we found evidence at the genomic level for at least 28 different myosin hc genes (Gerlach et al., 1990). The chicken also seems to have a larger number of myosin hc genes than mammalian species with at least 31 isogenes (Robbins et al., 1986).

Due to their large size, relatively few vertebrate myosin hc II genes have been sequenced completely. Those vertebrate myosin hc genes that have are very similar at the RNA level, where they are approximately 6000 nucleotides, and show a high degree of organizational and sequence homology. Both the human β cardiac (Jaenicke *et al.*, 1990) and the chicken embryonic (Molina *et al.*, 1987) myosin hc genes have a total of 40 exons while the rat embryonic myosin hc gene (Strehler *et al.*, 1986) has 41 exons. However, the exon positions differs slightly. The equivalent of exon 37 in the human β cardiac gene is split into two exons in the chicken and rat embryonic genes such that the position of the additional intron is identical in both of the latter two genes. Also, the intron separating the final two exons of the human β cardiac (exons 39 and 40) and the rat embryonic (exons 40 and 41) genes is absent in the chicken embryonic gene. Genomic clone analysis of the carp myosin hc gene which has been called FG2 (Ennion *et al.*, 1995, Gauvry *et al.*, 1996a) has demonstrated that while the exon/intron boundaries are very similar to the mammalian and chicken myosin hc genes, the gene itself is approximately half the size. The gene transcribes to an mRNA transcript of approximately 6000 nucleotides and the difference in size at the genomic level is solely due to shorter introns. Like the human β cardiac and rat embryonic genes, the carp FG2 gene has the final two exons (40 and 41) separated by an intron.

The determination of the deduced amino acid sequences of complete myosin hc proteins from genomic and cDNA sequences has allowed the various subdomains of the protein (reviewed by Lowey, 1986) to be assigned to specific regions within the protein sequence (Fig. 1). The globular S1 domain, which accounts for approximately 40% of the protein, begins at the first amino acid and ends at the

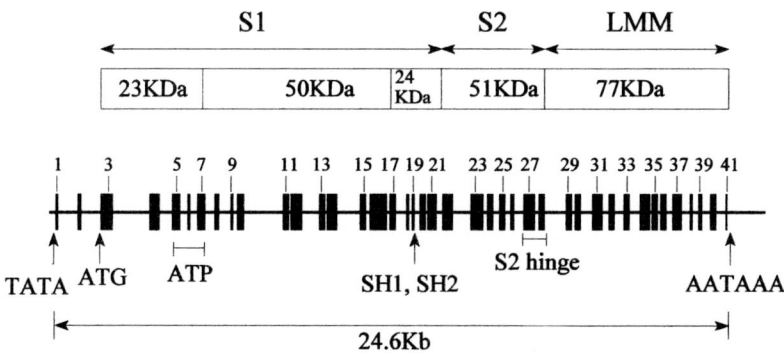

Fig. 1. Structure of the rat embryonic MyoHC gene. The 41 exons of the gene are numbered and represented as black boxes. Positions of the TATA box, ATG start site, and polyadenylation signal (AATAAA) are shown as arrows. The positions of the two reactive thiol groups (SH1 and SH2), the hinge region (S2 hinge), and the ATP binding site (ATP) are also shown. A schematic of the protein subdomains above the gene structure indicates the positions of various tryptic fragments. (Redrawn from Strehler *et al.*, 1986).

proline residue at position 838 in the rat embryonic myosin hc gene (Strehler *et al.*, 1986). The tryptic fragments of the S1 head of the rat embryonic myosin hc were also assigned to amino acid residues 1–206 (23-kDa fragment), 207–633 (50-kDa fragment), and 634–838 (24-kDa fragment). The S2 region is encoded by approximately 7 exons and covers positions 839–1279 and the rod (LMM) region covers positions 1280–1939.

Extensive sequence homology exists between myosin hc genes. This homology is present between species as diverse as nematode and rat and is especially noticeable between different isoforms within the same species. The sequence of the globular head region (S1) of the molecule is more highly conserved between species than the rod region (Karn *et al.*, 1983; Strehler *et al.*, 1986). The rod regions of both muscle and non-muscle myosin hc have a repetitive sequence organization characteristic of α helical coiled-coil proteins. All vertebrate myosin hc rod regions sequenced to date have two common features. They all contain 40 28-residue repeats, the first 39 of which are complete (termination of the rod region occurs within repeat 40; Stedman *et al.*, 1990), and they all contain an extra "skip" residue, the function of which is unknown, at the end of repeats 13, 20, 27, and 35.

IV. EXPRESSION OF MYOSIN ISOGENES DURING EARLY DEVELOPMENT AND MUSCLE REGENERATION

The embryonic formation of muscle results from the proliferation of mononucleated mesodermal cells which subsequently differentiate into mononucleated myoblasts. These fuse together forming multinucleated myotubes and begin to express muscle-specific structural and contractile proteins. In mammalian species, there are two major waves of muscle fiber formation with the first wave of fiber formation providing a framework of primary fibers along which subsequent myotubes longitudinally orientate themselves and fuse to form secondary fibers (Kelly and Zack, 1969). In fish species, this biphasic development of primary and secondary fibers is not present. The origins of different muscle fiber type are, however, both spatially and temporally segregated. The development of fast and slow fiber types in zebrafish axial cells has been mapped by labeling myogenic precursor cells with vital fluorescent dyes (Devoto *et al.*, 1996). Adaxial cells of the segmental plate were shown to migrate radially from adjacent to the notochord to the lateral surface of the myotome where they differentiate into slow muscle fibers. The fast muscle fibers of the myotomal muscle arise from lateral presomatic cells that do not contact the notochord. Subsequent to the formation of an inner white and superficial red zones of muscle, a second stage of muscle development occurs in the free swimming larvae when the adult red and pink fiber types form and a new phase of fiber hyperplasia begins in the white muscle zone. The newly formed small muscle fibers in these later stages of development have been shown to be immunohistochemically different from more mature fibers (Akster, 1983; Rowlerson et al., 1985).

A complex transition of myosin isoform expression occurs during the formation and development of muscle fibers and these have mainly been characterized in mammalian species. In the rat the initial primary fibers express high levels of the embryonic myosin hc and lower and variable levels of the neonatal and slow type I isoforms (Narusawa et al., 1987; Dhoot, 1986). At about day 16–17 these primary fibers differentiate into two distinct fiber types, one expressing the embryonic and slow type I isoforms and one expressing the embryonic and neonatal isoforms (Condon et al., 1990). The secondary fibers, however, only express the neonatal myosin hc isoform. Later in embryonic development, embryonic isoform expression decreases and shortly after birth the neonatal isoform also disappears. Concomitantly with the decrease in these developmental isoforms, the adult slow and fast myosin hc isoforms begin to be predominantly expressed as the muscle takes on the adult phenotype and some fibers cease to express the slow β cardiac isoform and begin to express the adult fast isoforms (Barbet et al., 1991). Chicken skeletal muscle development involves the expression of two distinct embryonic myosin hc isoforms (Hofmann et al., 1988) and *Xenopus laevis* has also been shown to express at least two embryonic isoforms named E3 and E19 (Radice and Malacinski, 1989). The E3 isoform is predominant in the fast larval fiber types, while the E19 isoform is predominant in the slow fibers. However, a significant proportion of slow and fast fibers co-express both isoforms (Radice, 1995).

The myosin hc isoform transitions occurring during embryonic development in mouse (Lyons et al., 1990a) and rabbit (McKoy et al., 1998) have been quantified at the mRNA level by *in situ* hybridization and RNase protection, respectively, and changes at the mRNA level are closely mirrored by changes in the corresponding proteins, suggesting that the control for isoform transitions occurs at the transcription level. Relatively little is known about developmental myosin hc isoforms expressed in fish, although muscle phenotype changes during post-hatching growth in the barbel have been demonstrated (Focant et al., 1992), the eel (Chanoine et al., 1992), and the herring (Johnston et al., 1997). Most studies have relied on protein separation by electrophoresis under dissociating conditions and the identification of individual myosin hc isoforms has been hampered by co-migration of closely related isoforms. Nevertheless, several studies at the protein electrophoretic level have indicated the existence of embryonic myosin isoforms in a variety of fish species. During ontogenesis in Arctic charr a minor protein band has been shown to be present only in charr embryos (Martinez et al., 1991) and in Atlantic salmon an isoform which is referred to as neonatal is expressed in the white muscle of parrs while another fast isoform emerges several months after smoltification (Martinez et al., 1993). Trout myotubes derived from primary cell cultures also exhibit a distinct myosin hc protein band which is not present in the adult (Gauvry et al., 1996b). However, at the gene level only one myosin hc containing clone could be isolated form cDNA libraries prepared from eyed embryos and subadult trout, and this gene was shown to be expressed throughout development (Gauvry and Fauconneau, 1996).

In order to investigate at the gene level myosin hc isoforms which are expressed during carp muscle development, we conducted RACE-PCR (rapid amplification of cDNA ends polymerase chain reaction) on cDNA prepared form unhatched carp embryos using an oligonucleotide from a conserved region of the gene. Two distinct clones for myosin hc were isolated (named Eggs22 and Eggs24). The coding sequence (37 amino acids) of these two clones were very similar and differed only having 3 amino acid substitutions and 1 amino acid addition, but had markedly different 3' untranslated regions. The latter strongly indicates they are derived from two distinct genes. The expression patterns of these two transcripts were determined over embryonic development by using specific 3' untranslated region probes in wholemount *in situ* and Northern hybridizations. The transcripts showed identical temporal patterns of expression with both commencing after 22 hr post-fertilization. This was coincident with the switch from exclusively β-actin to both β- and α-actin gene expression, and continued for two weeks after hatching (Ennion et al., 1999). Neither of these transcripts are detected in juvenile or adult carp (Fig. 2). Wholemount *in situ* hybrid-

Fig. 2. Northern hybridization of EGGS24 UTR probe with RNA from carp at different stages of development. Total RNA (30 µg) from carp was separated by electrophoresis, transferred to nylon membrane, and hybridized sequentially with the probes EGGS24 and FGA101 (carp actin Gerlach et al., 1990). Lanes contain RNA from the pooled muscle samples of a number of fish as indicated below. Panel A: The membrane was hybridized under high stringency conditions with the probe EGGS24 and the blot exposed to X-ray film (Fuji RX) at −70°C for one week. Panel B: Subsequently the membrane was stripped and reprobed under high stringency conditions with the carp actin probe FGA101 which recognizes both α- and β-actin (Gerlach et al., 1990). The blot was exposed to X-ray film (Fuji RX) at −70°C for 72 hr. Panel C: Ethidium bromide staining of the original RNA agarose gel. Lanes are as follows: (1) Red muscle from four carp acclimatized to 28°C for 5 weeks; (2) adult carp spleen; (3) white muscle from three carp acclimated to 10°C for 5 weeks; (4) white muscle from two carp acclimated to 10°C for 5 weeks; (5) white muscle from four carp acclimated to 28°C for 1 week; (6) blank lane no RNA; (7) white muscle from four carp acclimated to 28°C for 2 weeks; (8) white muscle from two carp acclimated to 28°C for 3 weeks; (9) white muscle from two carp acclimated to 28°C for 3 weeks; (10) white muscle from four carp acclimated to 28°C for 4 weeks; (11) white muscle from two carp acclimated to 28°C for 5 weeks; (12) white muscle from two carp acclimated to 28°C for 5 weeks; (13) white muscle from three juvenile (1-year-old) carp; (14) five whole carp fry, acclimatized to 28°C for 2 weeks; (15) five whole carp fry, acclimatized to 15°C for 2 weeks; and (16) unhatched carp eggs 20 hr after fertilization.

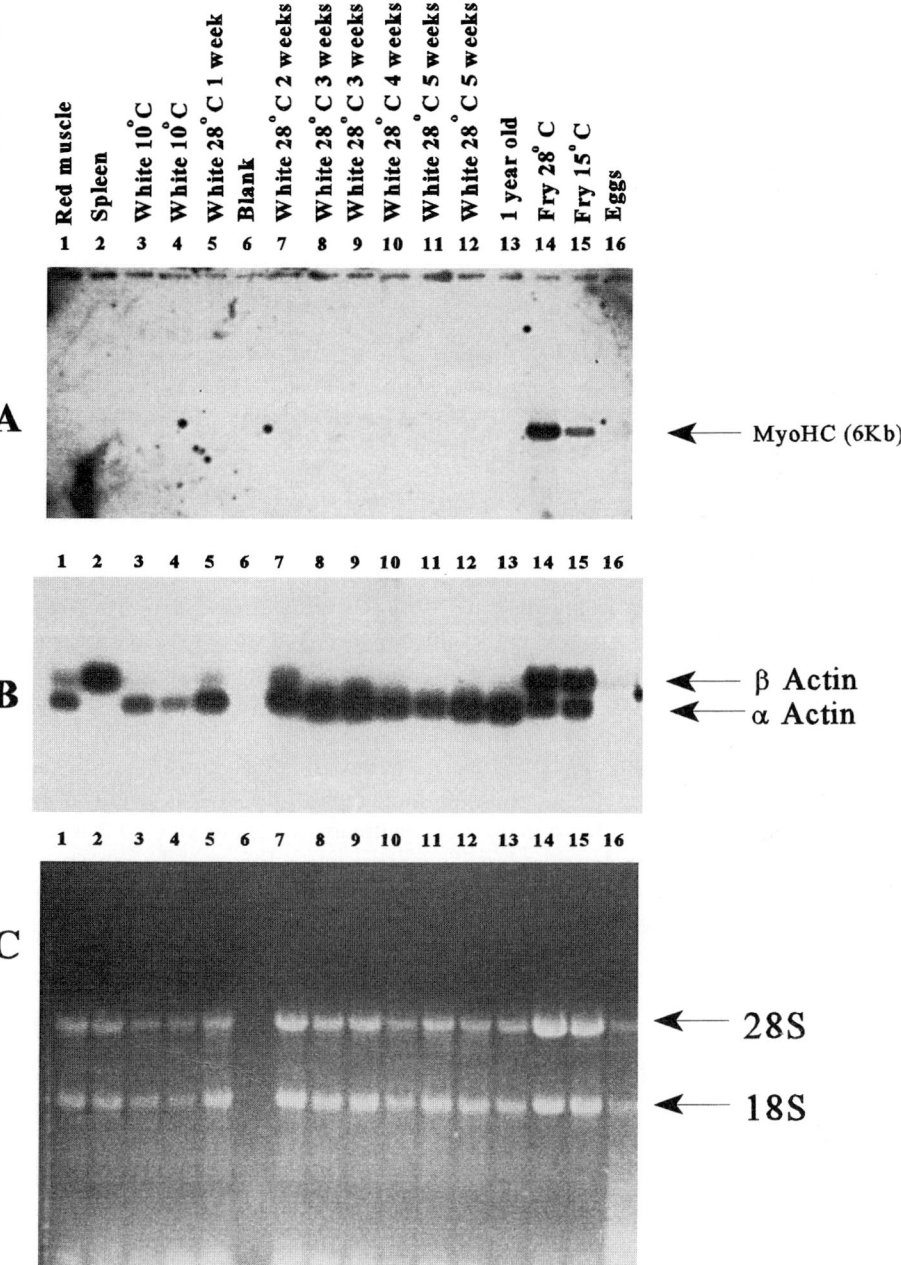

ization showed that both transcripts are expressed initially in the rostral region of the developing trunk and progress caudally. Both are expressed in the developing pectoral fin and protractor hyoideus muscles. However, the muscles of the lower jaw express the *Eggs22* transcript but not the *Eggs24* transcript. Interestingly, a distinct chevron pattern of expression was observed in the myotomal muscle and this is a result of the localization of mRNA to the myoseptal regions of the fibers; the sites of new sarcomere addition during muscle growth (Fig. 3, see color plate). This suggests transport of myosin hc mRNA transcripts to the sites where they will be required for protein synthesis or selective expression from nuclei near the ends of the fibers.

A recapitulation of the processes involved in myogenesis is triggered during muscle regeneration in response to injury. Again, mammalian species have been well studied in this respect with far fewer studies on fish species. However, those fish species which have been studied seem to follow the same general scheme of events. In response to muscle injury, satellite cells, which are remnants of embryonic myogenesis that have remained undifferentiated and "stored" between the basal lamina and the plasma membrane, proliferate and fuse to form new muscle fibers. These new fibers transiently express the developmental isoforms of myosin hc (d'Albis *et al.,* 1989; Dix and Eisenberg, 1991; Sartore *et al.,* 1982) in a way that resembles those isoforms which are expressed by myoblasts that give rise to secondary muscle fibers, i.e., the embryonic and neonatal isoforms are expressed in the early stages but not the slow type I myosin hc (Whalen *et al.,* 1990). Muscle regeneration in sea bream and zebrafish has been studied by Rowlerson and co-workers (Rowlerson *et al.,* 1997). Both species showed a vigorous regeneration giving rise to new muscle fibers with an initial myosin hc composition differing from the existing mature fibers. However, the nature of the myosin isoforms expressed in each species differed in that the sea bream myosin resembled that seen in the new fibers produced in post-larval white muscle, whereas in the zebrafish it resembled that of the primitive monolayer fibers formed during embryonic development, suggesting the type of satellite cell utilized for muscle regeneration may be species dependent.

V. A MYOSIN GENE THAT IS ASSOCIATED WITH ADULT MUSCLE HYPERPLASIA IN CARP

In fish, increases in fiber number (hyperplasia) as well as an increase in size of existing fibers (hypertrophy) contribute to adult myotomal muscle growth (Stickland, 1983; Greer-Walker, 1983; Weatherley and Gill, 1984). This is unlike the situation in mammalian and avian species where hyperplasia stops shortly

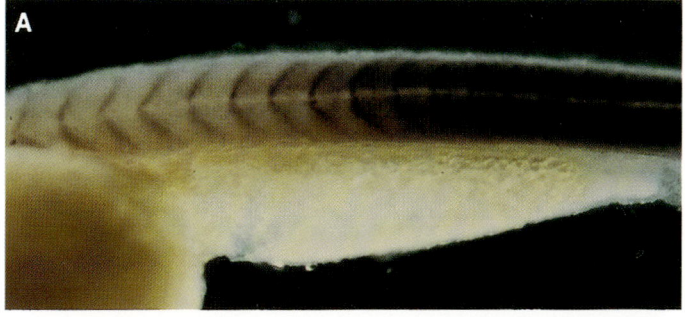
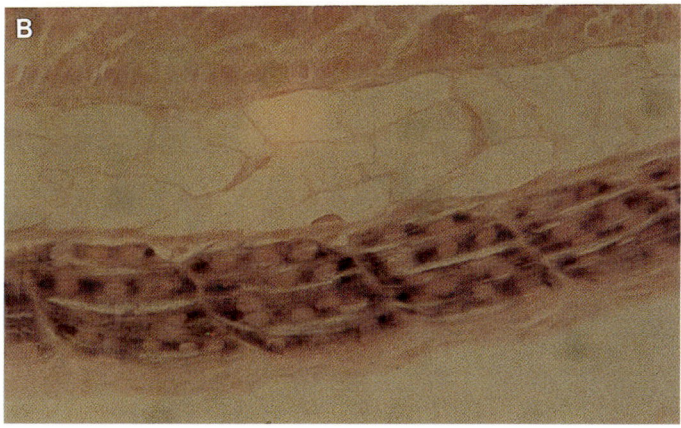

Figure 3. Localization of mRNA for myosin heavy chain to the sites of protein synthesis. (A) Wholemount *in situ* hybridization on 12-hr posthatching carp larvae with a digoxigenin labelled antisense cRNA probe corresponding to the 3' untranslated region of the carp EGGS22 myosin heavy chain gene (only trunk section shown). Hybridized probe was detected with an alkaline phosphatase conjugated anti-digoxigenin antibody with DAB staining (purple). Note the chevron pattern of gene expression through the myotomes. Longitudinal growth has been shown to involve the addition of sarcomeres in series onto the ends of the existing myofibrils (reviewed by Goldspink, 1996) and this is reflected in the higher concentration of myosin RNA at the ends of the fibers near the myosepta. (B) The above embryo was embedded and longitudinally sectioned. This panel shows an eosin-stained (pink) section through trunk musculature. Note the muscle fibers running from myosept to myosept and the stronger hybridization (purple) of the Eggs22 cRNA probe at the myoseptal ends of the fibers which explains the chevron pattern observed in (A).

after embryonic development (Goldspink, 1972). Furthermore, muscle growth rates in fish can vary significantly throughout life depending on factors such as nutrition or environmental temperature (Loughna and Goldspink, 1984). Myosatellite cells are involved in both hypertrophic and hyperplasic growth of fish muscle. In hypertrophic growth they fuse with the existing muscle fibers increasing the number of nuclei in the mature fibers, while in hyperplasic growth, they differentiate into new muscle fibers (Koumans et al., 1993a,b). The small diameter white muscle fibers of the gilthead sea bream, *Sparus aurata,* have a high mitotic activity suggesting that their origin is from myosatellite cells (Rowlerson et al., 1995). However, there may be species-dependent mechanisms of fiber hyperplasia since the sonic muscle fibers of the toadfish have been shown to increases in fiber number by splitting of existing fibers (Fine et al., 1993).

The small diameter fibers (<25 μm) present in the white myotomal muscle of adult carp (Rowlerson et al., 1985; Akster, 1983) are assumed to have arisen from myosatellite cells and hence are a consequence of the process of hyperplasia. Such fibers differ from the larger diameter fibers both histochemically and immunohistochemically, suggesting that their myosin hc content is also different (Rowlerson et al., 1985). We have isolated the gene for the specific myosin hc isoform which is expressed in these small diameter fibers, the carp FG2 myosin hc gene. By using *in situ* hybridization (Fig. 4) we have shown that expression of the FG2 gene is restricted to only the small diameter fibers of rapidly growing adult carp (Ennion et al., 1995). These *in situ* hybridization studies were performed with a highly specific 3' untranslated region cRNA probe labeled with digoxigenin (Fig. 4) and show that this gene is not expressed in the red muscle fibers or in the larger diameter more mature fibers of the white muscle. The carp FG2 myosin hc gene is transiently expressed during the differentiation of satellite cells into small diameter fibers during rapid growth. During the regeneration of damaged mammalian skeletal muscle fibers, myosatellite cells are also recruited; proliferating and fusing either into multinucleated myotubes or with the ends of damaged mature muscle fibers (Hinterberger and Barald, 1990). Recruitment of myosatellite cells in mammalian muscle regeneration is known to involve the expression of the embryonic and neonatal isoforms of the myosin hc (Sartore et al., 1982). However, the situation in carp satellite cell recruitment appears to be somewhat different in that the FG2 gene cannot be considered as an embryonic or neonatal myosin hc equivalent since Northern hybridization analysis shows that it is not expressed in juvenile carp, carp fry, or developing carp embryos. Thus, the carp FG2 myosin hc gene is best regarded as a "growth isoform gene" for which a mammalian or avian equivalent has not been isolated. The fact that there is a MyoHC gene which is expressed specifically during the growth phase of the small diameter white muscle fibers highlights another difference between muscle growth in fish compared to other species.

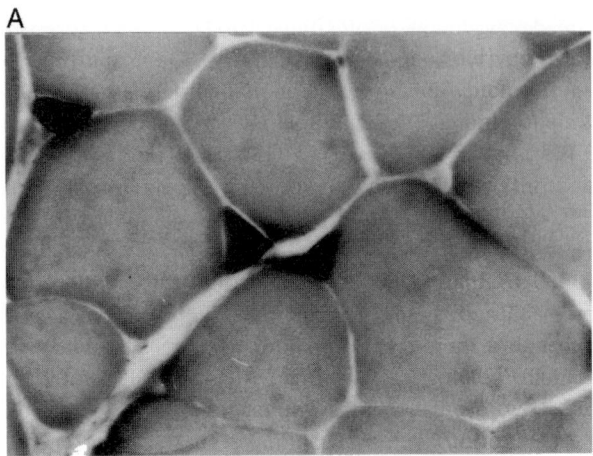

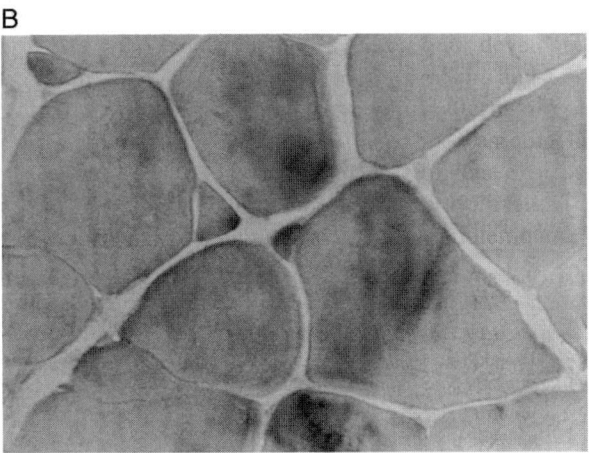

Fig. 4. Expression of the carp FG2 myosin heavy chain gene in small diameter fibers of adult carp. *In situ* hybridization on white muscle from an adult carp acclimated to a warm environmental temperature. The cRNA probe used corresponds to the 3' untranslated region of the carp FG2 myosin heavy chain gene and was labeled with digoxigenin. Hybridization was detected with an alkaline phosphatase anti-digoxigenein antibody and DAB staining. (A) Antisense labeled probe and (B) sense labeled probe. The same area of muscle fibers is shown in each photomicrograph.

VI. MOLECULAR SIGNALS INVOLVED IN MUSCLE GENE EXPRESSION DURING PRE- AND POST-HATCHING GROWTH

In vertebrates growth the speed and span of the growing period is for the most part controlled genetically. During evolution different species have evolved different optima for maximum growth, particularly with regard to temperature. Indeed, eurythermal species can even change their optima on a seasonal basis for muscle protein synthesis (Loughna and Goldspink, 1985). However, growth can also be influenced by other factors including availability of food, oxygen, salinity, and activity (Davison, 1989; Matschak et al., 1998; Christiansen et al., 1992) as well as environmental temperature (Martinez and Pettersen, 1992; Martinez et al., 1995; Johnston and Bennett, 1995). Because the environmental as well as the genetic influences are multifactorial and form a complex matrix there is a real need to investigate the molecular mechanisms as these should enable us to rationalize growth regulation in fish and other animals. For this reason we are studying the mechanical and chemical signal molecules that induce gene expression related to growth and adaptation. Some of the most important growth factors expressed during early development include insulin, IGF-I (insulin-like growth factor I), IGF-II, growth hormone (GH), and thyroid hormone. With regard to the latter, protein studies have shown that T_3 as well as environmental temperature result in myosin isoform transitions (Martinez et al., 1995). The myosin hc gene apparently has response elements in 5' flanking sequence to these growth factors and thus the prospects exist of establishing the molecular and cellular mechanisms involved in muscle differentiation and growth. The myogenic factors which include Myogenin and MyoD are important in determining the initial stages of muscle development *in ovo,* The expression of these during early *ontogeny* is discussed in other chapters of this book. Therefore, we restrict our comments mainly to the IGFs, the expression of which preceeds and continues beyond the other signal molecules

The members of the insulin family of peptides include insulin, IGF-I, and IGF-II. They are single-chain polypeptide hormones with a high degree of homology at the amino acid level and have overlapping activities due to the similarity of the "insulin-fold" tertiary structure. (Blundell and Humbel, 1980). While the importance of the IGFs in the regulation of postnatal growth and development in mammals has been established, the role of these peptides during embryogenesis and early growth has received little attention. The major source of IGFs is the liver although it is now understood that both IGF-I and IGF-II are produced in numerous extrahepatic tissues in fish (Berwert et al., 1995; Mack et al., 1995) as well as in mammals (Yang et al., 1996; McKoy et al., 1999).

The first IGF-I cloned from a fish species was by Cao et al. (1989) in coho salmon. Since then, IGF-I sequence has been isolated and determined in rainbow trout, carp, Atlantic salmon, and others (Duguay et al., 1992, 1996; Liang et al., 1996). Homology at the amino acid level is well conserved between these fish species and comparisons of the sequence with mammalian IGF-I shows a high degree of conservation during vertebrate evolution to an extent that human IGF-I is biologically active in teleost fish (Duan and Hirano, 1990). Analysis has shown that chum salmon have two non-allelic IGF-I genes (Kavsan et al., 1993, 1994), each of the genes are organized into four exons, with mature IGF-I encoded by exon 2 and 3 and the E peptide encoded by exons 3 and 4. As with many fish genes the genomic span of the IGF-I gene is more compact and simpler than in mammalian genes. The chum salmon IGF-I gene spans 20 kb compared with the human IGF-I gene which spans over 100 kb of chromosomal DNA. So it would seem that the IGF-I gene arrangement is simpler in lower vertebrates and additional exon sequences have been acquired during the evolution of the higher phyla. Alternative splicing also seems to be present in salmonoid fish with four different transcripts detected (Duan et al., 1994).

IGF-I has been detected in many juvenile and adult fish tissues (Berwert et al., 1995; Kagawa et al., 1995). Studies have shown that as in mammals GH regulates liver IGF-I expression in teleost fish (Marchant and Moroz, 1993; Moriyama, 1995). The importance of GH in post-hatching growth is clearly seen in transgenic fish overexpressing GH where increased size is observed (Du et al., 1992; Inoue et al., 1993; Devlin et al., 1994). Fauconneau et al. (1997) have shown that GH supplementation enhanced the percentage of small diameter fibers indicating a role of this hormone in the control of muscle hyperplastic growth.

Although injection of GH into fish increases hepatic IGF-1 mRNA, it does not have the same effect on other tissues, including heart, brain, and spleen (Duan et al., 1993). The difference in the response to GH indicates a somewhat different role of the non-liver IGF-1s, for example, the muscle IGF-1s which in mammals are produced in response to mechanical signals (Yang et al., 1996; McKoy et al., 1999). This provides a mechanism via which activity upregulates expression of the myosin and other structural genes.

IGF-I isoforms appear to be regulated in a tissue specific and temporal manner and are expressed in a wide number of tissues during fetal and postnatal stages of development. Funkenstein et al. (1996) reported detectable IGF-I transcripts in sea bream unfertilized eggs and embryos, and postulate that the IGF-I mRNA may be maternal in origin; what is still not clear, is whether there are biologically active IGF-I peptides at this early embryonic stage. Evidence that IGF-I plays a role in muscle growth comes from experiments where mammalian IGF-I stimulates protein synthesis in muscles of the gulf killifish (Negatu and Meier, 1995). Other studies have demonstrated that IGF-I is a potent stimulator of DNA synthesis and proteoglycan synthesis of skeletal tissues in fish (Duan and Hirano, 1992) and plasma IGF-I levels are connected with growth rate in juvenile coho salmon (Duan et al., 1995). IGF-I shows significant levels of mRNA expression in tissues other than liver, including muscle, and neutralization of endogenously secreted IGF-I inhibits cell proliferation (Jones and Clemmons, 1995), which implies that IGF-I in the development of muscle tissue in the fish. Although IGF-I originating from the liver influences growth and development in fish in an endocrine fashion, the role of IGF-I in an autocrine/paracrine manner remains unclear and requires further investigation. Free swimming embryos of teleost fish make them ideal subjects for investigation of the mechanism by which IGF-I act to regulate cell proliferation, differentiation, and development and will contribute to the to the growing knowledge of IGF-I action during early development.

VII. EXPRESSION OF MYOSIN ISOGENES IN RESPONSE TO TEMPERATURE

Both acute and chronic changes in temperature have profound effects on virtually all biological processes, not least the molecular events occurring during muscle contraction (Rome and Bennett, 1989). In order to minimize the biological effects of temperature change, many animals have evolved various physiological or behavioral means to maintain a constant body temperature. The body temperature of most species of fish, however, is strictly dependent on the environmental water temperature, although in some species muscle tissue has the ability to adapt to different temperature regimes. Thus environmental temperature determines not only the rate of muscle development but also what type of muscle is produced. In other words there are qualitative as well as quantitative changes in gene expression. Different species have different tolerances to temperature and one may therefore expect there may be different mechanisms. However, here we review adaptation in carp as only in this species have the events been elucidated at the gene level.

A major determinant of muscle function is the myofibrillar ATPase activity and much research over the years has focused on the effects of temperature acclimation on this enzyme activity in fish. The first evidence that the contractile properties of fish muscle could be modified by thermal acclimation came from Johnston et al. (1975). They acclimated goldsfish (*Carassius auratus L.*) to either 1 or 26°C and showed that the specific myofribrillar ATPase activity was 2.8 times greater in cold acclimated compared to warm acclimated fish. Similar changes in ATPase activity with thermal acclimation have also been reported for other members of the cyprinid family which include carp (*Cyprinus carpio*), roach (*Rutilus rutilus*), and tench (*Tinca tinca*); (Heap et al., 1985). The changes in myofibrillar ATPase activity associated with warm and cold temperature acclimation occur within four weeks and have been shown to be totally reversible (Heap et al., 1985). Interestingly, these changes were not observed in starved fish where rates of protein synthesis are low (Heap et al., 1986), which was one of the first indications that this type of adaption involved the expression of different genes and the synthesis of new and different proteins.

There are other mechanisms involved in the rebuilding of the contractile apparatus in eurythermal fish such as the carp during temperature acclimation. These affect the rate at which the contractile system is activated as well as the rate at which it generates force during each contractile cycle. These alterations have implications for not only the thermodynamic efficiency but also the mechanochemical efficiency of location at different environmental temperatures. With regard to the latter, the rate at which the contractile system uses energy is determined primarily by the specific myofibrillar ATPase, and we now know that this involves expressing different myosin genes and changing the molecular motor type. The rate at which the system is activated depends on the diffusion of Ca^{++} from the sarcoplasmic reticulum (SR). Penney and Goldspink (1981a) showed that the conformation and distribution of the SR changed during temperature acclimation in the carp in that myofibrils were smaller and surrounded by more small SR tubules in the cold acclimated fish. Thus the surface area and diffusion distances are greatly reduced.

Alterations in the Ca^{++} regulatory proteins, tropomyosin and troponin, are also apparently involved. Johnston and co-workers (Johnston, 1979) demonstrated that densensitized actomyosin from which the Ca^{++} regulatory proteins had been removed by high salt concentrations showed an ATPase activity which was almost identical in warm- and cold-acclimated goldfish (*Carassius auratus*). Furthermore, they showed that cross-hybridization of regulatory proteins from 2°C-acclimated goldfish to desensitized actomyosin from 31°C-acclimated fish altered the ATPase activity toward that of the 2°C-acclimated intact actomyosin. The converse of this was also demonstrated. These results were later confirmed by Penney and Goldspink (1981b), who also proposed that the Ca^{++} regulatory proteins play a more involved role in muscle contraction than a simple on/off switch. However, it might be expected that removal of the regulatory proteins will disrupt the orderly structure of the contractile apparatus and so these results may not be representative of physiological condtions within the myofibril. Guo and Watabe (1993) carried out a similar experiment in carp (*C. carpio*) but were unable to demonstrate any acclimation temperature dependent effect of regulatory proteins on the activity or thermostability of actomyosins from warm and cold acclimated fish.

Both the heavy and light chains of myosin have been investigated with regards to elucidating their role in the temperature acclimation process in fish. Crockford and Johnston (1990) showed that myofibrils from 8°C-acclimated carp contained an additional (MLC) light chain band on SDS polyacrylamide peptide gels which was not present in warm- (20°C) acclimated carp. Additionally, the MLC3/MLCI ratio was significantly lower in 8°C- than in 20°C-acclimated carp. The MLC3/MLCI ratio during temperature acclimation in carp has also been investigated by Northern blot analysis and the ratio at the mRNA level was shown to be significantly greater in 30°C-acclimated than in 10°C-acclimated carp (Hirayama *et al.*, 1997). Interestingly, this same study also demonstrated that carp MLC I and MLC3 are encoded by different genes. This is in marked contrast to mammalian

MLC I and MLC3 which are produced by alternative splicing from the same gene (Barton and Buckingham, 1985).

Investigations at the protein level regarding the role of the myosin hc in temperature acclimation are contradictory. Johnston and co-workers (Crockford and Johnston, 1990) purified myosin hc protein from the fast and slow myotomal muscle of both 8- and 20°C-acclimated carp and digested with the proteolytic enzymes V8 protease and chymotrypsin. Although different peptide maps were obtained for fast and slow muscle myosin hc, no differences were observed with thermal acclimation. However, the extensive protein studies by Watabe and co-workers did show changes in myosin hc proteins in response to temperature acclimation. By using chymotrypsin proteolytic digests of myosin hc protein, differences were clearly demonstrated between the peptide maps of myosin hc from warm- (30°C) and cold- (10°C) acclimated carp (Hwang et al., 1990). The difference in results obtained by Crockford and Johnston (1990) and Hwang et al., (1990) may possibly reflect differences in technique or the acclimation temperatures used in each experiment. The 20°C used by Johnston and co-workers may not be high enough to elicit the same acclimation response observed by Hwang and co-workers at 30°C. This group also demonstrated that the thermostability of myosin isolated from warm- and cold-acclimated carp differed such that the inactivation rate constant (K_p) of myosin from cold-acclimated fish is approximately three times greater than that of myosin from warm-acclimated fish. Watabe et al. (1994) peptide mapped SI myosin hc subfragments of warm- (30°C) and cold- (10°C) acclimated carp and demonstrated that cold-acclimated carp contain four different types of SIs (molecular motors) while warm-acclimated carp contain two SI types, which were distinct from the cold isoforms.

The above data at the protein level suggest that the changes observed in myofibrillar ATPase with temperature acclimation are due to the production of different isoforms of one or more of the contractile proteins rather than conformational changes in existing proteins. Such a change in isoforms would involve either the expression of different genes coding for different isoforms or alternative splicing of exons within the same gene. The first investigation of temperature acclimation at the level of myosin hc gene expression was conducted by Gerlach et al. (1990) by constructing a carp genomic library and isolating 28 different myosin hc clones. A fragment of one of these clones was shown to hybridize strongly at high stringency conditions to RNA isolated from the white muscle of warm acclimated carp but not to RNA from carp acclimated to 10°C or RNA from red muscle. At low stringency conditions the same probe bound equally to RNA from both warm- and cold-acclimated fish. From these data Gerlach and co-workers (1990) hypothesized that the temperature acclimation process in carp involves the differential expression of separate myosin hc genes. More recently, Imai et al. (1997) isolated cDNA clones encoding fast skeletal muscle myosin hc of carp acclimated to 10, 20, and 30°C identified genes which were the predominant types expressed in 10°C, 30°C, and an intermediate type present at all acclimation temperatures. Differences in primary structure were observed in two surface loops between the 10 and 30°C isoforms (Merati *et al.*, 1996). And 5 out of 16 amino acid sequences were different in loop I near the ATP-binding pocket, and 6 out of 20 were different in loop 2 on the actin-binding site, suggesting that the differential gene expression as a result of temperature acclimation produces myosin hc isoforms which differ in their functional characteristics. Indeed, experiments examining the temperature dependence of sliding velocity of fluorescent F-actin on myosins isolated from 10- and 30°C-acclimated carp demonstrate that activation energies for the sliding of F-actin differ between isoforms (Chaen *et al.*, 1996).

VIII. REGIONS DETERMINING FUNCTIONAL DIVERSITY IN MYOSIN hc ISOFORMS

The determination of the three-dimensional structure of the head region of a chicken pectoralis myosin hc (Rayment *et al.*, 1993a,b) by crystal X-ray diffraction highlighted regions of the molecule which could possibly be involved in determining the functional diversity between isoforms. This indicated that there are two flexible surface loops (residues 204–216 and residues 627–646) in the chicken myosin-head hc sequence. Cloning and sequencing studies have shown that although the myosin hc genes are highly conserved, these loops vary markedly between different myosin hc isoforms (Fig. 5). Therefore, it has been hypothesized that they determine functional diversity (Bobkov *et al.*, 1996; Spudich, 1994; Uyeda *et al.*, 1994; Rayment *et al.*, 1993a). Hypervariable loop 1 connects the 25 and 50-kDa domains and projects from the surface of the protein sitting over the ATP-binding pocket. Hypervariable loop 2 forms one of the actin-binding domains and is situated at the junction between the 50- and 20-kDa tryptic fragments. Tryptic digestions of myosin, to produce different percentages of molecules cleaved at either loop 1 or loop 2, have been used to characterize the roles of these regions by *in vitro* motility assays (Bobkov *et al.*, 1996). Selective cleavage of loop 2 produces a lower affinity for actin but does not change the rate of force development (sliding velocity), whereas cleavage of loop 1 decreases the mean sliding velocity by almost 50% but does not alter the affinity for actin. Loop 1 is thought to effect the sliding velocity by modulating the release of ADP (Spudich, 1994). The factors which may modulate the effect of loop 1 on the properties' contraction are unknown but are likely to involve both its length, its overall charge and the hydrophilic/hydrophobic properties of its constituent amino acids (Gauvry *et al.*, 1997).

Evidence for the role of loop 2 in determining ATPase activity has been provided by expressing chimeric myosin molecules in *Dictyostelium* (Uyeda *et al.*, 1994). Theses studies showed that substitution of the *Dictyostelium* loop 2 with

A

```
                  >  HYPERVARIABLE LOOP 1  <
10 C Type    VIQYFATI-AMAGPKKAEAVPGKMQSSLEDQIIAANPLLEAYGNAKT
Int-Type     .......V-..S....T.............................
30 C Type    .......VG..S....P.P..............V.............
Antrc A1     ........S-VS...R-----DASKG.................S.....
Antrc A4     ........A-V..G..M.QA-S..KG.................S.....
Trop 54      ........A.L-.A....PT.....G......V...............
                     >                        <
```

B

```
                  >HYPERVARIABLE LOOP 2<
10 C Type    YQKSALKVLALLYVA--VPEAEAAGKKGGKKKGGSFQTVSAVFRENLGK
Int-Type     ................--...-.GG...A..............L.....AG
30 C Type    ....S.....F....THGA...GG.G.K...............L......
Antrc A1     .M..SV.L.G...P----.VV.E................M....SQ......
Antrc A4     .M..SV.L....ASHNAA...--...AA..............L......
Trop 54      ....SN.L..M..A.HAGA.EA.G..................L......
                 >                        <
```

Fig. 5. Comparison of amino acid sequence over hypervariable loops 1(A) and 2(B) in myosin isoforms from fish species. The predominantly expressed myosin hc isoform of *Cyprinus Carpio* at 10°C, intermediate temperatures (Int), and 30°C, respectively (Hirayama and Watabe, 1997). Antrc A1 and AntrcA4 are white and red isoforms from the Antarctic species *Notothenia coreceps* and Trop54 is a white muscle isoform from the tropical species *Paracirricthyes forsterri*. Note the relative divergence of amino acid sequence within the loop region compared to the conservation in the flanking sequences. *N. coreceps* and Trop54 are white muscle isoforms from the tropical species *P. forsterri*. Note the relative divergence of amino acid sequence within the loop region compared to the conservation in the flanking sequences.

a rabbit fast skeletal myosin loop 2 produced a molecule with an ATPase activity fivefold higher than the original *Dictyostelium* myosin. Furthermore, loop 2 substitutions from a range of vertebrate isoforms produced molecules with ATPase activities which reflected that of the donor molecule, indicating that loop 2 is a major determinant of myosin ATPase activity. Thus although the muscle myosins show a very high degree of homology, the difference in their contractile characteristics appears to be due to subtle changes in the amino acid sequence of the two hypervariable loops, one of which is associated with the ATPase site and the other with the actin-binding site.

A. Regulatory Elements of the Myosin hc Genes

Muscle ontogeny requires different sets of genes, including the isoforms of myosin, to be precisely controlled at the transcriptional level. This control is not just a simple question of switching on or switching off genes but requires expression in the correct positional, temporal, and quantitative manner. All myosin hc genes have a regulatory or promoter region in the sequence flanking the ATG start site. This sequence is not transcribed into mRNA but it is needed to control the expression of the gene. The mechanisms of how gene expression is so precisely controlled during ontogeny are just beginning to be understood. Intracellular signaling cascades resulting from extracellular cues culminate in the binding of certain proteins or transcriptional factors at or near the promoter region which can either activate or repress expression of the gene. These transcriptional factors fall into several families with common features. At least three main families of transcription factors play a central role in the regulation of gene expression during muscle ontogeny (Ludolph and Konieczny, 1995).

The myogenic factors (MyoD, Myf-5, myogenin, and MRF4) contain a conserved central protein motif referred to as the basic helix-loop-helix domain. This family of transcription factors are muscle specific and have a well characterized role in determining the myogenic lineage during embryonic development and a less well understood role in adult muscle gene expression regulation. The MEF2 (myocyte enhancer) factor family (MEF2 A, B, C, and D) belongs to the MADS superfamily of transcription factors. Similar to the myogenic factors these factors play an important role in embryonic development but are also expressed in the adult (Black and Olson, 1998). However, unlike the myogenic factors the MEF2 factors are expressed in a broad range of tissue types. Members of the PAX (paired box) family of transcription factors, e.g., PAX-3, may also play a role in very early myogenic determination during embryonic development. Later, other transcriptional or growth factor signals became involved during pre- and postnatal hatching muscle growth. These may act in concert with the myogenic factors as it is necessary to maintain muscle tissue specificity as well as promote rapid increases in mass.

Cysteine-rich proteins (CRP) define a subclass of LIM-only proteins implicated in muscle differentiation. CRP proteins contain two tandemly arranged LIM domains linked to a glycine-rich motif. The LIM domains have been described as specific adapter elements that promote protein-protein interactions (Dawid *et al.*, 1995). Among the three CRPs that have been described in vertebrates, CRP3, also called MLP (muscle LIM protein), has been shown to interact physically with the myogenic factor MyoD via its first LIM domain, thereby increasing the transactivation activity of MyoD (Kong *et al.*, 1997). Recently a CRP cDNA, called TCRP, was isolated from rainbow trout (Delalande and Rescan, 1998). *In situ* hybridization of early trout embryos showed TCRP mRNA expression within the whole unsegmented pre-somitic mesoderm as well as in the newly formed somites. As the somites mature, the levels of TCRP mRNA fade. This pattern of expression overlaps with the expression of TMyoD. However, any functional interactions between TCRP and TMyoD in the early phases of the somitogenesis remains undetermined.

3. MYOSIN EXPRESSION DURING ONTOGENY

The myosin hc gene promoter regions of several mammalian (Chang et al., 1995; Takeda et al., 1992a,b; Rindt et al., 1993; Rindt et al., 1995; Lonn et al., 1993), avian (Gulick et al., 1985), and one fish species (Gauvry et al., 1996a) have been characterized. As with many other genes, a common feature is the presence of a "TATA box" which is usually 30–50 bases upstream of the transcription start site. This motif has been described as the "signpost" for RNA polymerase II and its function is to fix the location of the start site for transcription. In addition to the TATA box, most myosin hc promoter regions also contain a common upstream element known as the CCAAT box which is thought to be the binding site for transcriptional factors which regulate the level of gene transcription. This is not a universal feature with myosin hc genes, as the porcine fast skeletal myosin hc promoter does not contain this motif (Chang et al., 1995). As with many other muscle specific genes, another common feature in the myosin hc promoters is the presence of motifs known as E-boxes (CAANTG), which represent binding regions for members of the helix-loop-helix transcription factor family (Weintraub et al., 1991; Edmondson and Olson, 1989). Other regulatory sequences, such as enhancers, may also be involved in facilitating the binding of the transcriptional factor proteins to DNA. These may be at the 5' or 3' end of the gene and many bases from the start sequence. Enhancer sequences are also sometimes found in untranslated introns as well as exons. In the myosin hc genes these are found in introns 1 and 2 which are untranslated (Chang et al., 1995; Gauvry et al., 1996).

Thyroid hormone also affects the levels of gene expression of both cardiac and skeletal myosin hc genes in a complex manner with the effect being both tissue type and gene dependent, and the rat and human α cardiac myosin hc promoter regions are known to contain at least two thyroid responsive elements (TREs; Tsika et al., 1990; Izumo et al., 1986). In mammals there is a burst of thyroid hormone secretion just after birth and the expression of the α TH receptor is believed to be involved in the switch from expressing embryonic to adult type fast genes. In fish T_3 has been shown to be associated with changes in myosin type (Martinez et al., 1995) as detected by gel electrophoresis. Therefore similar response elements seem to exist in the promoter regions of piscine genes.

The only piscine myosin hc promoter to be characterized to any extent is the carp FG2 gene (Gauvry et al., 1996a). We have characterized the 5' regulatory sequence of this myosin gene in order to determine which regions are important for expression. Engineered plasmid constructs which consisted of different regions of the promoter sequence and attached to a reporter cDNA were introduced into cells in culture and into carp muscle in vivo by direct injection. In this way we were able to determine which upstream regulatory sequences were necessary for expression (Gauvry et al., 1996a). Analysis of 901 bp flanking the ATG start site revealed several common features with the mammalian and avian promoters suggesting that similar factors to those which control myosin hc expression in these species are also present in fish. A characteristic TATA box is located 30 bp upstream of the transcription start site, a CCAAT box is located at −74 bp, and several putative E-boxes are present. The carp FG2 promoter also contains a putative MEF 1 motif (C/G)N(G/A)(G/A)CA(C/G)(C/G)TG(C/T)(C/T)N(C/G). Mammalian type gene constructs under the control of mammalian promoters are not expressed or are weakly expressed in fish. However, the carp FG2 myosin promoter works well in zebrafish and catfish and we have used this promoter for producing transgenic fish (Muller et al., 1997).

The 3' flanking sequences of the myosin hc genes also appear to have a regulatory function, but this is not well understood. Although this region shows the greatest lack of homology between the myosin hc genes and hence is usually the region of choice for using as an isoform specific probe in hybridization studies, there are often small regions of conserved sequences between isoforms (Ennion et al., 1999). For example, the 10 nucleotide motif of "AAAATGTGAA" can be found in the 3' untranslated regions of myosin hc isogoenes from carp, chicken, human, pig, rabbit, rat, and *Xenopus*. This strongly indicates a muscle specific function for this sequence and there is some evidence that certain proteins bind to this region (A. Kiri, personal communication). There is also evidence that the 3' untranslated region determines the stability of the message as many nucleases degrade mRNA from the 3' end inward. Hence the specific sequence of the RNA may allow the formation of "hairpin" like structures which prevent or slow down the action on exonucleases from degrading the RNA. The stability of mRNA represents a method of regulating gene expression and it is particularly important in adaptive changes as this may determine the rate at which gene expression can be switched from expressing one isoform to another isoform of myosin. It is thought that the 3' untranslated region of several genes encodes a "ZIP" or mailing code for the mRNA which determines cellular location. It appears that within cells the mRNA is directed from the nucleus to different regions, e.g., perinuclear or submembrane locations (Wiseman et al., 1997). In muscle fibers that are syncytial units and in which during growth new contractile proteins are synthesized at a very rapid rate and assembled into sarcomeres at the end of the fibers, the distance from the more centrally located muscle fibers to the ends is thousands of kilometers on the molecular scale. Instead of transporting the proteins which are relatively large, it is more effective to transport the mRNA so that the proteins can be synthesized at the site of assembly.

IX. CONCLUSIONS

As is the case in mammalian, avian, and amphibian species, the myosin hc isoforms present in fish species exist as a multigene family. The fish species best studied to date in terms of myosin diversity is the carp with an estimated number of 28 separate myosin isoform genes for this species (Gerlach *et al.,* 1990). The size of the myosin hc family in other fish species has not been estimated, so it is not known whether the carp has an atypically large family due to its tetraploid genome and its acquisition of the ability to adapt to different environmental temperatures. The ability to adapt during development and growth and to changed environmental conditions clearly has great survival value as it is difficult to see how just one type of molecular motor could meet the diverse functional demands. This has apparently been achieved during evolution by multiplication of the myosin hc genes and slight modifications to the region that codes for the globular S1 head of the protein where differences in amino acid sequence affect the crossbridge cycle. Expression of different isoforms of myosin hc in response to temperature acclimation in carp (Hirayama and Watabe, 1997) is a particularly good example.

In addition to considering contractile properties, it must be borne in mind that the myosin hc genes differ significantly in their regulatory regions. The cellular environment is probably sufficiently different in the newly developing or growing muscle fibers to necessitate the existence of isogenes which differ in their regulatory regions. Thus it may be necessary to build the initial myofibrillar infrastructure using a myosin hc isoform that is transiently but strongly expressed. Thereafter, this myosin hc isoform can be readily exchanged for the molecules of the adult type genes by the law of mass action (Goldfine *et al.,* 1991).

An important unresolved aspect of muscle biology is elucidating the precise molecular mechanism be which myosin hydrolyzes ATP, interacts with actin, and causes movement of the thick and thin filaments, i.e., the molecular mechanism of muscle contraction. As well as looking at the mutations in zebrafish, there is probably a greater diversity in this phyla with respect to muscle function than any other in the animal kingdom. The study of myosin isoforms in different fish species may provide some very useful pointers in elucidating the details of the contractile mechanism by studying the way molecular motors have been adapted during evolution. This range not only includes molecular motors for different modes of swimming (burst vs. cruise speeds), electric organs development, and even high frequency sound production (Rome *et al.,* 1999), but also includes adaptation to extremes in pressure in deep sea species as well as adaptation to environmental temperature.

ACKNOWLEDGMENTS

The research described in this review was supported by grants from NERC, Wellcome Trust, and BBSRC to Professor Goldspink. Dr. Ennion was a Dowager Duchess Eleanor Peel Research Fellow and Dr. Wilkes received his Fellowship from the European Community Fair Programme.

REFERENCES

Akster, H. A. (1983). A comparative study of fibre type characteristics and terminal innervation in head and axial muscle of the carp (*Cyprinus carpio* L.): a histochemical and electronmicroscopial study. *Neth. J. Zool.* **33**, 164–188.

Barbet, J. P., Thornell, L. E., and Butler-Browne, G. S. (1991). Immunocytochemical characterisation of two generations of fibers during the development of the human quadriceps muscle. *Mech. Dev.* **35**, 3–11.

Barton, P. J. R., and Buckingham, M. E. (1985). The myosin alkali light chain proteins and their genes. *Biochem. J.* **231**, 249–261.

Berwert, L., Senger, H., and Reinecke, M. (1995). Ontogeny of IGF-I and the classical islet hormones in the turbot, *Scophthalmus maximus*. *Peptides* **16**, 113–122.

Black, B. L., and Olson, E. N. (1998). Transcriptional control of muscle development by myocyte enhancer factor-2 (MEF2) proteins. *Ann. Rev. Cell Dev. Biol.* **14**, 167–196.

Blundell, T. L., and Humbel, R. E. (1980). Hormone families: pancreatic hormones and homologous growth factors. *Nature* **287**, 781–787.

Bobkov, A. A., Bobkova, E. A., Lin, S. H., and Reisler, E. (1996). The role of surface loops (residues 204–216 and 627–646) in the motor function of the myosin head. *Proc. Natl. Acad. Sci. U.S.A.* **93**, 2285–2289.

Cao, Q.-P., Doguay, S. J., Plisetskaya, E. M., Steiner, D. F., and Chan, S. J. (1989). Nucleotide sequence and growth hormone-regulated expression of salmon insulin-like growth factor I mRNA. *Mol. Endocrinol.* **3**, 2005–2010.

Chaen, S., Nakaya, M., Guo, X. F., and Watabe, S. (1996). Lower activation energy for sliding of F-actin on a less thermostable isoform of carp myosin. *J. Biochem.* **120**, 788–791.

Chang, K. C., Fernandes, K., and Dauncey, M. J. (1995). Molecular characterization of a developmentally regulated porcine skeletal myosin heavy chain gene and its 5' regulatory region. *J. Cell. Sci.* **108**, 1779–1789.

Chanoine, C., Guyot-Lenfant, M., El Attari, A., Saadi, A., and Gallien, C. (1992). White muscle differentiation in the eel (*Anguilla anguilla*, L.). Changes in the myosin isoforms pattern and ATPase profile during post-metamorphic development, *Differentiation*, **49**, 69–75.

Christiansen, J., Martinez, I., Jobling, M., and Amin, A. B. (1992). Rapid somatic growth and muscle damage in a salmonid fish. *Bas. Appl. Myol.* **2**, 235–239.

Condon, K., Silberstein, L., Blau, H. M., and Thompson, W. J. (1990). Development of muscle fiber types in the prenatal rat hindlimb. *Dev. Biol.* **138**, 256–274.

Crockford, T., and Johnston, I. A. (1990). Temperature acclimation and the expression of contractile protein isoforms in the skeletal muscles of the common carp (*Cyprinus carpio*). *J. Comp. Physiol. B* **160**, 23–30.

d'Albis, A., Couteaux, R., Janmot, C., and Mira, J. C. (1989). Myosin isoform transitions in regeneration of fast and slow muscles during postnatal development of the rat. *Dev. Biol.* **135**, 320–325.

Davison, W. (1989). Training and its effects on teleost fish. *Comp. Biochem. Physiol.* **94A**, 1–10.

Dawid, I. B., Toyama, R., and Taira, M. (1995). LIM domain proteins. *C. R. Acad. Sci. (Paris) III* **318**, 295–306.

Delalande, J. M., and Rescan, P. Y. (1998). Expression of a cysteine-rich protein (CRP) encoding gene during early development of the trout. *Mech. Dev.* **76**, 179–183.

Deng, Z., Liu, P., Marlton, P., Claxton, D. F., Lane, S., Callen, D. F., Collins, F. S., and Siciliano, M. J. (1993). Smooth muscle myosin heavy chain locus (MYH11) maps to 16p13.13-p13.12 and establishes a new region of conserved synteny between human 16p and mouse 16. *Genomics* **18**, 156–195.

Devlin, R. H., Yesaki, T. Y., Blagi, C. A., Donaldson, E. M., Swanson, P., and Chan, W.-K. (1994). Extraordinary salmon growth. *Nature* **371**, 209–210.

Devoto, S. H., Melancon, E., Eisen, J. S., and Westerfield, M. (1996). Identification of separate slow and fast muscle precursor cells *in vivo*, prior to somite formation. *Development* **122**, 3371-3380.

Dhoot, G. K. (1986). Selective synthesis and degradation of slow skeletal myosin heavy chains in developing muscle fibers. *Muscle Nerve* **9**, 155-164.

Dix, D. J., and Eisenberg, B. R. (1991). Distribution of myosin mRNA during development and regeneration of skeletal muscle fibers. *Dev. Biol.* **143**, 422-426.

Du, S. J., Gong, Z., Fletcher, G. L., Sears, M. A., King, M. J., Idler, D. R., and Hew, C. L. (1992). Dramatic growth enhancement in transgenic Atlantic salmon: use of an all fish chimeric growth hormone gene construct. *Biotechnology* **10**, 176-181.

Duan, C., and Hirano, T. (1990). Stimulation of ^{35}S-sulphate uptake by mammalian insulin-like growth factor I and II in cultured cartilage of Japanese eel, *Anguilla japonica*. *J. Exp. Zool.* **256**, 347-350.

Duan, C., and Hirano, T. (1992). Effects of insulin-like growth factor I and insulin on the *in vitro* uptake of sulfate by eel branchial cartilage: evidence for the presence of independent hepatic and pancreatic sulfation factors. *J. Endocrinol.* **133**, 211-219.

Duan, C., Duguay, S. J., and Plisetskaya, E. M. (1993). Insulin-like growth factor I (IGF-I) mRNA expression in coho salmon, *Oncorhynchus kisutch:* tissue distribution and effects of growth hormone/prolactin family peptides. *Fish Physiol. Biochem.* **11**, 371-379.

Duan, C., Duguay, S. J., Swanson, P., Dickhoff, W. W., and Plisetskaya, E. M. (1994). Tissue-specific expression of mRNAs in salmonids: developmental, hormonal, and nutritional regulation. "Perspective in Comparative Endocrinology" (Davey, K. G., Tobe, S. S., Peter, D. E., eds.), pp. 365-372. National Research Council of Canada, Toronto, Canada.

Duan, C., Plisetskaya, E. M., and Dickhoff, W. W. (1995). Expression of insulin-like growth factor I in normally and abnormally developing coho salmon (*Oncorhynchus kisutch*). *Endocrinology* **136**, 446-452.

Duguay, S. L., Park, L. K., Samadpour, M., and Dickhoff, W. W. (1992). Nucleotide sequence and tissue distribution of three insulin-like growth factor I prohormones in salmon. *Mol. Endocrinol.* **6**, 1202-1210.

Duguay, S. L., Lai-Zhang, J., Steiner, D. F., Funkenstein, B., and Chan, S. J. (1996). Developmental and tissue regulated expression of insulin-like growth factor I and II mRNA in *Sparus aurata*. *J. Mol. Endocrinol.* **16**, 123-132.

Edmondson, D. G., and Olson, E. N. (1989). A gene with homology to the myc similarity region of MyoD1 is expressed during myogenesis and is sufficient to activate the muscle differentiation program [published erratum appears in *Genes Dev.* 1990 Aug., 4(8):1450]. *Genes Dev.* **3**, 628-640.

Ennion, S., Gauvry, L., Butterworth, P., and Goldspink, G. (1995). Small-diameter white myotomal muscle fibres associated with growth hyperplasia in the carp (*Cyprinus carpio*) express a distinct myosin heavy chain gene. *J. Exp. Biol.* **198**, 1603-1611.

Ennion, S., Wilkes, D., Gauvry, L., Alami-Durante, H., and Goldspink, G. (1999). Identification and expression analysis of two developmentally regulated myosin heavy chain gene transcripts in carp (*Cyprinus carpio*). *J. Exp. Biol.* **202**, 1081-1090.

Fauconneau, B., Andre, S., Chmaitilly, J., LeBail, P. Y., Krieg, F., and Kaushik, S. J. (1997). Control of skeletal muscle fibres and adipose cells size in the flesh of rainbow trout. *J. Fish. Biol.* **50**, 296-314.

Fine, M. L., Bernard, B., and Harris, T. M. (1993). Functional morphology of toadfish sonc muscle fibres—relationship to possible fiber division. *Can. J. Zool.* **71**, 2262-2274.

Focant, B., Huriaux, F., Vandewalle, P., Castelli, M., and Goessens, G. (1992) Myosin, parvalbumin and myofibril expression in barbel (*Barbus barbus* L.) lateral white muscle during development. *Fish Physiol. Biochem.* **10**, 133-143.

Funkenstein, B., Shermer, R., and Cohen, I. (1996). Nucleotide sequence of the promotor region of

Sparus aurata insulin-like growth factor I gene and expression of IGF-I in eggs and embryos. *Mol. Mar. Biol. Biotech.* **5**, 43–51.

Gauvry, L., Ennion, S., Hansen, E., Butterworth, P., and Goldspink, G. (1996a). The characterisation of the 5′ regulatory region of a temperature-induced myosin-heavy-chain gene associated with myotomal muscle growth in the carp. *Eur. J. Biochem.* **236**, 887–894.

Gauvry, L., Perez, C., and Fauconneau, B. (1996b). Rainbow trout myosin heavy chain polymorphism during development. *In* "Gene Expression and Manipulation in Aquatic Organisms" (Ennion, S., and Goldspink, G., eds.), pp. 149–164. Cambridge University Press, Cambridge.

Gauvry, L., and Fauconneau, B. (1996). Cloning of a trout fast skeletal myosin heavy chain expressed both in embryo and adult muscles and in myotubes neoformed *in vitro*. *Comp. Biochem. Physiol.* **115B**, 183–190.

Gauvry, L., Mohan-Ram, V., Ettelaie, S., Ennion, S., and Goldspink, G. (1997). Molecular motors designed for different tasks and to operate at different temperatures. *J. Therm. Biol.* **22**, 367–373.

Gerlach, G. F., Turay, L., Malik, K. T., Lida, J., Scutt, A., and Goldspink, G. (1990). Mechanisms of temperature acclimation in the carp: a molecular biology approach. *Am. J. Physiol.* **259**, R237–R244.

Goldfine, S. M., Einheber, S., and Fischman, D. A. (1991). Cell free incorporation of newly synthesized myosin subunits into thick filaments. *Muscle Res. Cell Motil.* **12**, 161–170.

Goldspink, G. (1972). Postembryonic growth and differentiation of striated skeletal muscle. *In* "The Structure and Function of Muscle" (Bourne, G. H., Ed.), pp. 179–236. Academic Press, New York.

Goldspink, G. (1996). Muscle growth and muscle function: a molecular biological perspective. *Res. Vet. Sci.* **60**, 193–204.

Greer-Walker, M. (1983). Growth and development of skeletal muscle fibres in the cod. *J. Cons. Int. Explor. Mer.* **33**, 228–244.

Gros, F., and Buckingham, M. E. (1987). Polymorphism of contractile proteins. *Biopolymers* **26**, S177–S192.

Gulick, J., Kropp, K., and Robbins, J. (1985). The structure of two fast-white myosin heavy chain promoters. A comparative study. *J. Biol. Chem.* **260**, 14513–14520.

Guo, X., and Watabe, S. (1993). ATPase activity and thermostability of actomyosins from thermally acclimated carp. *Nippon Suisan Gakkaishi* **59**, 363–369.

Heap, S. P., Watt, P. W., and Goldspink, G. (1985). Consequences of thermal change on the myofibrillar ATPase of five freshwater teleosts. *J. Fish Biol.* **26**, 733–738.

Heap, S. P., Watt, P. W., and Goldspink, G. (1986). Myofibrillar ATPase activity in the carp *Cyprinus carpio*: interactions between starvation and environmental temperature. *J. Exp. Biol.* **123**, 373–382.

Hinterberger, T. J., and Barald, K. F. (1990). Fusion between myoblasts and adult muscle fibres promotes remodelling of fibres into myotubes *in vitro*. *Development* **109**, 139–148.

Hirayama, Y., Kanoh, S., Nakaya, M., and Watabe, S. (1997). The two essential light chains of carp fast skeletal myosin, LC1 and LC3, are encoded by distinct genes and change their molar ratio following temperature acclimation. *J. Exp. Biol.* **200**, 693–701.

Hirayama, Y., and Watabe, S. (1997). Structural differences in the crossbridge head of temperature-associated myosin subfragment-1 isoforms from carp fast skeletal muscle. *Eur. J. Biochem.* **246**, 380–387.

Hofmann, S., Dusterhoft, S., and Pette, D. (1988). Six myosin heavy chain isoforms are expressed during chick breast muscle development. *FEBS Lett.* **238**, 245–248.

Hwang, G. C., Watabe, S., and Hashimoto, K., (1990). Changes in carp myosin ATPase induced by temperature acclimation. *J. Comp. Physiol. B* **160**, 233–239.

Inoue, K., Yamada, S., and Yamashita, S. (1993). Introduction, expression, and growth-enhancing effects of rainbow trout growth hormone cDNA fussed to an avian chimeric promotor in rainbow trout fry. *J. Mar. Biotechnol.* **1**, 131–134.

Imai, J., Hirayama, Y., Kikuchi, K., Kakinuma, M., and Watabe, S. (1997). cDNA cloning of myosin heavy chain isoforms from carp fast skeletal muscle and their gene expression associated with temperature acclimation. *J. Exp. Biol.* **200,** 27–34.

Izumo, S., Nadal-Ginard, B., and Mahdavi, V. (1986). All members of the MHC multigene family respond to thyroid hormone in a highly tissue-specific manner. *Science* **231,** 597–600.

Jaenicke, T., Diederich, K. W., Haas, W., Schleich, J., Lichter, P., Pfordt, M., and Vosberg, H. P. (1990). The complete sequence of the human beta-myosin heavy chain gene and a comparative analysis of its product. *Genomics* **8,** 194–206.

Jones, I. J., and Clemmonds, D. R. (1995). Insulin like growth factors and their binding proteins: biological actions. *Endocr. Rev.* **16,** 3–34.

Johnston, I. A., Frearson, N., and Goldspink, G. (1973). The effects of environmental temperature on the properties of myofibrillar adenosine triphosphatase from various species of fish. *Biochem. J.* **133,** 735–738.

Johnston, I. A., Davison, W., and Goldspink, G. (1975). Adaptations in Mg^{++}-activated myofibrillar ATPase induced by temperature acclimation. *FEBS Lett.* **50,** 293–295.

Johnston, I. A., Davison, W., and Goldspink G. (1977) Energy metabolism of carp swimming muscles. *J. Comp. Physiol.* **114,** 203–216.

Johnston, I. A. (1979). Calcium regulatory proteins and temperature acclimation of actomyosin ATPase from a eurythermal teleost (*Carassius auratus* L.). *J. Comp. Physiol. B* **129,** 163–167.

Johnston, I. A., and Lucking, M. (1983). Temperature induced variation in the distribution of different types of muscle fibre in the goldfish (*Carassius auratus*). *J. Dev. Physiol.* **124,** 111–116.

Johnston, I. A., Sidell, B. D., and Driedzic, W. R. (1985). Force-velocity characteristics and metabolism of carp muscle fibres following temperature acclimation. *J. Exp. Biol.* **119,** 239–249.

Johnston, I. A., Cole, N. J., Vieira, V. L. A., and Davidson, I. (1997). Temperature and developmental plasticity of muscle phenotype in herring larvae. *J. Exp. Biol.* **200,** 849–868.

Johnson, T. P., and Bennett, A. F. (1995). The thermal-acclimation of burst escape performance in fish—an integrated study of molecular and cellular physiology and organismal performance. *J. Exp. Biol.* **198,** 2165–2175.

Karn, J., Brenner, S., and Barnett, L. (1983). Protein structural domains in the *Caenorhabditis elegans* unc-54 myosin heavy chain gene are not separated by introns. *Proc. Natl. Acad. Sci. U.S.A.* **80,** 4253–4257.

Kagawa, H., Moriyama, S., and Kawauchi, H. (1995). Immunocytochemical localisation of IGF-I in the ovary of the red seabream, *Pagrus major. Gen. Comp. Endocrinol.* **99,** 307–315.

Kavsan, V. M., Gray, E. S., Siharath, K., Nicoll, C. S., and Bern, H. A. (1993). Structure of the chum salmon insulin-like growth factor I gene. *DNA Cell Biol.* **12,** 729–737.

Kavsan, V. M., Grebenjuk, V. A., Kova, A. P., Skorokhod, A. S., Roberts, C. T., and LeRoith, D. (1994). Isolation of a second nonallelic insulin-like growth factor I gene from the salmon genome. *DNA Cell Biol.* **13,** 555–559.

Kelly, A. M., and Zack, S. I. (1969) The histogenesis of rat intercostal muscle. *J. Cell Biol.* **42,** 135–153.

Kong, Y., Flick, M. J., Kudla, A. J., and Konieczny, S. F. (1997). Muscle LIM protein promotes myogenisis by enhancing the activity of MyoD. *Mol. Cell. Biol.* **17,** 4750–4760.

Koumans, J. T. M., Akster, H. A., Booms, G. H. R., and Osse, J. W. M. (1993a). Growth of carp (*Cyprinus carpio*) white axial muscle: hyperplasia and hypertrophy in relation to the myonucleus/sarcoplasm ratio and the occurence of different subclasses of myogenic cells. *J. Fish Biol.* **43,** 69–80.

Koumans, J. T. M., Akster, H. A., Booms, R. G. H., and Osse, J. W. M. (1993b). Influence of fish size on proliferation of cultured myosatellite cells of white axial muscle of carp (*Cyprinus carpio* L.). *Differentiation* **53,** 1–6.

Leinwand, L. A., Fournier, R. E., Nadal-Ginard, B., and Shows, T. B. (1983a). Multigene family for sarcomeric myosin heavy chain in mouse and human DNA: localization on a single chromosome. *Science* **221,** 766–769.

Leinwand, L. A., Saez, L., McNally, E., and Nadal-Ginard, B. (1983b). Isolation and characterization of human myosin heavy chain genes. *Proc. Natl. Acad. Sci. U.S.A.* **80,** 3716–3720.

Liang, Y. H., Cheng, C. H., and Chan, K. M. (1996). Insulin-like growth factor Iea2 is the predominantly expressed form of IGF in common carp (*Cyprinus carpio*). *Mol. Mar. Biol. Biotechnol.* **5,** 145–152.

Lonn, U., Lonn, S., and Stenkvist, B. (1993). Appearance of amplified thymidylate synthase or dihydrofolate reductase genes in stage-IV breast-cancer patients receiving endocrine treatment. *Int. J. Cancer* **54,** 237–242.

Loughna, P. T., and Goldspink, G. (1984). The effects of starvation upon protein turnover in red and white myotomal muscle of rainbow trout, *Salmo gairdneri* Richardson. *J. Fish Biol.* **25,** 223–230.

Loughna, P. T., and Goldspink, G. (1985). Muscle protein synthesis rates during temperature acclimation in a eurythermal (*Cyrinus carpio*) and a stenothermal (*Salmo gardneri*) species of teleost. *J. Exp. Biol.* **118,** 267–276.

Lowey, S. (1986). The structure of vertebrate muscle myosin. In "Myology, Basic and Clinical" (Engel, A. G., and Banker, B. Q., eds.), pp. 563–587. McGraw-Hill, New York.

Ludolph, D. C., and Konieczny, S. F. (1995). Transcription factor families: muscling in on the myogenic program (review). *FASEB J.* **9,** 1595–1604.

Lyons, G. E., Ontell, M., Cox, R., Sassoon, D., and Buckingham, M. (1990). The expression of myosin genes in developing skeletal muscle in the mouse embryo. *J. Cell Biol.* **111,** 1465–1476.

Mack, A. E., Balt S. L., and Fernald, R. D. (1995). Localization and expression of insulin-like growth factor in the teleost retina. *Visual Neurosci.* **12,** 457–461.

Marchant, T. A., and Moroz, B. M. (1993). The influence of insulin-like growth factor and related peptides on the uptake of [^{35}S]-sulphate into the cartilage from the goldfish (*Carassius auratus* L.). *Fish Physiol. Biochem.* **11,** 393–400.

Martinez, I., Christiansen, J. S., Ofstad, R., and Olsen, R. L. (1991). Comparison of myosin isoenzymes present in skeletal and cardiac muscles of the Arctic charr *Salvelinus alpinus* (L.). Sequential expression of different myosin heavy chains during development of the fast white skeletal muscle. *Eur. J. Biochem.* **195,** 743–753.

Martinez, I., and Pettersen, G. W. (1992). Temperature-induced precocious transitions of myosin heavy chain isoforms in the white muscle of the Arctic charr, *Salvelinus alpinus* (L.). *Bas. Appl. Myol.* **2,** 89–95.

Martinez, I., Bang, B., Hatlen, B., and Blix, P. (1993). Myofibrillar proteins in skeletal muscles of parr, smolt and adult atlantic salmon (salmo-salar l.)—comparison with another salmonid, the arctic charr salvelinus-alpinus (l.). *Comp. Biochem. Physiol. B* **106,** 1021–1028.

Martinez, I., Dreyer, B., Agersborg, A., Leroux, A., and Boeuf, G. (1995). Effects of T3 and rearing temperature on growth and skeletal myosin heavy chain isoform transition during early development in the salmonid *Salvelinus alpinus* (L.). *Comp. Biochem. Physiol.* **112B,** 717–725.

Matschak, T. W., Hopcroft, T., Mason, P. S., Crook, A. R., and Stickland, N. (1998). Temperature and oxygen tension influence the development of muscle cellularity in embryonic rainbow trout. *J. Fish Biol.* **53,** 581–590.

McKoy, G., Léger, M., Bacou, F., and Goldspink, G. (1998). Differential Expression of Myosin heavy chain mRNA and protein isoforms in four functionally diverse rabbit skeletal muscles during pre- and postnatal development. *Dev. Dynamics* **211,** 193–203.

McKoy, G., Ashley, W., Mander, J., Yang, S. Y., Williams, N., and Goldspink, G. (1999). Expression of IGF-1 splice variants and structural genes in rabbit skeletal muscle are induced by stretch and stimulation. *J. Physiol.* **516.2,** 583–592.

Merati, A. L., Bodine, S. C., Bennett, T., Jung, H. H., Furuta, H., and Ryan, A. F. (1996). Identification of a novel myosin heavy chain gene expressed in the rat larynx. *Bba. Gene. Struct. Express.* **1306,** 153–159.

Molina, M. I., Kropp, K. E., Gulick, J., and Robbins, J. (1987). The sequence of an embryonic myosin heavy chain gene and isolation of its corresponding cDNA. *J. Biol. Chem.* **262,** 6478–6488.

Moriyama, S. (1995). Increased plasma insulin-like growth factor I (IGF-I) following oral intraperitoneal administration of growth hormone to rainbow trout, *Oncorhynchus mykiss*. *Growth Regulat.* **5,** 164–167.

Muller, F., Williams, D. W., Kobolak, J., Gauvry, L., Goldspink, G., Orban, L., and Maclean, N. (1997). Activator effect of coinjected enhancers on the muscle-specific expression of promoters in zebrafish embryos. *Mol. Reprod. Dev.* **47,** 404–412.

Narusawa, M., Fitzsimons, R. B., Izumo, S., Nadal-Ginard, B., and Rubinstein, N. A. (1987). Slow myosin in developing rat skeletal muscle. *J. Cell Biol.* **104,** 447–459.

Negatu, Z., and Meier, A. H. (1995). *In vitro* incorporation of [^{14}C] glycine into muscle protein of gulf killifish (*Fundulus grandis*) in response to insulin-like growth factor I. *Gen. Comp. Endocrinol.* **98,** 193–201.

Newman, J. N., and Wu, T. Y.-T. (1975). Hydromechanical aspects of fish swimming. *In* "Swimming and Flying in Nature" (Wu, T.Y-T., Brokshaw, C. J., and Brennen, C., eds.), pp. 615–634. Plenum Press, New York.

Penney, R. K., and Goldspink, G. (1981a). Temperature adaptation by the myotomal muscle of fish. *J. Therm Biol.* **6,** 297–306.

Penney, R. K., and Goldspink, G. (1981b). Regulatory proteins and thermostability of myofibrillar ATPase in acclimated goldfish. *Comp. Biochem. Physiol. B* **69B,** 577–583.

Radice, G. P., and Malacinski, G. M. (1989). Expression of myosin heavy chain transcripts during *Xenopus laevis* development. *Dev. Biol.* **133,** 562–568.

Radice, G. P. (1995). Spatial expression of two tadpole stage specific myosin heavy chains in *Xenopus laevis*. *Acta Anat.* **153,** 254–262.

Rayment, I., Holden, H. M., Whittaker, C. B., Yohn, C. B., Lorenz, M., Holmes, K. C., and Milligan, R. A. (1993a). Structure of the Actin-Myosin complex and its implications for muscle contraction. *Science* **261,** 58–65.

Rayment, I., Rypniewski, W. R., Schmidt-Base, K., Smith, R., Tomchick, D. R., Benning, M. M., Winkelmann, D. A., Wessenberg, G., and Holden, H. M. (1993b). Three-dimensional structure of myosin subfragment-1: a molecular motor. *Science* **261,** 50–58.

Rindt, H., Gulick, J., Knotts, S., Neumann, J., and Robbins, J. (1993). *In vivo* analysis of the murine beta-myosin heavy chain gene promoter. *J. Biol. Chem.* **268,** 5332–5338.

Rindt, H., Subramaniam, A., and Robbins, J. (1995). An *in vivo* analysis of transcriptional elements in the mouse alpha-myosin heavy chain gene promoter. *Transgenic Res.* **4,** 397–405.

Robbins, J., Horan, T., Gulick, J., and Kropp, K. (1986). The chicken myosin heavy chain family. *J. Biol. Chem.* **261,** 6606–6612.

Rome, L. C., Loughna, P. T., and Goldspink, G. (1984). Muscle fiber activity in carp as a function of swimming speed and muscle temperature. *Am. J. Physiol.* **247,** R272–R279.

Rome, L. C., Funke, R. P., Alexander, R. M., Lutz, G., Aldridge, H., and Scott, F. (1988). Why animals have different muscle fibre types. *Nature* **335,** 824–827.

Rome, L. C., and Bennett, A. F. (1989). Influence of temperature on muscle and locomotor performance. *Am. J. Physiol.* **258,** R189–R190.

Rome, L. C. (1990). Influence of temperature on muscle recruitment and muscle function *in vivo*. *Am. J. Physiol.* **259,** R210–R222.

Rome, L. C., Cook, C., Syme, D. A., Connaughton, M. A., Ashley-Ross, M., Kilimov, A., Tikunov, B. M., and Goldman, Y. E. (1999). Trading forces for speed: why superfast crossbridge kinetics leads to superlow force. *PNAS,* **96,** 5826–5831.

Rowlerson, A., Scapolo, P. A., Mascarello, F., Carpene, E., and Veggetti, A. (1985). Comparative study of myosins present in the lateral muscle of some fish: species variations in myosin isoforms and their distribution in red, pink and white muscle. *J. Muscle Res. Cell Motil.* **6,** 601–640.

Rowlerson, A., Mascarello, F., Radaelli, G., and Veggetti, A. (1995). Differentiation and growth of muscle in the fish *Sparus aurata* (L): II. Hyperplastic and hypertrophic growth of lateral muscle from hatching to adult. *J. Muscle Res. Cell Motil.* **16,** 223–236.

Rowlerson, A., Radaelli, G., Mascarello, F., and Veggetti, A. (1997). Regeneration of skeletal muscle in two teleost fish: *Sparus aurata* and *Brachydanio rerio*. *Cell Tissue Res.* **289,** 311–322.

Sartore, S., Gorza, L., and Schiaffino, S. (1982). Fetal myosin heavy chains in regenerating muscle. *Nature* **298,** 294–296.

Schiaffino, S., and Reggiani, C. (1996). Molecular diversity of myofibrillar proteins: Gene regulation and functional significance. *Physiol. Rev.* **76,** 371–423.

Sellers, J. R., Wang, F., and Goodson, H. V. (1996). A myosin family reunion. *J. Muscle Res. Cell Motil.* **17,** 7–22.

Spudich, J. A. (1994). How molecular motors work. *Nature* **372,** 515–518.

Stedman, H. H., Eller, M., Jullian, E. H., Fertels, S. H., Sarkar, S., Sylvester, J. E., and Rubinstein, N. A. (1990). The human embryonic myosin heavy chain. Complete primary structure reveals evolutionary relationships with other developmental isoforms. *J. Biol. Chem.* **265,** 3568–3576.

Stickland, N. C. (1983). Growth and development of muscle fibres in the rainbow trout (*Salmo gairdneri*). *J. Anat.* **137,** 323–333.

Strehler, E. E., Strehler-Page, M. A., Perriard, J. C., Periasamy, M., and Nadal-Ginard, B. (1986). Complete nucleotide and encoded amino acid sequence of a mammalian myosin heavy chain gene. Evidence against intron-dependent evolution of the rod. *J. Mol. Biol.* **190,** 291–317.

Takeda, S., North, D. L., Lakich, M. M., Russell, S. D., Kahng, L. S., and Whalen, R. G. (1992a). Evolutionarily conserved promoter motifs and enhancer organization in the mouse gene encoding the IIB myosin heavy chain isoform expressed in adult fast skeletal muscle. *C. R. Acad. Sci. Ser. III* **315,** 467–472.

Takeda, S., North, D. L., Lakich, M. M., Russell, S. D., and Whalen, R. G. (1992b). A possible regulatory role for conserved promoter motifs in an adult-specific muscle myosin gene from mouse. *J. Biol. Chem.* **267,** 16957–16967.

Tsika, R. W., Bahl, J. J., Leinwand, L. A., and Morkin, E. (1990). Thyroid hormone regulates expression of a transfected human alpha-myosin heavy-chain fusion gene in fetal rat heart cells. *Proc. Natl. Acad. Sci. U.S.A.* **87,** 379–383.

Uyeda, T. Q. P., Ruppel, K. M., and Spudich, J. A. (1994). Enzymatic activities correlate with chimaeric substitutions at the actin-binding face of myosin. *Nature* **368,** 567–569.

Watabe, S., Guo, X. F., and Hwang, G. C. (1994). Carp express specific isoforms of the myosin cross-bridge head, subfragment-1, in association with cold and warm temperature acclimation. *J. Therm. Biol.* **19,** 261–268.

Weatherley, A. H., and Gill, H. S. (1984). Growth dynamics of white myotomal muscle fibres in the bluntnose minnow *Pimephales notatus* Gafinesque and comparison with rainbow trout *Salmo gairdneri* Richardson. *J. Fish Biol.* **25,** 13–24.

Weeds, A. G., and Lowey, S. (1971). Substructure of the myosin molecule II. The light chains of myosin. *J. Mol. Biol.* **61,** 701–725.

Weintraub, H., Davis, R., Tapscott, S., Thayer, M., Krause, M., Benezra, R., Blackwell, T. K., Turner, D., Rupp, R., Hollenberg, S., Zhuang, Y., and Lasser, A. (1991). The MyoD gene family: nodal point during specification of the muscle lineage. *Science* **251,** 761–766.

Weiss, A., and Leinwand, L. A. (1996). The mammalian myosin heavy chain gene family. *J. Therm. Biol.* **22,** 367–373.

Whalen, R. G., Harris, J. B., Butler-Browne, G. S., and Sesodia, S. (1990). Expression of myosin isoforms during notexin-induced regeneration of rat soleus muscles. *Dev. Biol.* **141,** 24–40.

Wiseman, J. W., Glover, L. A., and Hesketh, J. E. (1997). Evidence for a localization signal in the 3' untranslated region of myosin heavy chain messenger RNA. *Cell Biol. Int.* **21,** 243–248.

Xiong, Y. L., and Blanchard, S. P. (1994). Dynamic gelling properties of myofibrillar protein from skeletal muscles of different chicken parts. *J. Agr. Food Chem.* **42,** 670–674.

Yang, S. Y., Alnaqeeb, M., Simpson, H., and Goldspink, G. (1996). Cloning and characterisation of an IGF-1 isoform expressed in skeletal muscle subjected to stretch. *J. Muscle Res. Cell Motil.* **17,** 487–495.

4

MUSCLE SATELLITE CELLS IN FISH

BENOIT FAUCONNEAU
GILLES PABOEUF

    I. Abstract
   II. Introduction
  III. Characteristics of Satellite Cells *In Vivo*
  IV. Characteristics of Satellite Cells *In Vitro*
      A. Satellite Cell Isolation
      B. Satellite Cell Enrichment
      C. Methods of Culture
      D. Proliferation
      E. Differentiation
   V. Origin of Satellite Cells and Related Questions
  VI. Contribution of Satellite Cells to Muscle Growth
 VII. Effect of Intrinsic Factors
      A. Changes with Age/Body Weight
      B. Effect of Muscle Type
      C. Effect of Genetic Origin
      D. Hormonal Control
VIII. Effect of Extrinsic Factors
      A. Effect of Temperature
      B. Effect of Feeding
      C. Other Environmental Factors
  IX. Conclusions
     Acknowledgments
     References

I. ABSTRACT

The characteristics of fish satellite cells *in situ* and *in vitro* are described. The *in situ* proliferation of satellite cells appears to be very low and therefore not easy to quantify. Differences in the *in vitro* proliferation capacities of satellite cells have been observed between species, but a large part of these differences could be related to methodology. Most studies on fish satellite cells have been carried out on cyprinids and salmonids. The relationship between proliferation capacity and muscle growth is discussed in these fish. The differentiation of fish satellite cells is partially similar to that of myoblasts. However, there is a great heterogeneity in satellite cell differentiation that needs to be characterized.

The effects of intrinsic and extrinsic factors on satellite cells are reviewed. It appears that aging, genetic origin, and muscle type affect either the number or the proliferation capacity of satellite cells. The factors which are involved in these intrinsic differences in proliferation capacities are not yet known. Among the extrinsic factors, feeding dramatically affects the state of satellite cells and in particular their initial proliferation capacities. This has never been suspected in mammals and birds. It is not yet known if other environmental factors could also modulate satellite cell proliferation. Temperature certainly seems to affect differentiation rate.

The effects of various intrinsic and extrinsic factors on satellite cells suggest that there are different populations of satellite cells which contribute to both hyperplastic and hypertrophic growth of muscle.

II. INTRODUCTION

Satellite cells have only been described recently in mammalian and avian muscle (Mauro, 1961). These cells are quiescent cells closely attached to existing muscle fibers and are located under the basal lamina (Campion, 1984). The contribution of these cells to muscle growth by hypertrophy of existing fibers and to muscle regeneration by rebuilding damaged fibers is now well known in vertebrates (Stockdale, 1992; Shultz, 1996). However, although the physiological context of the initiation of satellite cells is known, the factors which are involved in their initiation are not fully understood and are subject to debate (Quinn *et al.*, 1990). Due to the anatomical location of the satellite cells, factors affecting the muscle fibers will have secondary effects on the cells. In addition, factors affecting the attachment of satellite cells to the fibers or the integrity of the basal lamina will have direct effects on the cells. Satellite cells originate from a pool of precursor cells thought to be produced at the end of embryonic development (Stockdale, 1992) and although these cells are considered to be pluripotent (Quinn *et al.*, 1988), their number of divisions are limited and their division capacity is probably progressively less.

Compared with other vertebrate groups, less attention has been paid to satellite cells in fish. Satellite cells have only been recently described *in situ* (Nag and Nursall, 1972; Krivy and Eide, 1977; Willemse and van den Berg, 1978; O'Connell, 1981; Akster, 1983). Unlike other vertebrates, myogenesis in fish still occurs after hatching and throughout most of the life cycle through the production of new fibers (hyperplasia). Measurement of changes in fiber number demonstrates clearly the process (Greer-Walker, 1970; Stickland, 1983; Koumans *et al.*,

1993b). Satellite cells will account both for hypertrophy of existing fibers and for myogenesis of new fibers. Involvement of satellite cells is, however, not easy to demonstrate because *in situ* direct observations of the process is difficult. In addition they are not very easy to extract and observe *in vitro*. Many attempts to develop fish muscle cell lines have not been successful (Hightower and Renfro, 1988) and most of the characteristics of these cells have been described on primary cell culture *in vitro*.

This review aims to describe characteristics of fish satellite cells and to emphasize factors which are involved in the initiation of these cells and their contribution to muscle growth.

III. CHARACTERISTICS OF SATELLITE CELLS *IN VIVO*

A few papers have described satellite cells *in situ* (Nag and Nursall, 1972; Willemse and van den Berg, 1978; Akster *et al.*, 1995; Koumans *et al.*, 1995; Stoiber and Sänger, 1996). Satellite cells are small (less than 5 μm) and in transverse sections of muscle can be seen at the periphery of fibers under the basal lamina (Fig. 1a) but separated from the fibers by cell membranes. Satellite cells are characterized by a heterochromatic nucleus and reduced cytoplasm. These criteria have been used to identify and quantify myosatellite cells *in situ*. Cells with such characteristics are often termed quiescent stem cells and appear very different from mesenchymal cells and myoblasts observed in early and late embryogenesis (Nag and Nursall, 1972; Stoiber and Sänger, 1996; Johnston *et al.*, 1998). In mammals, satellite cells lie near myofiber nuclei on transverse sections and are uniformly distributed along fibers in longitudinal sections (Campion, 1984). In fish, satellite cells are generally located at the multiple points of existing fibers and more specifically at angles of the polygonal shape formed by muscle fibers in transverse sections (Willemse and van den Berg, 1978; Romanello *et al.*, 1987; Koumans *et al.*, 1991). In longitudinal section these cells appear as elongated spindle-shaped cells distributed uniformly along the fibers (Johnston *et al.*, 1998).

The number of muscle satellite cells observed *in situ* in rainbow trout accounts for 3–7% of the 100–580 total myonuclei per 100 fibers quantified depending on size (Alfei *et al.*, 1989). Similar percentages of myosatellite cells were found in shark and carp (6–8%, respectively, Krivy and Eide, 1977 and 1–8%, respectively, Koumans *et al.*, 1991). In carp, the total number of cells was estimated to lie between 10^3 and 10^7 cells.g^{-1} muscle (Koumans *et al.*, 1991). Krivy and Eide (1977) found the percentage of satellite cells to be similar between red and white muscle.

Satellite cell can also be characterized by their ability to enter the cell cycle (Table I). It has been demonstrated by cytofluorimetry (Alfei *et al.*, 1989) and by BrdU uptake/PCNA cyclin expression (Koumans *et al.*, 1991; Alfei *et al.*, 1993,

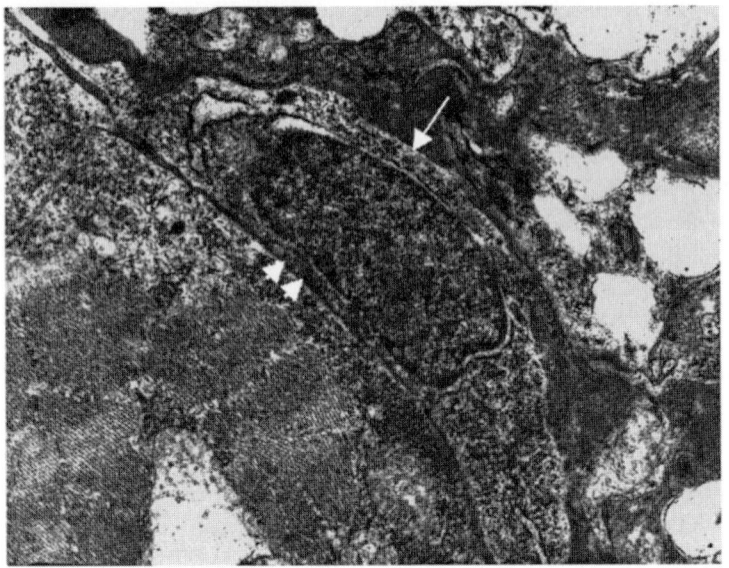

a

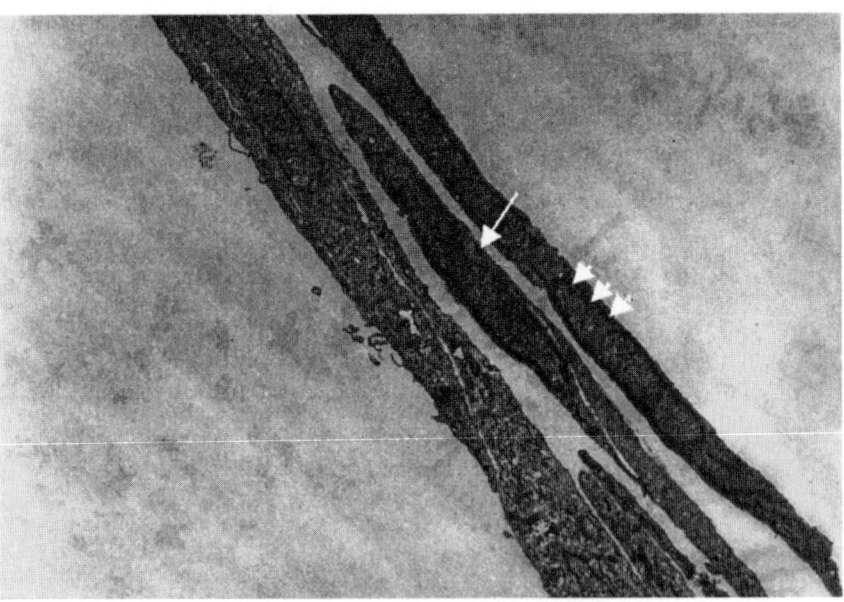

b

Fig. 1. Satellite cells (a) observed *in situ* in the white deep muscle of carp larvae at hatching (magnification × 30,000) and (b) satellite cells extracted from white muscle of rainbow trout and cultured *in vitro* (magnification ×10,000). Single arrow: satellite cells; double head: absence of basal lamina between satellite cells and fiber; triple arrow: myotubes with two nuclei.

Table I.

Number and *In Situ* Proliferation of Satellite Cells in Skeletal Muscle of Different Fish Species (FL: fork length; no.g^{-1} number of cells/g muscle)

Species	Stages	Muscle	Investigation method	Localization	Satellite cells (% muscle cells)	Satellite cells (no.g^{-1} muscle)	Proliferation	Factor tested	Authors
Shark *Etmopterus spinax*	10–23 cm FL	Caudal white & red	^{3}H-Thymidine	White muscle Red muscle	6–7% SC 7–8% SC		0.5% MN 2% MN		Kryvi and Eide, 1977
Rainbow trout *Salmo gairdneri*	Different stages	Caudal white	DNA cytofluorometry LM		3–7% SC	nd	2–7% MN	Age	Alfei et al., 1989
Carp *Cyprinus carpio*	4–18 cm FL	White	TEM/LM and DNA dissociation		54% MC 1–8% SC	10^3–10^7·g^{-1}	Not observed	Age	Koumans et al., 1991
Carp *Cyprinus carpio* L.	Different stages	Caudal white	BrdU uptake DNA cytofluorometry	Inner	5–6%	nd	5–6% MN	Age	Alfei et al., 1993
Carp *Cyprinus carpio* L.	Different stages	Caudal white	PCNA expression BrdU uptake	Inner	2–7% SC	nd	2–7% MN	Age	Alfei et al., 1995
Carp *Cyprinus carpio* L.	Juvenile	Anterior white and red	^{3}H-Thymidine uptake	Inner		nd	<1% MN 1.5–2% MN		Akster et al., 1995
Pearlfish *Rutilis frisii meidingeri*	Larvae	Anal myotomes	TEM	Dorsal and ventral EC and Inner SC	nd	nd			Stoiber and Sänger, 1996
Roach *Rutilis rutilis*	Swimm-up larvae	—	—	Inner SC	nd	nd			
Rainbow trout *Oncorhynchus mykiss*	Pre-hatch embryo	—	—	Dorsal-ventral EC/SC and inner SC	nd	nd	Suspected		
Herring *Culpea harengus* L.	Larvae		BrdU uptake	Dorsal-ventral EC and inner SC	nd	nd	nd	Temperature	Johnston et al., 1998
Rainbow trout *Oncorhynchus mykiss*	Juvenile	Caudal white	PCNA expression	Inner muscle			0–10% MN	Feeding	Fauconneau and Paboeuf, unpublished data

Note: Cells: EC, external cell; SC, Satellite cells; MC, muscle cells; MN, muscle nuclei. Methods: TEM, transmission electron microscopy; LM, light microscopy; DNA Quantification.

1995; Rowlerson et al., 1995, Fauconneau and Paboeuf, unpublished data) that muscle cells proliferate *in situ*. The amount of proliferation measured by these methods is not very high (less than 10% of muscle nuclei). It is not known, however, which categories of cells—muscle cells, fibroblasts, adipose cells—account for the proliferation observed.

However, further analysis of BrdU uptake or ^{3}H-thymidine uptake, combined with observation of the basal lamina at the ultrastructural level, clearly demonstrate the proliferation of satellite cells. In carp, proliferative muscle cell nuclei measured by BrdU uptake for 2–3 days vary between 2 and 7% of the total nuclei (Alfei et al., 1995). The percentage of muscle cell proliferation measured in carp by ^{3}H-thymidine labeling is however low: less than 1% in white muscle and 1.5–2% in red muscle. Additionally, in red muscle, only 14–15% of these nuclei were myosatellite cells. This is similar to previous findings by Krivy and Eide (1977) who demonstrated that in shark the incorporation of ^{3}H-thymidine in muscle nuclei cells is low in both white and red muscle (0.5 and 2% of muscle nuclei, respectively). Thus direct observation of proliferation could be complicated by the method used and the low proliferation rate of satellite cells.

Using transverse sections of muscle, some authors have observed muscle contractile proteins and even myofibrils in satellite cells (Willemse and van den Berg, 1978, Stoiber and Sänger, 1996). It is suggested that these cells are differentiated satellite cells or even myotubes.

IV. CHARACTERISTICS OF SATELLITE CELLS *IN VITRO*

Satellite cells and myoblasts have been extracted from various fish species: salmon, trout, carp, catfish, and even zebrafish. The common features and the differences in the extraction and culture methods are described, respectively, in Tables II and III. The characteristics and phenotypes of the cells extracted seem to be very similar among the different species, but the proliferation and differentiation programs expressed by these cells *in vitro* are very different, thus a critical analysis in the methods used is required.

A. Satellite Cell Isolation

Satellite cells have only been isolated from white muscle and no data on satellite cells from red muscle are available. However, as careful dissection is not always specified, part of the variability observed in the phenotype of satellite cells could be related to the extraction of a mixture of cells from white, red, and intermediate muscle.

Table II.
Methods for the Extraction of Satellite Cells from Skeletal Muscle of Different Fish Species

Species	Muscle	Extraction medium	Enzymatic digestion I	Enzymatic digestion II	Yield (10^6 cells·g^{-1} muscle)	Authors
Rainbow trout *Oncorhynchus mykiss*	Skeletal	PBS Ca-Mg-Free	Pronase E 0.1% $2 \times 1hr/20°C$		nd	Powell et al., 1989
Carp *Cyprinus carpio*	White	DMEM 90% HS 15%	Collagenase 0.2% 45 min/room T^{ure}	Trypsin 0.1% 2×15 min/room T^{ure}	$\sim 10^6 \cdot g^{-1}$	Koumans et al., 1990
Zebrafish *Brachydanio rerio*	Embryos	Ca-free Ringer 1 mM EDTA	Trypsin 0.25%			Sepich et al., 1994
Rainbow trout *Oncorhynchus mykiss*	Trunk	HBSS Ca-Mg-Free	Collagenase 1 mg/ml 1 h/room T^{ure}	Trypsin 1 mg/ml 45 min/room T^{ure}	$2 \times 10^6 \cdot g^{-1}$	Greenlee et al., 1995a,b
Rainbow trout *Oncorhynchus mykiss*	White	DMEM 90% HS 15%	Collagenase 0.2% 1 h/15°C	Trypsin 0.1% 2×20 min/15°C	2 to $4 \times 10^6 \cdot g^{-1}$	Rescan et al., 1995a
Salmon *Salmo salar*	White	DMEM 90% HS 15%	Collagenase 0.2% 90 min/11°C	Trypsin 0.1% 30 min/11°C	$3 \times 10^6 \cdot g^{-1}$	Matschak and Stickland, 1995

Note: Medium: DMEM Dulbecco modified Eagle medium; HBSS Hanks balanced salt solution, L15 Leibovitz. Supplement: HS, horse serum.

Table III.
Methods for Culture of Satellite Cells Extracted from Skeletal Muscle of Different Fish Species

Species	Purification of cell extract	Substrate	Cells density ($\times 10^5 \cdot cm^2$)	Culture medium	Proliferation	Differentiation	Markers tested	Authors
Rainbow trout *Oncorhynchus mykiss*	No purification	Gelatin	nd	L15; FBS 10–15%	No proliferation	Small myotubes less than 5 nuclei		Powell et al., 1989
Carp *Cyprinus carpio*	Adhesion on laminin 15 min	Laminin 20 µg/ml on PLL 100 µg/ml	$\sim 10^5 \cdot cm^2$	DMEM 90%; HS 15%	No proliferation (once seeded)	Large myotubes (cross-striation)	Desmin	Koumans et al., 1990
Zebrafish *Brachydanio rerio*	No purification	Laminin 10 µg/ml		L15–3% FCS; 10% embryo extract	Not assessed	Small myotubes (cross-striation)	Acetylcholine receptors	Sepich et al., 1994
Rainbow trout *Oncorhynchus mykiss*	No purification	Laminin 20 µg/ml on PLL 200 µg/ml	0.4 to $4.0 \times 10^5 \cdot cm^2$	L15; FBS 10%	30–60%·day^{-1} BrdU incorporation	Small myotubes	Desmin; myosin MF20	Greenlee et al., 1995a,b
Rainbow trout *Oncorhynchus mykiss*	Adhesion on laminin 20 min	Laminin 20 µg/ml on PLL 100 µg/ml	1 to $5 \times 10^5 \cdot cm^2$	DMEM 90%; FCS 10%	10–60%·day^{-1} BrdU incorporation	Large myotubes (cross-striation)	Desmin; myosin	Rescan et al., 1995a Fauconneau and Paboeuf, 1999
Salmon *Salmo salar*	Adhesion on laminin 20 min 11°C	Laminin 20 µg/ml on PLL 100 µg/ml	$3.2 \times 10^5 \cdot cm^2$	DMEM 90%; HS 15%	No proliferation BrdU incorporation	Small myotubes (cross-striation)	Myosin 83b6	Matschak and Stickland, 1995

Note: Antibodies: Myosin MF20 anti-myosin heavy chain chicken antibody; myosin 83b6 anti-myosin heavy chain mouse antibody.
Supplement: HS, horse serum; FBS, fetal bovine serum; FCS, fetal calf serum.
Substrates: PLL, Poly-L-lysine.

Muscle cells are isolated by a two-step enzymatic digestion. The collagenase-trypsin method of Yablonka-Reuveni *et al.* (1987) has been successfully developed in carp by Koumans *et al.* (1990) and applied to rainbow trout (Greenlee *et al.*, 1995a; Rescan *et al.*, 1995a) and salmon (Matschak *et al.*, 1995). The duration of enzymatic digestion has to be extended at low temperature, although this is species specific. The yeast collagenase generally used to extract fish cells is not used at its optimum temperature and thus tissue digestion requires more time as the temperature is decreased. Enzymatic digestion must be accompanied by mechanical dissociation by gentle trituration through pipettes. Cells extracted by this method, including salmon cells reared at a temperature of $5°C$ (Matschak *et al.*, 1995), differentiate very rapidly into small myotubes within a few days. A yield of several millions of cells per gram of muscle tissue is observed in the different species (Koumans *et al.*, 1990; Greenlee *et al.*, 1995a; Rescan *et al.*, 1995a).

The collagenase-trypsin method is considered by some authors to be too strong for muscle cell extraction and is often associated with the appearance of large syncitia such as those observed in carp (Koumans *et al.*, 1990) and rainbow trout (Rescan *et al.*, 1995a). Thus other gentle extraction processes, such as the classical method used in mammals which includes a pronase digestion, are sometimes recommended. This method has been tested in fish, but the few cells extracted were not in a very good state at the end of extraction process and were seeded directly on gelatin and not on a specific substrate (Powell *et al.*, 1989). Further investigations are thus required to ascertain which method of extraction is the most suitable for fish.

B. Satellite Cell Enrichment

Mononucleated muscle cells are composed of a mixture of fibroblast cells, satellite cells, blood cells, and even adipose cells. Only half the cells present *in situ* in muscle tissue are satellite cells (Koumans *et al.*, 1995) and the proportion seems to be similar after extraction (Koumans *et al.*, 1990). Enrichment is therefore needed to increase the number of myosatellite cells in the muscle cell extract.

Two methods are generally used in mammals, pre-plating or purification by sedimentation. Pre-plating of the muscle cell extract on a simple substrate (gelatin or plastic) enables fibroblast cells to adhere to the substrate, leaving the remaining extract enriched in satellite cells (Yafee, 1968). However, this method has been tested unsuccessfully in fish (Koumans *et al.*, 1990, Fauconneau and Paboeuf, unpublished data). It is thought that perhaps more time is needed for the adhesion of fish cells at the temperature tested (even room temperature) than for cells from warm blooded animals. The purification of muscle cells on a Percoll gradient has been tested with carp (Koumans *et al.*, 1990) and rainbow trout muscle cells (Greenlee *et al.*, 1995a). However, the cells appear to be too fragile to support this supplemental step and thus the percentage of myosatellite cells does not increase.

The purification of muscle cells by selective centrifugation has been tested in trout, but no differences in the percentage of myosatellite cells are observed between low density and high density extract (Fauconneau and Paboeuf, unpublished data).

The method developed in fish (Table III) is based on the relative affinity of satellite cells to a component of the basal lamina—laminin (Foster *et al.*, 1987). Muscle cells are seeded on a laminin layer which is deposited on a pre-coated poly-L-lysine substratum (Koumans *et al.*, 1990). The muscle cell extract is left on the substrate for 20–30 min and non-adherent cells are removed by washing. The remaining cells are enriched by up to 90% in myosatellite cells (Koumans *et al.*, 1990; Rescan *et al.*, 1995a). This enrichment step is not always implemented. In some studies all the cells are left adhered on the laminin substrate (Greenlee *et al.*, 1995a,b) and it could be suspected that the percentage of myosatellite cells is low.

C. Methods of Culture

As mentioned above, muscle cells are cultured on a laminin substrate, although other substrates have been tested in trout like fibronectin and matrigel (a commercial product of the extracellular matrix produced by hepatoma cells; Greenlee *et al.,*, 1995a). Matrigel is composed mainly of laminin, fibronectin, and entectin, which are three components of basal lamina. Matrigel can also contain various growth factors including those which inhibit proliferation of satellite cells. Compared with laminin, plating of cells on matrigel results in low efficiency of plating and low proliferation, but a high rate of differentiation (Greenlee *et al.*, 1995a). Fibronectin and laminin, which are the two main components of basal lamina, give similar results in terms of plating efficiency and proliferation.

The culture medium used for muscle cells is generally a classical minimum medium (Dubbelco's DMEM, Leibovitz-L15, RPMI) which is sometimes supplemented with a specific metabolic substrate such as glutamine or glucose. As for all fish cells, successful culturing of muscle cells depends on the osmolarity of the culture medium being the same as that found *in vivo*. The medium used for the extraction of cells is generally the same as the culture medium.

Mammal satellite cells are generally cultured in a controlled atmosphere containing 3–5% CO_2 (Campion *et al.*, 1984). This has been applied successfully to carp cells Koumans *et al.*, 1990). However, these conditions are very different from those *in situ*. Thus, muscle cells are generally cultured in a buffered medium (hepes, carbonate, or phosphate; Matschak and Stickland, 1995; Greenlee *et al.*, 1995a; Rescan *et al.*, 1995a).

The medium is generally supplemented with 10–15% horse serum or 10–15% fetal calf/bovine serum (FCS/FBS; Table III). Other supplements have been tested (Koumans *et al.*, 1990) including carp serum but the best results have been observed with horse serum. We have also tested different sera (Fig. 2a) including that of trout (Fig. 2b), but the proliferation rate was significantly higher with FCS.

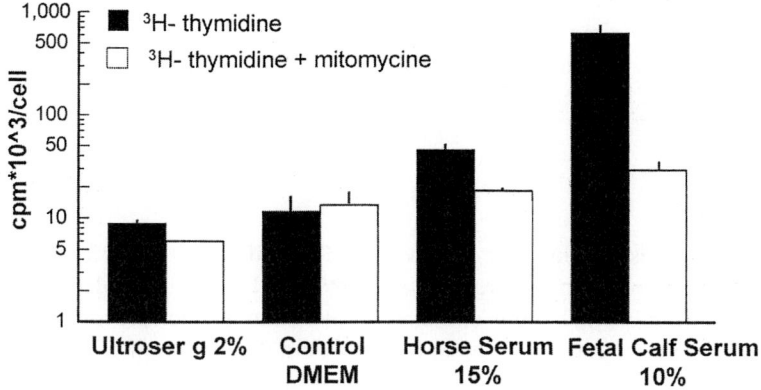

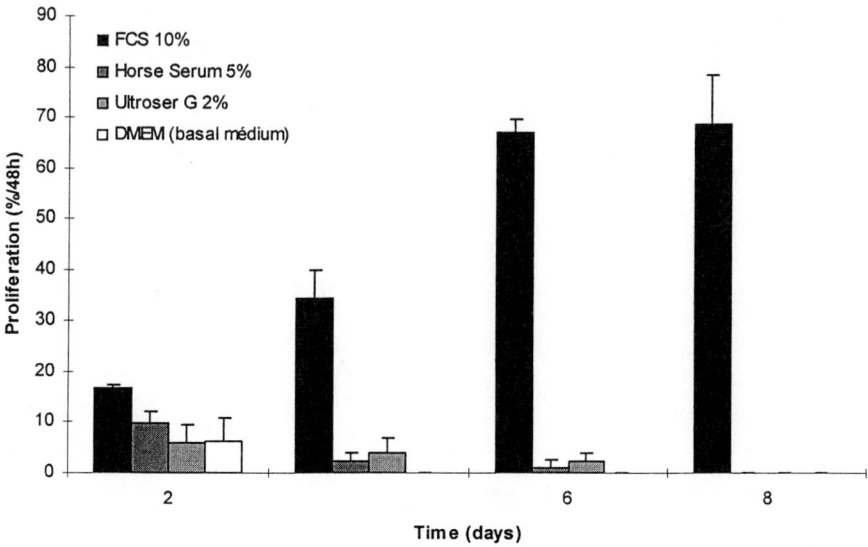

Fig. 2. The effect of different supplements to DMEM medium on the proliferation of satellite cells extracted from juvenile rainbow trout (2–3 g) and cultured on a laminin substrate at 15°C. Proliferation was measured (a) by incorporation of ^{3}H-thymidine during a 24-hr period (in cpm, counts per min) or (b) by incorporation of BrdU during a 24-hr period. FCS: fetal calf serum; HS: horse serum; TS: trout serum extracted from 1 kg trout either fed or fasted since 3 months; Ultroser G: semi-synthetic serum deprived of steroids.

A low concentration of the supplement has also been tested (Koumans *et al.*, 1990; Greenlee *et al.*, 1995a; Fauconneau and Paboeuf, unpublished data) and it appears that satellite cells can be cultured in a medium supplemented with only 1% FCS/FBS. Below 1%, proliferation seems to cease (Fauconneau and Paboeuf, unpublished data).

D. Proliferation

Recent studies on rainbow trout have found satellite cells to proliferate (Fig. 3) at a rate of 20–60% day^{-1} (Greenlee *et al.*, 1995a; Rescan *et al.*, 1995a). It has been demonstrated that laminin stimulates proliferation, which probably accounts for the proliferation rate observed in rainbow trout. Proliferation of satellite cells has not, however, been observed in carp (Koumans *et al.*, 1990) or salmon (Matschak and Stickland, 1995) and in the very first study in rainbow trout (Powells *et al.*, 1989). In carp, satellite cells demonstrate a capacity for proliferation as they incorporate BrdU during the extraction process (Koumans *et al.*, 1993a) but they do not proliferate once they are seeded (Koumans *et al.*, 1993a). This may be due to species-specific differences in the specificity of laminin or to the 5% CO_2 that is used in the controlled atmosphere during the culture of carp satellite cells. However, salmon myosatellite cells have been cultured in a buffered medium, like that used for rainbow trout, and no proliferation was observed (Matschak and Stickland, 1995).

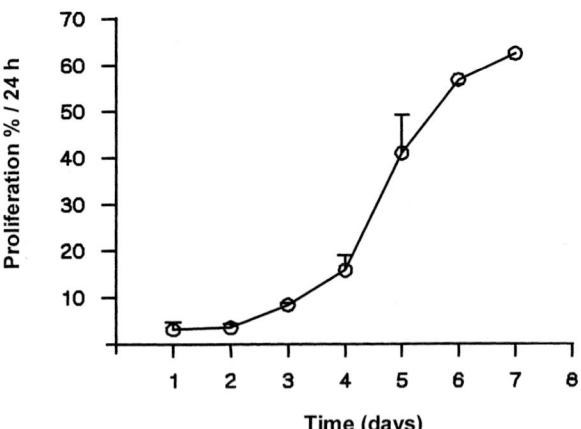

Fig. 3. The time-course of proliferation measured by BrdU uptake during a 24-hr period of satellite cells extracted from the white skeletal muscle of juvenile rainbow trout (2–3 g body weight) and cultured on a laminin substrate with a DMEM 90%/FCS 10% medium at 15°C (error bars are standard deviation).

For salmon, the culture is carried out at low temperatures (5 and 11°C) and the medium, which contains growth factors, is changed every 10 days, unlike rainbow trout where the culture is realized at 15–20°C and the medium is changed at least every 2 days (Greenlee *et al.*, 1995a; Rescan *et al.*, 1995a). Thus the conditions of culture may explain differences in cell behavior. However, intrinsic differences in the characteristics of satellite cells for different species cannot be ruled out.

E. Differentiation

Satellite cells actively differentiate first into almond- then spindle-shaped cells (Figs. 1b and 4). These satellite cells have a great capacity to migrate (direct observation with video recording) and form small or very large bundles of aligned cells which start to fuse within a few days. Most cells cultured *in vitro* differentiate into small myotubes (Fig. 1b) rather than large myotubes (Fig. 4, Table III). The differentiation of satellite cells into myotubes with spontaneous contractions, as has been generally observed in birds and mammals, has never been seen in fish. In birds and mammals, the morphology and degree of differentiation of the myotubes formed *in vitro* by skeletal myoblasts or satellite cells depends on the stage of development (fetal myoblasts vs. embryonic myoblasts vs. satellite cells; Cossu *et al.*, 1992; Stockdale, 1992). The medium requirement of muscle cells for *in vitro* growth is also stage dependent (Cossu *et al.*, 1992). It is proposed that fish satellite cells express a phenotype similar to that of myoblasts rather than that of classical satellite cells, and this may be related to the maintenance of myogenic capacities in fish muscle late after the embryonic stages.

It has been demonstrated that satellite cells differentiate very rapidly as within a few hours they express myoD (Rescan *et al.*, 1994) and then myogenin (Rescan *et al.*, 1995b). In all studies, the small myotubes express desmin and myosin (Table III). In zebrafish, small myotubes also express acetylcholine receptors (Sepich *et al.*, 1994). Depending on the model, desmin is either observed very early on in individual cells (carp) or late on small myotubes (rainbow trout; Fig. 5, see color plate). In small myotubes, myosin expression is observed, using antibodies against myosin heavy chain (MHC). In addition, cross-striation (light microscopy) and myofibrils (ultrastructure) are also observed (Table III). Myosin expression has also been observed in individual satellite cells in rainbow trout (Fig. 6), particularly when using a monoclonal antibody raised against embryonic MHC (Bobe and Fauconneau, 1999).

There is a large variability in the *in vitro* myotube phenotype. A few hours after seeding satellite cells, small myotubes of two to three nuclei have been observed (Koumans *et al.*, 1990; Rescan *et al.*, 1995a; Greenlee *et al.*, 1995a). Following a wave of proliferation lasting a few days, large myotubes (more than 10 nuclei) and very large myotubes (more than 20 nuclei) have been observed

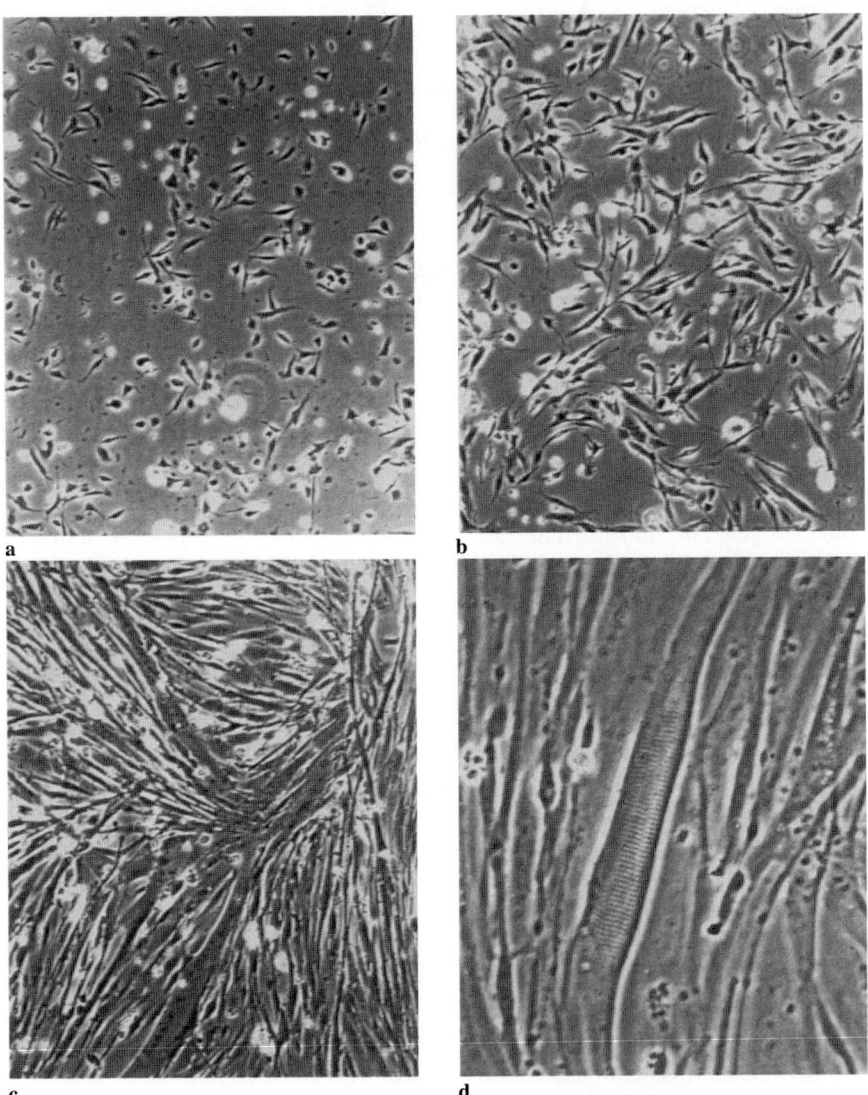

Fig. 4. The differentiation of satellite cells extracted from white skeletal muscle of juvenile rainbow trout (2 to 3 g) and cultured on laminin substrate with a DMEM 90%/ FCS 10% medium at 15°C for (a) a few hours, (b) 2, (c) 6, and (d) 10 days.

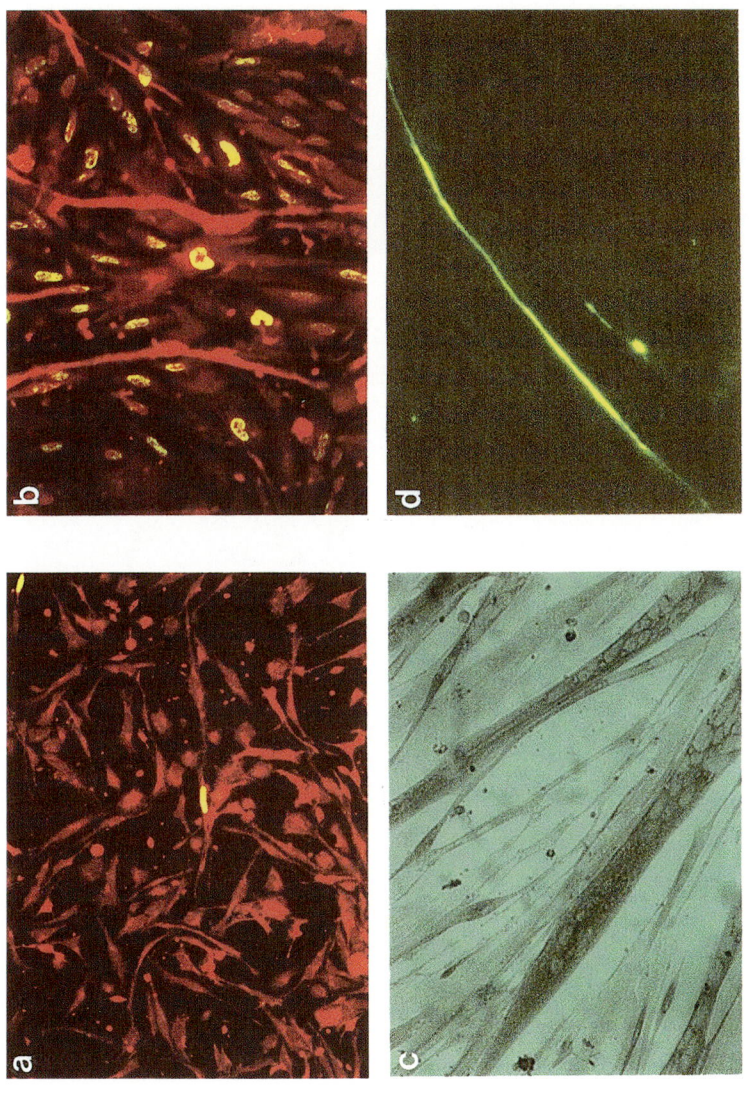

Figure 5. The expression of desmin (antibody against chicken gizzard desmin) at (a) 2 and (b) 6 days of culture and myosin using (c) a polyclonal antibody against skeletal MHC and (d) a monoclonal antibody against embryonic *Xenopus* MHC in myotubes built *in vitro* by satellite cells extracted from white skeletal muscle of juvenile rainbow trout (2–3 g) and cultured on a laminin substrate with a DMEM 90%/FCS 10% medium at 15°C.

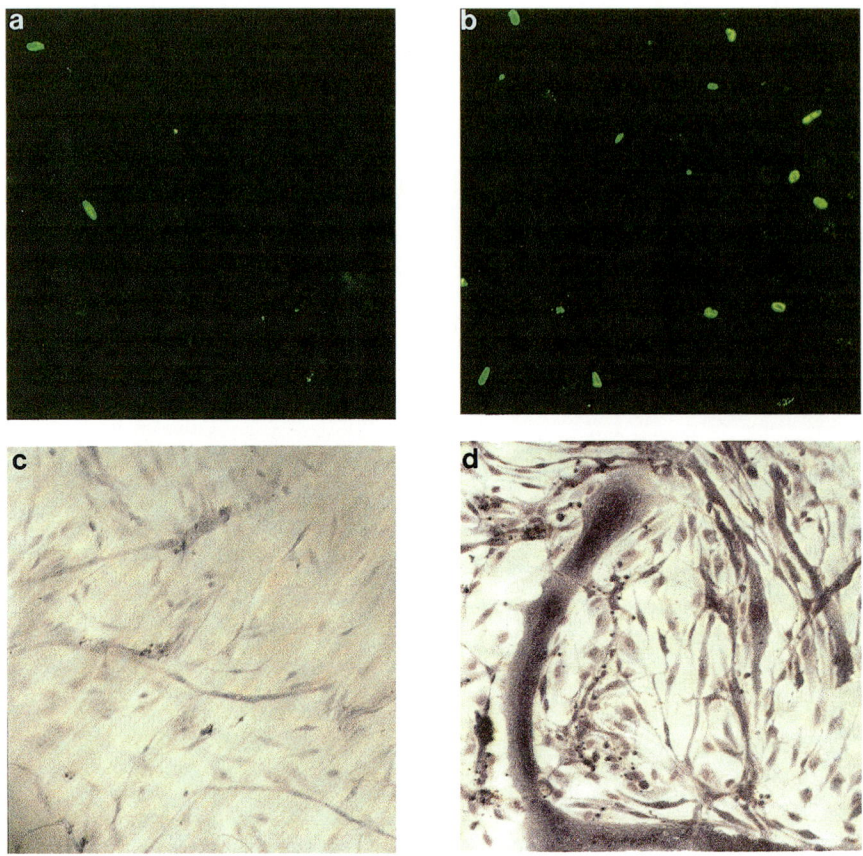

Figure 8. The effect of human recombinant IGF1 on short term satellite cell proliferation; (a) 15 ng IGF1/ml applied during 24 h compared to (b) control basal medium and long term differentiation into myotubes; (c) 30 ng IGF1/ml applied for 4 to 5 days compared to (d) control basal medium. Satellite cells are extracted from white skeletal muscle of juvenile rainbow trout and cultured on a laminin substrate with only DMEM medium at 15°C.

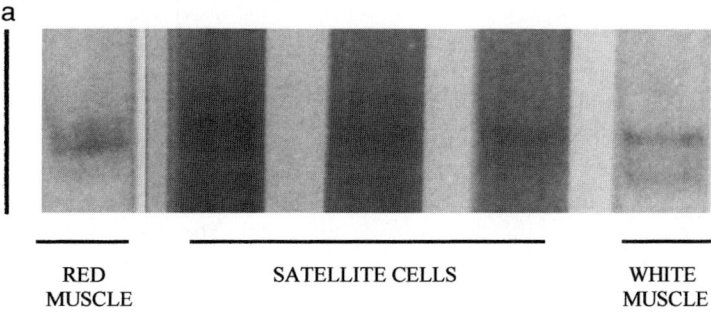

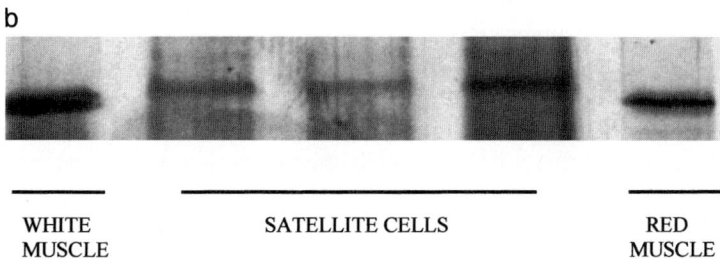

Fig. 6. The characteristics of (a) 5% acrylamide rod gel of native myosin and (b) SDS Page 5% acrylamide gels of MHC from satellite cells of the white skeletal muscle of juvenile rainbow trout cultured *in vitro* for 8 days and in white and red muscle from adult rainbow trout (1.5 kg).

(Koumans *et al.,* 1990; Rescan *et al.,* 1995a; Fauconneau and Paboeuf, 1999). A heterogeneity in the expression of myosin has also been demonstrated, as although all the myotubes expressed MHC (general antibody against skeletal MHC), specific monoclonal antibodies against slow or fast mammalian and trout MHC react only in a few dispersed myotubes (Fauconneau and Paboeuf, 1998).

V. ORIGIN OF SATELLITE CELLS AND RELATED QUESTIONS

Satellite cells are thought to be formed during the late stages of embryonic development. Depending on the species, a proliferative area is observed both dorsally and laterally. This occurs before hatching in cold water and large egg species (rainbow trout, salmon), around hatching in some small egg species (herring, carp), or during the larval stages in warm water and other small egg species (sea bass, sea bream; see Rowlerson and Vegetti, Chapter 5). Just after this period, small fibers appear in the deep white muscle giving it the classical mosaic appearance which is observed in most species studied (see Rowlerson and Vegetti, Chapter 5) with the exception of zebrafish. These small fibers are recruited from the very first satellite cells and therefore the pool of satellite cells is thus probably formed during this specific period of development. The first satellite cells observed during early development are not included beneath the basal lamina (Koumans et al., 1993b) as it is still developing. During the post-larval growth, satellite cells adjacent to fibers participate in both the hypertrophy of existing fibers and fiber hyperplasia.

The origin of muscle satellite cells has not been studied. There are various hypotheses all stating that a pool of embryonic stem cells is formed at the end of embryonic development (see Rowlerson and Vegetti, Chapter 5). Recent studies suggest that the precursor cells have a high capacity for migration both during early muscle development (Devoto et al., 1996; Stoiber and Sänger, 1996) and during late muscle regeneration (Rowlerson et al., 1997). Pioneer cells near the notochord, for example, migrate within the somite to form the embryonic red cells. We proposed that a few stem cells migrate within deep white muscle and then attach to existing fibers to form the pool of satellite cells. The source of stem cells could be located either in the external part of the somite under the layer of superficial red muscle fibers where proliferation has been observed in many species (Rowlerson et al., 1995), deep within the somite where the mosaic appearance of the white muscle is first seen in most species, (Koumans et al., 1995) or in the myosepta which contain pluripotent cells (Stoiber et al., 1997). Further data are required to conclusively ascertain the origin of satellite cells although the last hypothesis seems to be the most interesting.

The contribution of existing myotubes or fibers to the initial number of satellite cells is another important question. Does the number of fibers present in a somite at the end of embryonic development determine the size of the pool of precursor cells and thus the capacity for growth? Recent work on the effect of temperature during early development seems to demonstrate that the number of fibers at the end of embryonic development is associated with an altered capacity for hyperplastic and hypertrophic growth by juveniles (Nathanailides et al., 1995; Johnston et al., 1998). We propose that the construction of myotubes in the somites during early development and the recruitment of satellite cells are not independent. Consequently, the number of precursor cells formed would be dependent not only on the duration of the period of proliferation of stem cells but also on the surface of existing fibers at this stage.

VI. CONTRIBUTION OF SATELLITE CELLS TO MUSCLE GROWTH

During post-larval growth, the initial pool of satellite cells contribute during both to the synthesis of new fibers and the hypertrophy of existing fibers. From the data available it is not easy to conclude whether or not these two processes are connected with two populations of satellite cells. The existence of two different populations could be suspected *in vitro* both from the distinction between initial and stimulated proliferation of satellite cells in rainbow trout (Fauconneau and Paboeuf, 1999) and from the changes in the proportion of differentiated vs. proliferating myosatellite cells in carp with aging (Koumans *et al.*, 1993b). Furthermore, there is a great heterogeneity in the differentiation of satellite cells extracted from fish of the same development stage, particularly in the characteristics of the myotubes (size and myosin expression). The analysis of such heterogeneity will probably help to identify different populations. Attempts have been made to culture satellite cells under clonal conditions for this purpose, but fish satellite cells are difficult to raise under such conditions (Rescan and Paboeuf, personal communication). It is possible that the same population may contribute to the two processes and that, after a proliferative period, there may be specific paracrine and autocrine factors, which help the cells to distinguish between fusion with related satellite cells (hyperplasia) and fusion with adjacent fibers (hypertrophy).

In birds and in mammals, satellite cells which are recruited at the end of embryonic or fetal development, have a limited and fixed number of cell divisions. Furthermore, myoblasts have the capacity to undergo a given number of symmetrical cell divisions when satellite cells can only undergo a given number of asymmetrical divisions (Stockdale, 1992). It would be interesting to see if the number of fibers observed in juvenile fish and even in large fish of a commercial size could be related to the proliferation capacity of the satellite cells.

VII. EFFECT OF INTRINSIC FACTORS

A. Changes with Age/Body Weight

The relative number of muscle nuclei measured *in situ* seems to be very stable during ontogeny in carp (Koumans *et al.*, 1991), but a decrease in the percentage of myosatellite cells measured by two different methods is observed with aging. The percentage of proliferating muscle cells observed *in situ* with BrdU incorporation and PCNA expression (Alfei *et al.*, 1995) or observed *in vitro* with BrdU incorporation (Koumans *et al.*, 1993a) was also found to be very stable during ontogeny in carp and to even increase with aging. This is not in close agreement with the decrease in the contribution of hyperplasia to muscle growth during aging generally associated with a decrease in the percentage of small fibers (Koumans et al., 1993b; Alfei et al., 1993, 1995). It is, however, suspected from changes in DNA/protein ratio in white muscle (Koumans et al., 1993b) that myosatellite cells comprise different populations of myogenic cells: those involved in *de novo* myogenesis whose contribution decreases with aging and those involved in the hypertrophy of existing fibers whose contribution increases with aging (Koumans et al., 1995).

In rainbow trout, however, these features seem to be slightly different. We observed a decrease in the yield of satellite cells extracted in juvenile of rainbow trout (Fig. 7). However, the number of satellite cells extracted from large fish (>15 cm) was found to be lower than that extracted from small fish (<15 cm; Greenlee et al., 1995a). The percentage of myosatellite cell nuclei and the rate of proliferation of satellite cells measured by DNA flow cytometry also decreased *in situ* with aging (Alfei et al., 1989). This is in agreement with the decrease in the relative contribution of hyperplasia to muscle growth and the corresponding decrease in the percentage of small fibers (Stickland et al., 1983; Alfei et al., 1989).

Thus, carp and trout demonstrate differences in both the *in vitro* and *in vivo* proliferation of satellite cells with aging. This may be related not only to differences in the timing of muscle development in each species, but also to differences in the growth process related to divergent developmental pathways earlier in the development (Stoiber et al., 1998). It should be noted that, in all the species of fish studied, the percentage of small diameter fibers decreases with aging and fibers reach a maximum size of 200 μm at the standard maximum size of the

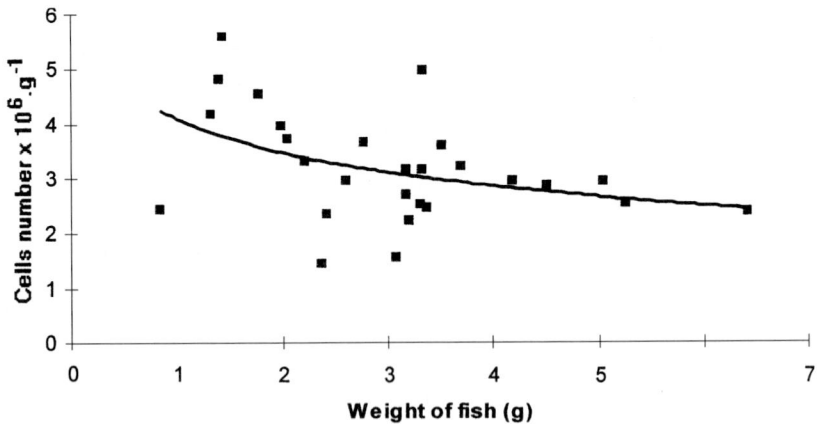

Fig. 7. The relationship between the yield of extraction of satellite cells ($10^6 \cdot g^{-1}$) from white skeletal muscle and body size of juvenile rainbow trout.

species. (Weatherley *et al.*, 1988). Thus, the two muscle growth processes determined by satellite cells are limited and this is consistent with a decrease in proliferation rate with aging.

B. Effect of Muscle Type

In mammals, it has been demonstrated that differentiation of satellite cells extracted from slow oxidative muscle produces different phenotypes *in vitro* compared to cells extracted from fast-glycolytic muscle (Feldman and Stockdale, 1991). There is, however, no such data available on satellite cells from red and white muscle in fish. It has only been found that the proliferation rate of muscle cells (including satellite cells) *in situ* is probably more active in red muscle than in white muscle (Krivy and Eide, 1977).

C. Effect of Genetic Origin

In fish, the differences in growth rate between individuals or between strains have never been related to differences in the relative contributions of hyperplasia and hypertrophy to muscle growth. In most studies, it has been concluded that the changes observed are only related to the stimulation of normal muscle growth, i.e., with a speed up of changes with aging of both hyperplasia and hypertrophy (Weatherley, 1990; Alami-Durante *et al.*, 1997; Fauconneau *et al.*, 1997). However, it has been demonstrated that there are differences between strains in the process of hyperplasia either during early (Johnston and MacLay, 1997) or late stages of development (Greer-Walker *et al.*, 1972; Fauconneau *et al.*, 1997; Valente *et al.*, 1999). Up to now differences between strains on satellite cells during initiation, proliferation, or differentiation has not been demonstrated.

The *in vitro* proliferation of satellite cells extracted from two strains differing in growth rate has been analyzed by Valente *et al.* (2000). Few differences were observed in proliferation rates of satellite cells extracted from fed fish of the two strains (Table IV). Proliferation rates of satellite cells extracted from fasted fish were, however, lower in slow-growing strains than in fast-growing strains. The characteristics of satellite cells from fed fish (especially initial proliferation rate) are supposed to be related to cells already activated *"in vivo"* and thus dependent on environmental factors including rearing conditions. On the contrary satellite cells extracted from fasted fish are supposed to be "quiescent" stem cells (see Section VIII.B, Effect of feeding) and they express rather intrinsic characteristics. Thus, the intrinsic proliferation capacity of satellite cells is different between fast- and slow-growing strains. The same has been observed in mammals where satellite cells extracted from fast-growing strains show higher proliferation rate than cells extracted from slow-growing strains (Quinn *et al.*, 1990).

Satellite cells from diploid and triploid rainbow trout have been also studied

(Greenlee et al., 1995b; Valente et al., 2000). The yield of extraction of satellite cells from triploid was significantly lower (around 33% less) than that of diploid. This is consistent with previous findings demonstrating that DNA content in muscle is similar in diploid and triploid when the amount of DNA present in one nucleus of triploid cell is 33% higher than that in diploid cells (Greenlee et al., 1995b). No proliferation is observed in satellite cells from diploid and triploid fish cultured on a matrigel substrate (Greenlee et al., 1995b). The proliferation of satellite cells cultured on alaminin substrate was not significantly different between diploid and triploid fed fish (Table IV). It was, however, significantly lower in satellite cells of fasted triploid fish than in that of fasted diploid fish (Valente et al., 2000). The proliferation capacity of satellite cells was thus proved to be different between diploid and triploid fish. These alterations in the characteristics of satellite cells could help to explain the delay in growth often observed in juvenile stages of triploid fish compared to diploid fish.

The differences in proliferation rate of satellite cells *in vitro* suggest either genetic differences in the duration of cell cycle or differences in the response of

Table IV.
Effect of Fasting on Yield of Extraction and Proliferation of Satellite Cells in Rainbow Trout; Interaction with Genetic Origin

	Yield ($\times 10^6 \cdot g^{-1}$)	Initial (%/24 hr)	Time for half (Max, days)	Max. proliferation (%/24 hr)	Authors
Fed	3.7	11.2	3.0	59	Fauconneau and
Fasting	4.0	0.3	5.5	56	Paboeuf, 2000
Fed	2.4	3	4.5	65	
Fasting	2.9	—	6.5	68	
Fast and refed	2.9	18	4.5	>65	
	$\times 10^6 \cdot g^{-1}$	%/48 hr	Days	%/48 hr	
Fed 2n	1,9	15.0	2.5	67	Valente et al., 2000
Fast 2n	2,0	0.8	5.0	64	
Fed 3n	1,5	10.8	3,0	67	
Fast 3n	1,6	3.2	5.5	75	
	$\times 10^6 \cdot g^{-1}$	%/24 hr	Days	%/24 hr	
Fed HG	1.3	3.4	3.5	48	Valente et al., 2000
Fast HG	1.3	1.4	5.5	48	
Fed LG	1.2	3.3	3.5	46	
Fast LG	1.4	0.8	6.0	39	

Note: 2n, 3n: diploid fish, triploid fish; HG, LG: high growth strain, low growth strain.

quiescent satellite cells to growth factors of the culture medium (FCS). If the duration of cell cycle varies, the proliferation rate of both fed fish and fasted fish would be affected. Therefore the differences observed are related rather to the response of the satellite cells to growth factors. If such observations *in vitro* could be extrapolated to satellite cells *in vivo*, it would explain to some extent differences in the intrinsic growth capacities of muscle. This has to be investigated further to characterize the growth factors which are involved in the initiation of satellite cells and to demonstrate variations in binding capacity of these growth factors to satellite cells.

D. Hormonal Control

The hormones which participate in the control of muscle growth *in vivo*—growth hormone and IGFs, thyroid and steroid hormones—would affect satellite cell proliferation and differentiation *in vitro* both directly and indirectly. However, there is no published data in this field, which is probably related to the fact that proliferation of satellite cells has only been observed in a few fish species and that quantification of differentiation is not very easy (see Chapter 5).

Among the different hormones mentioned we have not observed any effect of growth hormone and estradiol on the proliferation of satellite cells of juvenile rainbow trout cultured with the DMEM medium either supplemented or not with FCS (Fauconneau and Paboeuf, unpublished data). The same result was observed with T_3 and T_4 tested on satellite cells cultured in a DMEM medium supplemented with FCS partially deprived of thyroid hormones (by adsorption on an ion-exchange resin). These different hormones seemed to have stimulating effects on differentiation, but this has not been quantified (Fauconneau and Paboeuf, personal communication).

As growth hormone is suspected to contribute to the growth of muscle (Weatherley and Gill, 1982; Fauconneau *et al.*, 1997) through a stimulating effect of IGFs on hyperplasia, the effect of IGFs on proliferation of satellite cells has been examined. The binding of IGFs to satellite cells and to satellite cell membrane have been observed in rainbow trout satellite cells (Fauconneau *et al.*, 1998) but the specificity and affinity of binding needs further analysis to be characterized. The effect of IGFs on proliferation was tested with only mammalian IGFs in DMEM medium (Fig. 8, see color plate). The maximum proliferation rate observed with IGFs accounts for less than 10% of the proliferation observed with FCS (Fig. 9), thus IGFs are not the only factors involved. In the same conditions, a dose response of proliferation to IGFs was observed with IGF-I and IGF-II (IGF-II being less potent than IGF-I), but as the proliferation rate induced by IGF is low, it is difficult to conclude on the maximum effect which in some experiments was observed at a low concentration of 7–15 μ/ml and in another experiment at a high concentration of 30–50 μ/ml. The effect of IGFs will have to be tested with

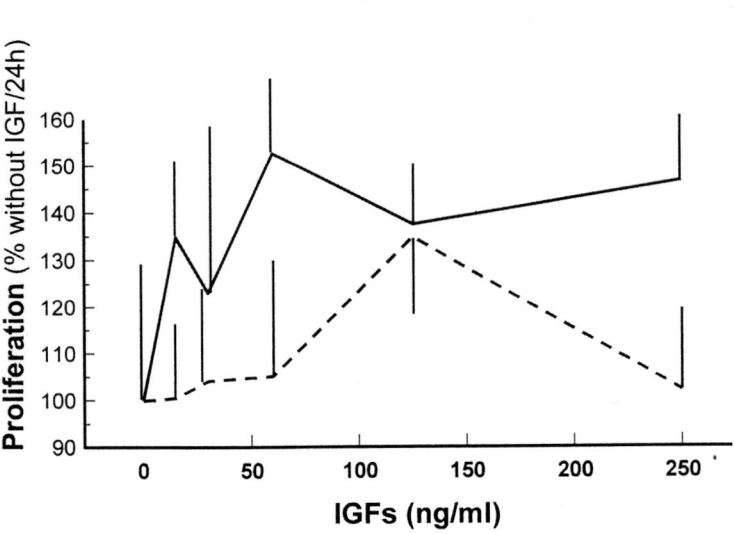
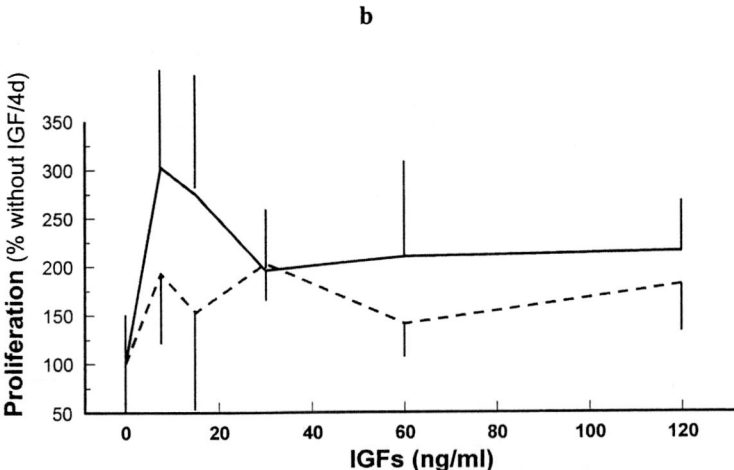

Fig. 9. Effect of human recombinant IGF1 (straight line) and IGF2 (dotted line) on satellite cells proliferation measured either (a) during a 24-hr period or (b) during a 4-day period by BrdU uptake of satellite cells extracted from white skeletal muscle of juvenile rainbow trout (2–3 g) and cultured on a laminin substrate with only DMEM medium at 15°C (error bars are standard deviation).

a minimum medium containing other stimulatory growth factors (FGF, EGF, PDGF; Quinn *et al.,* 1990) and also inhibitory factors (TGFs), as it has been demonstrated that the proliferation of muscle precursor cells *in vivo* is probably regulated not only positively but also negatively by growth factors (myostatine; McPherron *et al.,* 1997).

VIII. EFFECT OF EXTRINSIC FACTORS

A. Effect of Temperature

Fish growth and consequently muscle growth are dependent on temperature, but the changes observed in muscle cellularity are generally included in the normal allometric relationship of muscle cellularity with fish size or fish body weight (Weatherley, 1990; Alami-Durante *et al.,* 1997; Fauconneau *et al.,* 1997; Valente *et al.,* 1999). Temperature has been found to cause alterations in the cellularity of skeletal muscle (number and size of fibers) at the end of embryonic and larval development (Stickland *et al.,* 1988; Johnston, 1993; Johnston *et al.,* 1998).

The characteristics of myoblast and presumptive satellite cells from fish raised at different temperatures during early development have only been analyzed in a few species. The characteristics of salmon muscle satellite cells cultured *in vitro* at different temperatures have been studied (Matshack and Stickland, 1995). These cells did not demonstrate any proliferation but some differences in the state of differentiation of satellite cells depending on temperature is, however, suspected (Matshack and Stickland, 1995). In herring the absolute and relative number of presumptive myosatellite cells analyzed *in situ* at hatching are significantly higher at intermediate temperature (8°C compared to 5 and 12°C; Johnston, 1993). It means that rearing temperature during early development affects the number of fiber at hatching and also the population of precursor muscle cells involved in post-larval growth of muscle. The various hypotheses proposed to explain such an effect of temperature are related either to the duration of the cell cycle or to the length of the recruitment period of the precursor cells (Stickland *et al.,* 1988; Johnston, 1998).

B. Effect of Feeding

The large variability in the initial proliferation rate of satellite cells *in vitro* indicates a potential effect of feeding status. A comparison of the proliferation rates of satellite cells extracted from either fed or fasted fish clearly demonstrated this.

An initial proliferation rate was observed in satellite cells extracted from normally fed fish, but no proliferation rate was observed in satellite cells from fasted fish over a period of few days (Fauconneau and Paboeuf, 1999). The initial proliferation rate of satellite cells was also negatively affected by food restriction and positively affected by refeeding following a fasting period (Table IV). The same difference between fed and fasted fish was observed on proliferation of muscle cells *in situ* as analyzed by PCNA expression.

The proliferation rate of satellite cells from fed fish follows a normal exponential curve starting at the first day of culture. It takes a few days, however, for the proliferation of satellite cells from fasted fish to begin, which is consistent with the appearance of *in situ* PCNA expression in muscle after 4 days of refeeding in previously fasted fish (Fauconneau and Paboeuf, unpublished results). The time-course for the changes in proliferation rate of satellite cells from fasted fish is similar to that observed in satellite cells from fed fish (Fauconneau and Paboeuf, 1999).

This finding suggests that there are certainly two different populations of satellite cells in fish muscle. The initial proliferation observed *in vitro* accounts for a population which is already induced *in vivo* and actively proliferates once it is extracted (Koumans *et al.,* 1993a) and cultured *in vitro* (Fauconneau and Paboeuf, 1999). The existence of this population is largely dependent on feeding status as these cells are not found in fasted or restrained fish. Furthermore, if the number of divisions is limited, these satellite cells will probably also participate in the fast differentiation into small myotubes found in many different species (Table III). This process has generally been associated with another cell population known as post-mitotic cells, although no evidence for this population has been found *in situ* (Stoiber and Sänger, 1996).

A second population could be related to quiescent cells. These cells actively proliferate once they are induced by the extraction process and by the different growth factors found in the culture medium (Greenlee *et al.,* 1995a, Rescan *et al.,* 1995a) regardless of the feeding status of fish (Fauconneau and Paboeuf, 1999). The initiation of proliferation, however, takes more time for quiescent cells extracted from fasted fish than from fed fish. The presence of actively proliferative cells in fed fish is certainly associated with the synthesis of growth factors which are not present in the culture medium.

The composition of the food may be the important factors determining proliferation since the feed contains components which can either inhibit (for instance, oxidized fatty acid) or stimulate (for instance, vitamin C) the initiation and the maintenance of proliferation of satellite cells. Although the effect of some of these components on muscle satellite cell proliferation are currently being investigated (Fauconneau, Paboeuf, Bugeon, personal communication), this will require more investigation in the future.

C. Other Environmental Factors

The effect of different pollutants present in the environment could probably affect satellite cell status. Pollutants could theoretically cause depression of feeding which in turn would affect satellite cell characteristics as described above. Furthermore, some pollutants can have direct effects on the satellite cell's environment *in vivo* (synthesis of basal lamina, attachment to basal lamina. . .) which in turn would also affect satellite cell characteristics *in vitro* (Fauconneau and Paboeuf, unpublished data). Such research will help to demonstrate that some pollutants could directly affect muscle growth. Moreover, the pollutants could affect proliferation rates by different means and this helps also to identify some cellular and molecular targets for understanding the effect of growth factors on initiation or recruitment of these cells.

IX. CONCLUSIONS

The hyperplastic potential of fish muscle has been known for many years and is suspected to contribute to the intrinsic variability in muscle growth and its response to various external factors. Compared to the role of the hyperplastic growth, however, little data is still available on the cells which are thought to be the origins of this process. This may be due to the methodological difficulties in extracting and raising satellite cells from fish muscle and to the fact that no fish muscle cell lines have been developed. Part of the methodological difficulty is due to the fragility of these cells.

Furthermore, fish muscle cells express *in vitro* proliferation and differentiation patterns which are related more to myoblasts than to satellite cells. Such myoblast phenotypes are certainly associated with the capacity of these cells to contribute to hyperplasic growth.

The relative contributions of hyperplasia and hypertrophy to muscle growth seem to be very species specific and dependent on developmental stage and external factors. This probably explains why such large differences in proliferation capacities are observed *in situ* and *in vitro* between different species and, especially, between the two models which have been extensively studied, cyprinids and salmonids. Furthermore, all these studies have demonstrated greater numbers of satellite cells and higher proliferation rates in juveniles compared to adult fish and this is consistent with the decrease in growth capacity with aging. It has, however, been noted that the proliferation capacity of muscle cells is relatively low, even in juvenile fish, compared to birds and mammals. Most of the information could be gained from *in vitro* analysis of satellite cell characteristics.

Within a species the proliferation capacity of satellite cells seems to fit well with muscle growth variations depending on intrinsic and extrinsic factors. Furthermore, within the proliferation capacity, two different processes are observed *in vitro*. First, an initial proliferation rate is related to the instantaneous muscle growth rate and varies under the effects of controlling factors (nutritional state in particular but also temperature). Second, a stimulated proliferation rate is related to the intrinsic capacity of muscle growth and varies under the effects of determinant factors (muscle type, genetic origin. . .). The effects of different factors on the characteristics of initial and stimulated proliferation rates certainly need to be analyzed more systematically.

The satellite cells *in vitro* seem to express a great heterogeneity which has not been examined further. This must be a priority in future research as it will help to dissociate the different populations which are involved in the two muscle growth processes. If such populations can be dissociated, a systematic analysis of such cell populations based, in particular, on specific markers would enable a better understanding of muscle growth and regeneration process not only in fish but also in mammals and in man.

ACKNOWLEDGMENTS

We are very grateful to Dr. J. T. M. Koumans and to the late Dr. H. A. Akster for their kind help, advice, and encouragement in the development of satellite cell culture.

REFERENCES

Akster, H. A. (1983). A comparative study of fiber type characteristics and terminal innervation in head and axial muscle of carp (*Cyprinus carpio* L.). A histochemical and electron-microscopical study. *Neth. J. Zool.*, **33**, 164–188.

Akster, H. A., Koumans, J. T. M., Cuelenaere, J., and Osse, J. W. M. (1995). Uptake of tritiated thymidine in muscle of juvenile carp. *J. Fish Biol.*, **47**, 165–167.

Alami-Durante , H., Fauconneau, B., Rouel, M., Escaffre, A. M., and Bergot, P. (1997). Growth and multiplication of white skeletal muscle fibres in larval carp (*Cyprinus carpio* L.) in relation to somatic growth rate. *J. Fish Biol.*, **50**, 1285–1302.

Alfei, L., Maggi, F., Parvopassu, F., Bertoncello, G., and de Vita, R. (1989). Post-larval muscle growth in fish: DNA flow cytometric and morphometric analysis. *Bas. Appl. Histochem.*, **33**, 147–158.

Alfei, L., Colombari, P. T., Cavallo, D., Eleuteri, P., and de Vita, R. (1993). Use of 5′-bromodeoxyuridine immunohistochemistry to examine proliferative activity of fish tissues. *Eur. J. Histochem.*, **37**, 183–189.

Alfei, L., Onali, A., Spano, L., Colombari, P. T., Altavista, P. L., and de Vita, R. (1995). PCNA/Cyclin expression and BrdU uptake define proliferating myosatellite cells during hyperplastic muscle growth of fish (*Cryprinus carpio* L.). *Eur. J. Histochem.*, **38**, 151–162.

Baroffio, A., Bochaton-Piallat, M. L., Gabbiani, G., and Bader, C. R. (1995). Heterogeneity in the progeny of single human muscle satellite cells. *Differentiation*, **59**, 259–268.

Bobe, J., and Fauconneau, B. (1999). An ultrastructural and histoimmunological description of early myogenesis in rainbow trout embryos. *J. Exp. Biol.* (in press).

Campion, D. R. (1984). The muscle satellite cells: A review. *Int Rev. Cytol.*, **87**, 225–251.

Cossu, G., Cusella-De Angelis, M. G., De Angelis, L., Messogiorno, A., Murphy, P., Coletta, M., Vivarelli, E., Bouché, M., and Molinaro, M. (1992). Multiple myogenic cell precursors and their possible role in muscle histogenesis. *In* "Neuromuscular Development and Disease Edit." pp. 183–194. (Kelly, A. M., and Blau, H. M., eds). Raven Press, New York.

Devoto, S. H., Melançon, E., Eisen, J. S., and Westerfield, M. (1966). Identification of separate slow and fast muscle precursor cells *in vivo*, prior to somite formation. *Development,* **122,** 3371–3380.

Fauconneau, B., Andre, S., Chmaitilly, J., LeBail, P. Y., Krieg, F., and Kaushik, S. J. (1997). Control of skeletal muscle fibres and adipose cells size in the flesh of rainbow trout. *J. Fish Biol.,* **50,** 296–314.

Fauconneau, B., and Paboeuf, G. (1998). A cytoimmunochemical characterisation of myosin heavy chain expression in skeletal muscle of rainbow trout. *Prod. Anim.,* **11,** 154–156.

Fauconneau, B., and Paboeuf, G., 1999. Effect of fasting and refeeding on satellite cells status *in vitro* of rainbow trout. *Cell Tissue Res.* (in press).

Feldman, J. L., and Stockdale, F. E. (1991). Skeletal muscle satellite cells diversity: satellite cells form fibers of different types in cell culture. *Dev. Biol.,* **143,** 320–334.

Foster, R. F., Thompson, J. M., and Kaufman, S. J. (1987). A laminin substrated promotes myogenesis in rat skeletal muscle cultures: analysis of replication and development using anti-desmin and anti BrdU monoclonal antibodies. *Dev. Biol.,* **122,** 11–20.

Greenlee, A. R., Dodson, M. V., Yablonka-Reuveni, Z., Kersten, C. A., and Cloud, J. G. (1995a). In vitro differentiation of myoblast from skeletal muscle of rainbow trout. *J. Fish Biol.,* **46,** 731–747.

Greenlee, A. R., Kersten, C. A., and Cloud, J. G. (1995b). Effects of triploidiy on rainbow trout myogenesis *in vitro*. *J. Fish Biol.,* **46,** 381–388.

Greer-Walker, M. (1970). Growth and development of the skeletal muscle fibres of the cod (*Gradus morhua* L.). *J. Cons. Int. Expl. Mer.,* **33,** 228–244.

Greer-Walker, M., Burd, A. C., and Pull, G. A. (1972). The total numbers of white skeletal muscle fibres in a cross-section as a character for stock separation in North sea herring (*Clupea harengus* L.). *J. Cons. Int. Expl. Mer.,* **34,** 238–243.

Hightower, L. E., and Renfro, J. L. (1988). Recent applications of fish cell culture to biomedical research *J. Exp. Zool.,* **248,** 290–302.

Johnston, I. A. (1993). Temperature influences muscle differentiation and the relative timing of organogenesis in herring (*Clupea harengus*) larvae. *Mar. Biol.,* **116,** 363–379.

Johnston, I. A., and MacLay, H. A. (1997). Temperature and family effect on muscle cellularity at hatch and first feeding in Atlantic salmon (*Salmo salar* L.). *Can. J. Zool.,* **75,** 64–74.

Johnston, I. A., Cole, N. J., Abercromby, M., and Viera, V. L. A. (1998). Embryonnic temperature modulates muscle growth characteristics in larval and juvenile herring. *J. Exp. Biol.,* **201,** 623–646.

Johnston, I. A. (1998). Muscle development and growth: potential implications for flesh quality in fish. *Aquaculture,* **177,** 99–115.

Koumans, J. T. M., Akster, H. A., Dulos, G. J., and Osse, J. W. M. (1990). Myosatellite cells of *Cyprinus carpio* (Teleostei) *in vitro*: isolation, recognition and differentiation. *Cell Tissue Res.,* **261,** 173–181.

Koumans, J. T. M., Akster, H. A., Booms, G. H. R., Lemmens, C. J. J., and Osse, J. W. (1991). Numbers of myosatellite cells in white axial muscle of growing fish: *Cyprinus carpio* L. (Teleostei). *Am. J. Anat.,* **192,** 418–424.

Koumans, J. T. M., Akster, H. A., Booms, G. H. R., and Osse, J. W. (1993a). Influence of fish size on proliferation and differentiation of cultured myosatellite cells of white axial muscle of carp (*Cyprinus carpio* L.). *Differentiation,* **53,** 1–6.

Koumans, J. T. M., Akster, H. A., Booms, G. H. R., and Osse, J. W. (1993b). Growth of carp (*Cyprinus carpio*) white axial muscle; hyperplasia and hypertrophy in relation to the myonucleus/sarcoplasm ratio and the occurrence of different subclasses of myogenic cells. *J. Fish Biol.,* **43,** 69–80.

Koumans, J. T. M., and Akster, H. A. (1995). Myogenic cells in development and growth of fish. *Comp. Biochem. Physiol.,* **110A,** 3–20.

Krivy, H. and Eide, A. (1977). Morphometric and autoradiographic studies on the growth of red and white axial muscle fibres in the shark *Etmopterus spinax*. *Anat Embryol.,* **151,** 17–28.

McPherron, A. C., Lawler, A. M., and Lee, S. J. (1997). Regulation of skeletal muscle mass in mice by a new TGF-β superfamily member. *Nature*, **387**, 83–90.
Matschak, T. W., and Stickland, N. C. (1995). The growth of Atlantic salmon (*Salmo salar* L.) myosatellite cells in culture at two different temperatures. *Experientia.*, **51**, 260–266.
Mauro, A. (1961). Satellite cells of skeletal muscle fibers. *J. Biophys. Biochem. Cytol.*, **9**, 493–495.
Mulvaney, D. R., and Cyrino, J. E. (1995). Establishment of Channel catfish satellite cells cultures. *Bas. Appl. Myo.*, **5**, 65–70.
Nag, A. C., and Nursall, J. R. (1972). Histogenesis of white and red muscle fibres of trunk muscle of a fish *Salmo gairdneri*. *Cytobios*, **6**, 227–246.
Nathanailides, C., Lopez-Albors, O., and Stickland, N. C. (1995). Influence of prehatch temperature on the development of muscle cellularity in posthatch Atlantic salmon (*Salmo salar*). *Can. J. Fish Aquat. Sci.*, **52**, 675–680.
Ocalan, M., Goodman, S. L., Kühl, U., Hauschka, S. D., and von der Mark, K. (1988). Laminin alters cells shape and stimulated motility and proliferation of murine skeletal myoblasts. *Dev. Biol.*, **125**, 158–167.
O'Connell, C. P. (1981). Development of organ systems in the Northern anchovy *Engraulis mordax* and other teleosts. *Am. Zool.*, **21**, 429–446.
Powell, R. L., Dodson, M. V., and Cloud, J. G. (1989). Cultivation and differentiation of satellite cells from skeletal muscle of the rainbow trout. *J. Exp. Biol.*, **250**, 333–338.
Quinn, L. S., Norwood, T. H., and Nameroff, M. (1988). Myogenic stem cells commitment probability remains constant as a function of organismal and mitotic age. *J. Cell. Physiol.*, **134**, 324–336.
Quinn, L. S., Ong., L. D., and Roeder, R. A. (1990). Paracrine control of myoblast proliferation and differentiation by fibroblasts. *Dev. Biol.*, **40**, 8–19.
Rescan, P. Y., Gauvry, L., Paboeuf, G., and Fauconneau, B. (1994). Identification of a muscle factor related to MyoD in a fish species. *Biochem. Biophys. Acta*, **1218**, 202–206.
Rescan, P. Y., Paboeuf, G., and Fauconneau, B. (1995a). Myosatellite cells of *Oncorhynchus mykiss*: culture and myogenesis on laminin substrates. In "Biology of Protozoa Invertebrates and Fishes: *In Vitro* Experimental Models and Applications." IFREMER Editions n° 18, 63–68.
Rescan, P. Y., Gauvry, L., and Paboeuf, G. (1995b). A gene with homology to myogenin is expressed in developing myotomal musculature of the rainbow trout and *in vitro* during conversion of myosatellite cells to myotubes. *FEBS Lett.* **362**, 89–92.
Romanello, M. G., Scapolo, P. A., Luprano, S., and Mascarello, F. (1987). Post larval growth in the lateral white muscle of the eel *Anguilla anguilla*. *J. Fish Biol.*, **30**, 161–172.
Rowlerson, A., Mascarello, F., Radaelli, G., and Veggetti, A. (1995). Differentiation and growth of muscle in the fish *Sparus auratus* (L.) II. Hyperplastic and hypertrophic growth of lateral muscle from hatching to adult. *J. Muscle Res. Cell Motil.*, **16**, 223–236.
Rowlerson, A., Radaelli, G., Mascarello, F., and Veggetti, A. (1997). Regeneration of skeletal muscle in two teleost fish: *Sparus auratus* and *Brachydanio rerio*. *Cell Tissue Res.*, **289**, 311–322.
Sepich, D. S., Ho, R. K., and Westerfield, M. (1994). Autonomous expression of nic-1 acetylcholine receptor mutation in zebrafish muscle cells. *Dev. Biol.*, **161**, 84–90.
Schultz, E. (1996). Satellite cell proliferative compartments in growing skeletal muscle. *Dev. Biol.*, **175**, 84–94.
Stickland, N. C. (1983). Growth and development of muscle fibres in the rainbow trout. (*Salmo gairdneri*). *J. Anat.*, **137**, 323–333.
Stickland, N. C., White, R. N., Mescall, P. E., Crook, A. R., and Thorpe, J. E. (1988). The effect of temperature on myogenesis in embryonnic development of the Atlantic salmon. (*Salmo salar* L.). *Anat. Embryol.*, **178**, 253–257.
Stockdale, F. E. (1992). Myogenic cell lineages. *Dev. Biol.*, **154**, 284–298.
Stoiber, W., and Sänger, A. M. (1996). An electron microscopic investigation into the possible source of new muscle fibres in teleost fish. *Anat. Embryol.*, **194**, 569–579.

Stoiber, W., Haslett, J. R., Goldschmid, A., and Sänger, A. M. (1998). Patterns of superficial fibre formation in the European pearlfish (*Rutilus frisii meidingeri*) provide a general template for slow muscle development in teleost fish. *Anat. Embryol.,* **197,** 485–496.

Valente, L. M. P., Rocha, E., Gomes, E. F. S., Silva, M. W., Olivereira, M. H., Monteiro, R. A. F., and Fauconneau, B. (1999). Growth dynamics of white and red muscle fibres in fast- and slow-growing strains of rainbow trout. *J. Fish Biol.,* **55,** 675–691.

Valente, L. M. P., Paboeuf, G., Gomes, E. F. S., and Fauconneau, B. (2000). *In vitro* proliferation of muscle myosatellite cells in fast vs. slow growing strains and in diploid vs. triploid of rainbow trout (*Oncorhynchus mykiss*) in response to nutritional state. *Cell Tissue Res.* (submitted)

Vegetti, A., Mascarello, F., Scapolo, P. A., and Rowlerson, A. (1990). Hyperplastic and hypertrophic growth of lateral muscle in *Dicentrarchus labrax* (L.). *Anat. Embryol.,* **182,** 1–10.

Weatherley, A. H., and Gill, H. S. (1982). Influence of bovine growth hormone on the growth dynamics of mosaic muscle in relation to somatic growth of rainbow trout *Salmo gairdneri* Richardson. *J. Fish Biol.,* **20,** 165–172.

Weatherley, A. H., Gill, H. S., and Lobo, A. F. (1988). Recruitment and maximal diameter of axial muscle fibres in teleosts and their relationship to somatic growth and ultimate size. *J. Fish Biol.,* **33,** 851–859.

Weatherley, A. H. (1990). Approaches to understanding fish growth. *Trans. Am. Fish. Soc.,* **119,** 662–672.

Willemse, J. J., and van den Berg, P. G. (1978). Growth of striated muscle fibres in the muscle lateralis of the European eel *Anguilla anguilla* (L.) (Pisces Teleostei). *J. Anat.,* **125,** 447–460.

Yablonka-Reuveni, Z., Quinn, L. S., and Nameroff, M. (1987). Isolation and clonal analysis of satellite cells from chicken pectoralis muscle. *Dev. Biol.,* **119,** 252–259.

Yafee, D. (1968). Retention of differentiation potentialities during prolonged cultivation of myogenic cells. *Proc. Natl. Acad. Sci. U.S.A.,* **61,** 477–483.

5

CELLULAR MECHANISMS OF POST-EMBRYONIC MUSCLE GROWTH IN AQUACULTURE SPECIES

A. ROWLERSON
A. VEGGETTI

I. Abstract
II. Introduction
   A. Outline
   B. Methods of Investigation
III. Muscle Growth Mechanisms
   A. Hypertrophy
   B. Hyperplasia
   C. Relative Contributions of Hyperplasia and Hypertrophy to Post-Embryonic Muscle Growth
IV. Factors Affecting Muscle Growth
   A. Diet: Ration, Composition, Feeding Behavior, and Social Hierarchy Effects
   B. Hormonal Manipulations
   C. Genetic Effects: Strains, Sex, Triploids and Hybrids
   D. Season
   E. Exercise
   F. "Growth History" Effects
V. Trends for Future Research on Muscle Growth in Fish, Especially Aquaculture Species
   References

I. ABSTRACT

In this chapter we are concerned with the processes of muscle fiber hypertrophy (increase in size) and hyperplasia (increase in number) during post-embryonic growth, i.e., after the initial phase of myogenesis. A brief outline is given of methods used to measure these processes. We then discuss the main features of hypertrophic growth, and distinguish between two quite distinct phases of hyperplastic growth. The first phase of hyperplastic growth generally takes place during at least part of larval life and completes the formation of the main muscle layers which were initiated during embryonic myogenesis. As this process generates new fibers along a distinct germinal layer we have named it "stratified" hyperplasia. In fish which grow to a large final size this is followed by a second and quite different hyperplastic process. As new fiber production is now disseminated across the whole myotome this results in a mosaic of fiber diameters when the muscle is cut in transverse section and we have, therefore, called this process "mosaic" hyperplasia. Mosaic hyperplasia results in a large increase in total fiber number during juvenile growth, and is therefore very important for commercial aquaculture species; it is lacking in species which remain small. We also discuss various biotic and abiotic factors which can affect muscle growth, especially in the context of intensive aquaculture, and have attempted to identify some areas where new investigative approaches are needed.

II. INTRODUCTION

A. Outline

In this chapter we are concerned mainly with teleost fish which are important aquaculture species, and especially those which are raised by intensive methods. Freshwater species (especially cyprinids) are by far the most important aquaculture species in production terms (FAO, 1998), but their herbivorous or omnivorous habits make them well suited to traditional extensive and semi-intensive rearing methods. By contrast, intensive aquaculture generally concentrates on carnivorous, diadromous, and marine species of higher commercial value such as salmonids, sea bream, bass etc., and consequently much of the information specifically referring to muscle growth in fish is derived from these species.

Both hypertrophy (increase in fiber size) and hyperplasia (genesis of new fibers) contribute to muscle growth, but as their relative importance varies at different life stages we will initially treat them separately (Sections III.A and III.B), before discussing their relative importance at different life stages (Section III.C). In Section IV we consider possible interactions between these processes and a number of biotic and abiotic factors which are known to influence somatic growth. Our citation policy is to give as many citations as space permits to original work carried out on teleost fish muscle (inevitably biased toward aquaculture species), and to restrict references to more general aspects to review articles wherever possible. We have also excluded discussion of the earliest events in embryonic myogenesis, because they are addressed in Chapter 1.

B. Methods of Investigation

1. MORPHOMETRY

Although somatic growth can be easily measured in the form of body weight (or carcass weight or length and/or condition factor), this gives only an indirect measure of muscle growth. A long-established method, which provides useful quantitative data, is measurement of muscle fiber diameters (or cross-sectional areas) in a representative area of lateral (trunk) muscle in fish of different ages, sizes, or conditions (e.g., Willemse and van den Berg, 1978; Weatherley et al., 1979, 1980a,b, 1988; Stickland 1983; Veggetti et al., 1990; Kiessling et al., 1991; Meyer-Rochow and Ingram, 1993; Rowlerson et al., 1995; Alami-Durante et al., 1997; Fauconneau et al., 1997; Johnston et al., 1998; Galloway et al., 1999a,b; Radaelli et al., 1999; Valente et al., 1999). The diameters of the larger fibers provide an index of hypertrophic growth which continues until they reach the functional maximum value characteristic of the species. Fibers also grow in length, but as measurement of this form of hypertrophy requires a more complex sampling technique, it is less often used (see Kiessling et al., 1991; Alami-Durante et al., 1997).

The distribution of fiber diameters (or areas), and especially the presence of very small diameter fibers, is often used as a measure of the appearance of new fibers and thus of hyperplasia (examples are shown in Fig. 1 and a method for quantitative analysis described by Johnston et al., 1999). Strictly, however, the presence of small fibers does not necessarily indicate fast growth because they are typical of fish size rather than growth rate (Weatherley and Gill, 1982, 1987a; Weatherley et al., 1988), and even some slow-growing fish have muscle containing small diameter fibers (Weatherly and Gill, 1987b). In longitudinal studies it is desirable to count the total fiber number in a transverse section of the trunk at each time point, or if that is not possible, at least to derive an estimate from the whole area occupied by muscle and the mean fiber diameter. Ideally, this estimation should use an appropriately weighted value for diameter if there is a zonal distribution of different diameters within the myotome. Fortunately, zonal differences are largest in the smaller fish (in which total fiber counts can be made relatively easily), and are less marked (although not insignificant, see Kiessling et al., 1991) in larger fish for which total fiber number counts are not practicable.

Some authors use a combination of fiber number and size to estimate "cellularity" (fiber number in relation to fiber area and/or whole muscle cross-sectional area, e.g., Stickland et al., 1988; Nathanailides et al., 1995a; Johnston and McLay, 1997; Matschak et al., 1997, 1998; Johnston, 1999), and counts of nuclear numbers in relation to fiber numbers or areas can also provide useful information (Koumans et al., 1991, 1993a; Johnston, 1993a; Alfei et al., 1994; Nathanailides et al., 1996; Johnston and McLay, 1997; Alami-Durante et al., 1997).

2. BIOCHEMISTRY

Various biochemical measures of muscle growth have been investigated. Some enzyme activities (ornithine decarboxylase, citrate synthase, cytochrome oxidase) have been shown to correlate with overall growth rates (Arndt et al., 1994; Benfey et al., 1994; Pelletier et al., 1993; Houlihan et al., 1993). More

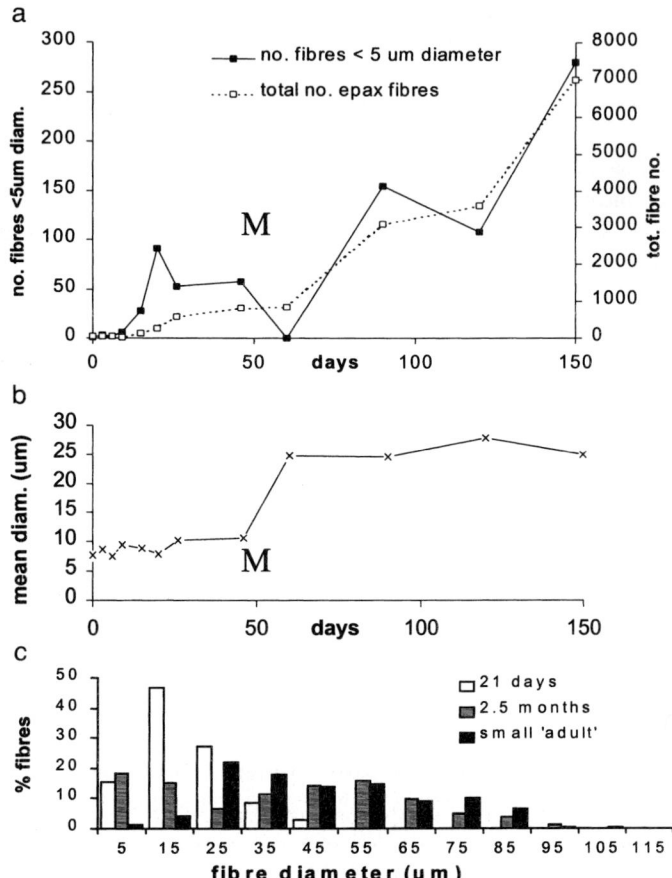

Fig. 1. Morphometric data illustrating the growth of fast-white fibers in an epaxial quadrant of lateral muscle in the sea bream, *Sparus aurata*, from hatching to 5 months (a,b) and in the sole, *Solea solea*, at the ages shown (c). Data refer to one representative large subject at each age, and are from studies by Rowlerson et al., 1995 and Veggetti et al., 1999. (a) Hyperplasia (represented by the number of fibers with diameter less than 5 μm) occurs in two phases: an early stratified phase which peaks in midlarval life followed by a later, mosaic phase which starts in the month following metamorphosis (at the time indicated M) and continues well into juvenile life (beyond the period shown here). (b) As hyperplasia ceased toward the end of larval life, mean diameter increased. As hyperplasia increased again after 60 days the mean diameter remained steady again (and only increased substantially after 150 days when the proportion of very small diameter fibers decreased again). The mosaic phase generates a far larger number of fibers than was formed in the earlier phase, and together with hypertrophic growth brings the fish to commercial size. (c) The histogram illustrates the left-skewed unimodal distribution of fiber diameters during the stratified hyperplastic phase in a 21-day-old sole, the typical bimodal distribution of fiber diameters during active mosaic hyperplasia (2.5 months), and the wide range of fiber diameters (but very few under 10 μm) in a small subject at one year when hyperplastic growth had ceased, but most fibers were still relatively small in diameter.

commonly, however, the relative contents or rates of synthesis of protein and nucleic acids in muscle are used. For example, RNA concentration generally correlates well with the rate of protein growth (Houlihan et al., 1993; Mathers et al., 1993), the RNA:DNA ratio has also been found to correlate with (protein) growth (Lone and Ince, 1983; Grant, 1996), and protein:DNA ratios have been used as a measure of hypertrophic growth (Pelletier et al., 1995). Conversely, an increase in DNA concentration can be taken to indicate an increase in hyperplastic growth (Luquet and Durand, 1970; Valente et al., 1998). However, it is not easy to separate the relative contributions of hyperplasia and hypertrophy in muscle from these data, doubts have been expressed about the use of the RNA:DNA ratio (Houlihan et al., 1993; Suresh and Sheehan, 1998b) and interpretation may be complicated by effects of recent nutritional status (Miglavs and Jobling, 1989; McLaughlin et al., 1995).

3. PHENOTYPES

Changes in muscle fiber phenotype (especially in contractile protein expression) can also be indicators of muscle growth. In mammals, new fibers express developmental (embryonic, neonatal/fetal) isoforms of myosin which reflect their lineage and time of appearance, and are relatively easily detected by biochemical and immunostaining methods. Furthermore, phenotype differences in the expression of myogenic regulatory factors are now being discovered (Rescan, 1997; Delalande and Rescan, 1999). Some developmental myosin isoforms have been identified in muscle of several fish species (e.g., Carpenè and Veggetti, 1981; Scapolo et al., 1988; Martinez et al., 1991; Chanoine et al., 1992; Brooks and Johnston, 1993; Johnston, 1993b; Johnston and Horne, 1994; Ennion et al., 1995, 1999; Johnston et al., 1998; Bobe et al., 2000; see also Fig. 2 and discussion in Chapter 3). In post-embryonic trout the presumed new fibers contain myosin isoforms which could not be distinguished from those in mature fibers (Higgins, 1990; Kiessling et al., 1995; Gauvry and Fauconneau, 1996), although a first indication that immunostaining might resolve this issue in rainbow trout has been obtained by Picard et al., 1998. As the expression of developmental isoforms in new fibers is seen throughout the vertebrates, including fish, it seems most likely that the lack of evidence for such forms in trout is due to the lack of suitable methods, rather than an absence of developmental myosins. This is a relatively neglected area of research which deserves more attention; development of a reliable (and preferably histological) species-independent method for detection of developmental isoforms of myosin in fish muscle would be extremely helpful.

4. MITOTIC ACTIVITY

Labeling methods which reveal the site(s) and intensity of mitotic activity in cells, e.g., incorporation of tritiated thymidine or 5BrdU (5-bromo-deoxy-uridine), or by detection of cell-cycle-dependent expression of nuclear proteins such

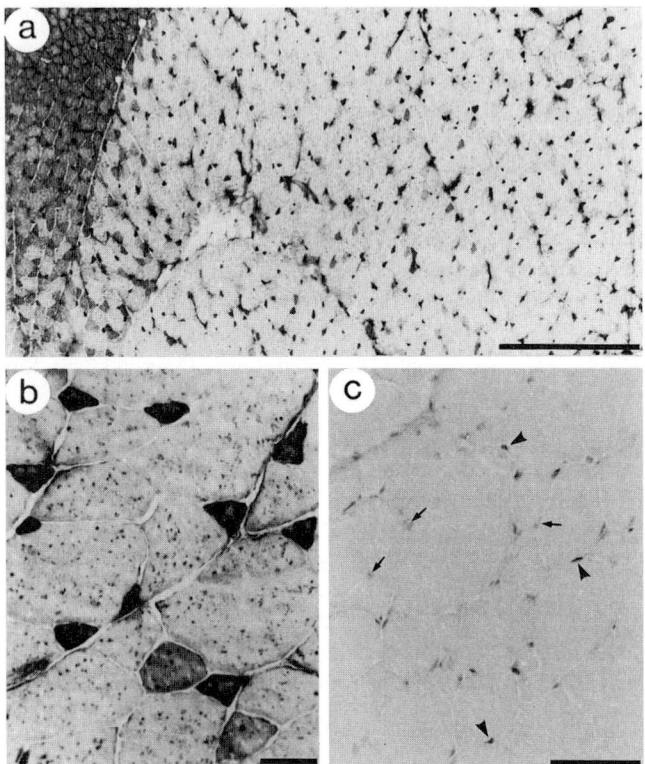

Fig. 2. (a,b) Histochemical staining for myosin-ATPase activity in a transverse section of mullet muscle. Low power view in (a) shows part of the slow-red muscle layer at the top left but the rest of the area is fast-white muscle in the mosaic phase, containing many scattered new (darkly stained) fibers, and (b) shows a higher power view of a small area of fast-white muscle. (c) Immunostaining for PCNA (followed by counterstaining with hematoxylin) in fast-white lateral muscle of the sole aged 2.5 months (post-metamorphosis). Some of the labeled nuclei are indicated with arrowheads; examples of PCNA-negative nuclei (hematoxylin-stained only) are indicated with small arrows. This muscle was already in the mosaic hyperplastic phase (see Section III.B.3.). Scale bars = 500 μm (a), 50 μm (b,c).

as proliferating cell nuclear antigen (PCNA), have all been applied to fish muscle to identify hyperplastic growth processes (Alfei et al., 1993, 1994; Akster et al., 1995; Johnston et al., 1995; Rowlerson et al., 1995; Veggetti et al., 1999). The advantage of PCNA as a marker is that no manipulation of fish is needed (PCNA is an intrinsic protein), and potentially, immunostaining for PCNA (illustrated in Fig. 2c) in combination with the use of small biopsy samples (Grant, 1996) might provide a useful way of monitoring muscle growth in live fish. Ideally, immunostaining for PCNA would be combined with a marker for myogenic lineage (cf. Johnston et al., 1999), but at least during rapid growth the vast majority of the PCNA-positive nuclei do belong to myogenic cells. Practically, however, the use of BrdU incorporation and PCNA immunostaining may be limited to phases of rapid hyperplastic growth because of the relatively low frequency of mititic events in muscle at other times. The value of these methods in the quantification of hyperplastic growth *in vivo* has not yet been fully explored.

III. MUSCLE GROWTH MECHANISMS

A. Hypertrophy

Muscle fibers grow by hypertrophy throughout post-embryonic life until they reach a functional maximum diameter that is in the range 100–300 μm for fast-white fibers in most fish, but rather smaller for slow-red fibers which are much more dependent upon the oxygen supply from adjacent capillaries (e.g., Johnston, 1982; Egginton and Johnston, 1982; Sänger, 1993).

Nascent fibers examined in late embryonic and larval life have sparse myofibrils, but the volume fraction of myofibrils soon increases rapidly so that even small diameter fibers are filled with closely packed myofibrillar material (e.g., Veggetti *et al.,* 1990; Brooks *et al.,* 1995; Ayala *et al.,* 1999; Vieira and Johnston, 1999). During subsequent hypertrophic growth the formation of new myofibrils keeps pace with the increase in fiber size.

The rate of hypertrophic growth will vary with somatic growth rate and at different life stages. For example, in newly hatched sea bream larvae, hypertrophic growth was very slow for the first few days after hatching (maximum diameter about 11 μm), then (following the transition to exogenous feeding) increased rapidly for the next 2–3 weeks (to maximum about 30 μm) before reaching an intermediate growth rate which continued through juvenile life up to a body weight of about 200 g (when maximal diameter reached 140 μm), and during the period of initial sexual maturation as males (to maximal diameter 170 μm at 600 g), before finally reaching a maximum value of 200 μm after the subsequent transformation to female (2800 g body weight; Rowlerson *et al.,* 1995). This last stage was also accompanied by a change to a right-skewed distribution of diameters and a substantial increase in mean diameter. Much larger diameters can be found in Antarctic fish like *Notothenia neglecta,* in which juvenile growth from the onset of the demersal stage is by hypertrophy only (Battram and Johnston, 1991).

Hypertrophic growth during larval life has been described for several species, e.g., carp (Alami-Durante *et al.,* 1997), cod (Galloway *et al.,* 1999a), flatfish (Gibson and Johnston, 1995; Galloway, 1999b; Veggetti *et al.,* 1999), herring (Johnston, 1993a; Johnston *et al.,* 1998) salmon (Nathanailides *et al.,* 1995a), sea bass (Veggetti *et al.,* 1990), and sea bream (Rowlerson *et al.,* 1995). The persistence of hypertrophic growth throughout juvenile life up to adult stages, even after hyperplastic growth has ceased, has also been described for a variety of fish, including trout (Weatherley et al., 1980a; Stickland, 1983; Weatherley and Gill, 1987a; Alfei et al., 1989; Kiessling et al., 1991), carp (Koumans et al., 1993a), hake (Calvo, 1989), sea bass (Veggetti et al., 1990), sea bream (Rowlerson et al., 1995), eel (Willemse and van den Berg, 1978; Romanello et al., 1987), and many others (Weatherley et al., 1988; Kundu and Mansuri, 1990; Meyer-Rochow and Ingram, 1993; Zimmerman and Lowery, 1999).

Counts of myofiber nuclei with respect to fiber area or volume indicate that fibers acquire additional nuclei as they grow (Alfei et al., 1989; Johnston, 1993a; Usher et al., 1994; Nathanailides et al., 1996; Alami-Durante et al., 1997). This process, which was first described in mammalian muscle (Moss and Leblond, 1971; Cardasis and Cooper, 1975), maintains a fairly constant ratio of nuclear to cytoplasmic volume as the fibers grow in both length and diameter. The new nuclei are supplied by a population of myogenic cells (already present in the muscle), which fuse with existing muscle fibers to provide the additonal nuclei. To supply the number of nuclei required during growth, this population must be capable of proliferation. A careful study of growing carp muscle by Koumans et al. (1990, 1991, 1993a,b), using a variety of techniques, showed that although classical satellite cells (see Section III.B.4) were present, they were unlikely to fulfill this role and a morphologically undifferentiated population of cells present in the muscle was tentatively identified as the proliferation-competent cells required (Koumans and Akster, 1995). This issue is considered further in Section III.B.

Although addition of myonuclei to growing fibers is important, there are clearly many other factors involved in regulating hypertrophic growth; for example, in the mouse, over-expression of the proto-oncogene *ski* results in a striking hypertrophy of type IIB and IIX muscle fibers (roughly equivalent to the fast-white type in fish) without nuclear addition (and therefore resulting in a fall in the nuclear to volume ratio; Sutrave et al., 1990). These factors have yet to be investigated in fish.

B. Hyperplasia

1. DISTINCT PHASES OF HYPERPLASIA

"Hyperplastic" growth of muscle refers to the increase in muscle fiber number due to formation of new fibers. It does not include the proliferation of myogenic cells which is necessary to support hypertrophic growth (Section III.A). After the initial two muscle layers have been formed during embryonic life (see Chapter 1), hyperplastic growth continues in two successive and distinct phases. The first phase is in some respects a continuation of embryonic myogenesis, since it occurs by apposition along a growth zone and completes the formation of the definitive muscle layers (slow-red, pink, and fast-white; see Section III.B.2 and

Fig. 4). Generally, in fish which grow to a large final size (e.g., aquaculture species such as trout, salmon, sea bream, sea bass, carp, etc.), this is followed by a second and quite different hyperplastic process resulting in a large increase in the total number of fibers in all muscle layers, especially in the fast-white layer which acquires a typical mosaic appearance (see Section III.B.3). Various aspects of phenotypic differentiation of the various fiber types are addressed in Chapters 3 and 7.

The relative timing of the main hyperplastic growth processes in relation to the life cycle varies between species (see Table I, and Koumans and Akster, 1995). In the sea bream, for example, the first hyperplastic phase is completed during larval life, and the second starts after metamorphosis has been completed (Figs. 1 and 4). In other fish the first phase may finish earlier and the second phase starts before metamorphosis. Thus, as description of these two phases as "larval" and

Table I.
Characteristics of Mosaic Hyperplasia in Different Species

Species	Age/size of fish at onset[a] of mosaic hyperplasia	Difference in myosin between new and mature fibers of mosaic	Authors
Anguilla anguilla	10 mm (?post-met.)	Yes	Romanello et al., 1987
Clupea harengus	22 mm (onset met.)	[b]	Johnston et al., 1998
Cyprinus carpio	6 weeks (post-met.)	Yes	Nathanailides et al., 1996
Dicentrarchus labrax	80 days (end-met.)	Yes	Veggetti et al., 1990; Scapolo et al., 1988
Engraulis mordax	21 mm (pre-met.)	[b]	O'Connell, 1981
Pleuronectes platessa	>30 mm (post-met.)	No	Brooks and Johnston, 1993
Oncorhynchus mykiss (Salmo gairdneri)	Before 22 mm (post-met.)	No	Weatherley and Gill, 1981
Rutilis rutilis and other cyprinids	ca 14 mm (pre-met.)	Yes	Stoiber and Sänger, 1996; Kilarski, 1990
Sardinops melanostictus	22 mm (onset met.)	[b]	Matsuoka, 1998
Salmo salar	ca 0.2 g (first feeding)	No	Johnston and McLay, 1997; Higgins, 1990
Salmo trutta	ca 22 mm (first feeding)	No	Killeen, 1999
Solea solea	ca 40 days (late met.)[c]	[b]	Veggetti et al., 1999
Sparus aurata	60–90 days (post-met.)	Yes	Mascarello et al., 1995

[a]Onset refers to appearance of new fibers, not the presence of presumptive myogenic cells.
[b]No information.
[c]Approximate date (onset is during second half of metamorphosis).
Note: met., metamorphosis.

"post-larval" is not valid for all species, we have named these two phases as "stratified" and "mosaic" according to their primary morphological feature (see below).

2. STRATIFIED HYPERPLASTIC GROWTH

a. The Main Growth Zone and Presumptive Fast-White Muscle. Initial myogenesis in the embryo results in the formation of a superficial monolayer of mitochondria-rich ("red") fibers and the underlying presumptive fast-white fibers (reviewed in Chapter 1). In many species, new presumptive fast-white fibers continue to be added throughout embryonic life and into larval life, and they appear in a germinal layer or proliferation zone which lies just under the superficial monolayer and extends dorsally from the horizontal septum into the apex of the myotome. This process, which is reminiscent of events in trunk muscle growth in embryonic chickens (Amthor *et al.,* 1999 and references therein), is illustrated in Fig. 3 and summarized in Fig. 4a. Although the superficial monolayer is initially a complete layer, covering the entire surface of the myotome, it subsequently recedes from the apical areas because the expansion of the presumptive white muscle is so much faster here. This germinal layer is the principal source of new fibers added in late embryonic and larval life; it gives rise to characteristic gradients in muscle fiber diameter from superficial to deep and from apical to central areas of the myotome, and has been described in many species (anchovy: O'Connell, 1981; cod: Galloway *et al.,* 1999a; cyprinids: Stoiber and Sänger, 1996; halibut: Galloway, 1999b; plaice: Brooks and Johnston, 1993; salmon: Usher *et al.,* 1994; Johnston and McLay, 1997; sea bream: Rowlerson *et al.,* 1995; sea bass: Veggetti *et al.,* 1990; turbot: Gibson and Johnston, 1995; zebrafish: Waterman, 1969). In the herring there is complete cessation of muscle hyperplasia for many days between inital myogenesis and mid-larval life, but this is followed later by some addition of new fibers in the apical zones (Johnston, 1993a; Johnston *et al.,* 1998). A distinct pause in hyperplasia also occurs in the first few days after hatching in some other species (Gibson and Johnston, 1995; Galloway *et al.,* 1999a,b; Veggetti *et al.,* 1999).

b. Superficial Monolayer and "External" Cells. The number of superficial monolayer cells also increases after initial myogenesis, because this layer eventually covers the entire lateral surface of the myotome again after the phase of apical hyperplasia (e.g., as illustrated for the salmon, by Johnston and McLay, 1997). However, it is not clear if the new fibers are derived at this early (larval) stage from the main germinal layer or also from a separate source such as the external cells described in some species (zebrafish: Waterman, 1969; sea bass: Veggetti *et al.,* 1990; herring: Johnston, 1993a; sea bream: Ramírez-Zarzosa *et al.,* 1995, Patruno *et al.,* 1998; pearlfish: Stoiber and Sänger, 1996, Stoiber *et al.,* 1998).

5. CELLULAR MECHANISMS OF POST-EMBRYONIC MUSCLE GROWTH

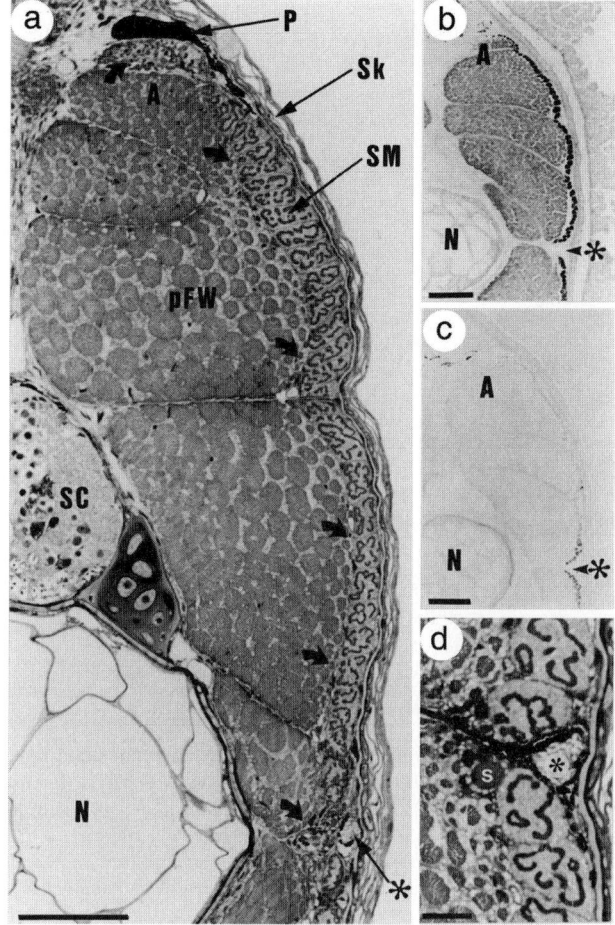

Fig. 3. Photomicrograph of (a) a transverse section of the epaxial quadrant of a sea bream larva aged 35 days, toluidine blue stain; (b) and (c) serial sections of an epaxial quadrant of a stage 31 salmon embryo, immunostained to show the distribution of a developmental myosin (b) in *all* superficial monolayer fibers, and slow myosin (c) which is strongly expressed only in those superficial monolayer fibers located adjacent to the lateral line nerve; and (d) a small area immediately surrounding the lateral line nerve in a near adjacent section to that shown in (a), illustrating the position of newly produced slow-fibers immediately under the superficial monolayer fibers close to the nerve in sea bream. A = apical region, SM = superficial monolayer, pFW = presumptive fast-white (larval) muscle, S = slow fiber, N = notochord, P = pigment, SC = spinal cord, * = lateral line nerve, and Sk = skin. Bold arrows point to areas of new fiber formation along the germinal zone which is located immediately underneath the superficial monolayer and extends from the horizontal septum to the apex of the myotome. External cells are too small to be seen clearly at this magnification. In the sea bream SM and S fibers are clearly distinct, not only in size, shape, and myofibrillar packing, as shown here, but also in myosin expression (see Section III.B.2., and Mascarello *et al.*, 1995). Scale bars = 50 μm (a), 100 μm (b,c), and 10 μm (d).

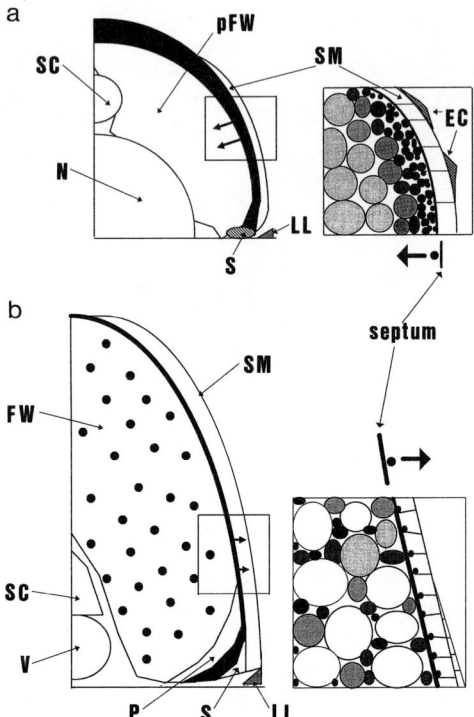

Fig. 4. Schematic representation of post-hatching myogenic proliferation zones in an epaxial quadrant of lateral muscle. This is based principally on observations made on sea bream and sea bass (Veggetti et al., 1990; Rowlerson et al., 1995), but the main features apply to many other species (see Sections III.B.2 and 3). The "stratified" stage shown in (a) can be considered to be an extension of embryonic myogenesis in that it continues to add to the two initial muscle layers (a superficial monolayer and an underlying presumptive fast-white muscle, separated by an initially very thin connective tissue septum), during at least part of larval life in most species. During this stage, hyperplasia occurs in the restricted proliferation zone of the myotome indicated in black (left panel), and most of the new fibers produced at this stage contribute to growth of the white muscle (see detail in right panel). On the superficial surface of the superficial monolayer, "external cells" (EC) may be present, and may contribute to the formation of an additional layer of superficial monolayer fibers toward the end of this stage and in the following stage (below right). Phenotypically "slow" fibers (S) start to appear during this phase in the part of the proliferation zone closest to the lateral line nerve. (b) shows the subsequent "mosaic" hyperplastic phase, when new fibers are generated throughout the fast-white muscle (black dots in left panel), giving rise to the characteristic mosaic appearance in transverse section. This phase is typically post-larval, and most intense in the first few months when it gives rise to very rapid growth. The previous proliferation zone also persists for a while, but now gives rise mainly to new slow-red fibers on its lateral surface (see detail in right panel), and (probably) to pink (P) fibers medially in the area close to the lateral line. SC = spinal cord, N = notochord, LL = lateral line nerve, SM = superficial monolayer, pFW = presumptive FW muscle layer, FW = fast-white muscle layer, and V = vertebral column. (Adapted from Fig. 7, Rowlerson et al. (1995) with kind permission from Kluwer Academic Publishers.)

Veggetti *et al.* (1990) and Patruno *et al.*, (1998) have suggested that external cells may contribute to the addition of new fibers to the superficial monolayer close to the lateral line as it thickens during late larval life. However, other authors working with other species take a different view because external cells appeared too briefly during development (Johnston, 1993a) or were too dissimilar in their fine structure (Stoiber and Sänger, 1996). It is still not known if external cells are non-myogenic (probably fibroblasts) or myoblasts which contribute to muscle fiber formation either in the superficial layer or elsewhere (by migration, cf. Stoiber and Sänger, 1996). This issue could now be resolved with the detection of a suitable myogenic marker, e.g., c-met receptor tyrosine kinase or myf-6 (cf. Johnston *et al.*, 1999).

c. Slow-Red Muscle Layer. Although the superficial monolayer fibers will eventually transform phenotype into slow fibers (Scapolo *et al.*, 1988; Mascarello *et al.*, 1995; El-Fiky and Wieser, 1988; Johnston and McLay, 1997), they are not the main source of slow fibers, even in zebrafish (van Raamsdonk *et al.*, 1982). Phenotypically and morphologically distinct slow fibers first appear *de novo* during larval life in some species (sea bass: Scapolo *et al.*, 1988; herring: Johnston and Horne, 1994; sea bream: Mascarello *et al.*, 1995 and Ramírez-Zarzosa *et al.*, 1995; catfish: Koumans and Akster, 1995) in a position close to the lateral line nerve (see Figs. 3d and 4). It is not known if these fibers have a distinct origin (as implied by Fig. 1c of Koumans and Akster, 1995) or are simply one of the many products of the main germinal layer. In the sea bream, the main germinal layer clearly also gives rise to new slow fibers along its full extent at a later period (early post-larval life: Mascarello *et al.*, 1995), as summarized in Fig. 4b. In other species such as salmonids, slow myosin expression appears initially in mature superficial monolayer fibers adjacent to the lateral line (by phenotypic transformation, see Fig. 3b,c and also Bobe *et al.*, 2000), and the slow muscle layer later increases by hyperplasia. The development of the slow-red muscle layer is associated with the onset of "cruising" type swimming activity (Koumans and Akster, 1995).

d. Pink Muscle. Finally, in most species the first appearance of pink (intermediate) muscle fibers occurs close to the horizontal septum during the stratified hyperplastic growth phase (Fig. 4; van Raamsdonk *et al.*, 1982; Scapolo *et al.*, 1988; Koumans and Akster, 1995; Mascarello *et al.*, 1995; Ramírez-Zarzosa *et al.*, 1995; Nathanailides *et al.*, 1995b; Stoiber and Sänger, 1996; Matsuoka, 1998). From their position and very small diameter when first seen, these pink fibers are presumably derived from the main germinal layer, although at later stages phenotypic transformation of nearby fast-white fibers may contribute to the pink fiber population (Stoiber *et al.*, 1999).

3. MOSAIC HYPERPLASTIC GROWTH

In most fish which grow to a large final size, the vast majority of muscle fibers will be formed in a long lasting hyperplastic process disseminated thoughout the entire myotome. This process gives rise to a typical mosaic appearance of muscle cut in transverse section, with fibers of different age (and therefore diameter) intermingled as shown in Figs. 2 and 4b. Mosaic hyperplastic muscle growth, which occurs principally during juvenile life, is of the greatest interest in commercial aquaculture because it brings the fish to market size and has therefore been extensively studied over many years (reviewed in Weatherley and Gill, 1987a; see also Kiessling *et al.*, 1991). It is this phase which is greatly reduced or entirely lacking in species such as zebrafish, guppies, and other fish which remain small (van Raamsdonk *et al.*, 1983; Weatherley and Gill, 1984, 1985; Weatherley *et al.*, 1988; Veggetti *et al.*, 1993; Koumans and Akster, 1995), and in species of intermediate size (such as the Antarctic fish *N. neglecta;* Battram and Johnston, 1991) which are capable of extreme hypertrophic growth.

Early work on mosaic muscle growth used principally morphometric analysis to show that this was a hyperplastic process (reviewed in Weatherley and Gill, 1987a), but the introduction of a variety of histochemical and immunostaining methods for contractile protein expression and electronmicroscopic studies of ultrastructure has enabled this to be confirmed in other ways in some fish (e.g., carp: Koumans *et al.*, 1991, 1993a,b and Alfei *et al.*, 1994; other cyprinids: Stoiber and Sänger, 1996; eel: Romanello *et al.*, 1987; herring: Johnston *et al.*, 1998; mullet: Carpenè and Veggetti, 1981; sea bass: Scapolo *et al.*, 1988; sea bream: Mascarello *et al.*, 1995, and Rowlerson *et al.*, 1995; trout: Alfei *et al.*, 1989).

Both phenotypic properties of the (presumed) new fibers and the age of onset of the mosaic hyperplastic phase vary widely between species (see Table I). Unfortunately, because many authors concentrate on either only early growth (embryonic and larval) or post-larval growth (from fingerling size to adult), there are relatively few studies of muscle growth right through from embryonic stages to adult life; as a result the age/stage of onset of mosaic hyperplasia is reported for only a few species.

In some species (e.g., sea bream, Fig. 1) mosaic hyperplasia starts after the previous phase has ceased, but in others it may initially overlap with stratified growth (Stoiber and Sänger, 1996; Johnston *et al.*, 1998). The intensity of mosaic hyperplastic growth is greatest in early juvenile life; later it wanes gradually until the fish reaches a characteristic fraction of body size (about 44% of final length in many fish, but see Zimmerman and Lowery, 1999, for an exception) after which further growth occurs by hypertrophy only (e.g., Weatherley *et al.*, 1980a, 1988; Stickland, 1983; Weatherley and Gill, 1987a; Veggetti *et al.*, 1990; Meyer-Rochow and Ingram, 1993; Koumans *et al.*, 1993b; Rowlerson *et al.*, 1995). The distribution of fiber diameters in the mosaic, and proportion of smallest diameter (most recently produced) fibers, is therefore characteristic of fish size, and this needs to be taken into account when comparing hyperplastic growth in fish which have reached very different sizes (Kiessling et al., 1991; Fauconneau et al., 1997).

Although this has received much less attention, the slow-red and pink muscle layers also undergo further growth in juveniles by hyperplasia (e.g., Willemse and van den Berg, 1978; Stickland, 1983; Kundu and Mansuri, 1990; Higgins and Thorpe, 1990; Meyer-Rochow and Ingram, 1993; Johnston et al., 1998), and at least in sea bream this appears to be a mosaic hyperplasia (Mascarello et al., 1995). Satellite cells have also been found in carp red muscle during the juvenile growth period, and may be the source of new fibers in this case (Akster et al., 1995).

Various factors which influence the extent and rate of mosaic hyperplastic growth are reviewed in Section IV.

4. SOURCE OF NEW FIBERS PRODUCED
DURING MOSAIC HYPERPLASTIC GROWTH

Fast-white muscle taken from juvenile fish of various ages yields myogenic precursor cells capable of giving rise to new myotubes when cultured *in vitro* (Koumans et al., 1990; Powell et al., 1989; Greenlee et al., 1995a; Matschak and Stickland, 1995; Chapter 4), although they were unable to proliferate in these conditions (Matschak and Stickland, 1995). Indirect evidence for the existence of a distinct population of myogenic cells supporting mosaic hyperplastic growth has also been provided by the observation that growth hormone treatment cannot provoke hyperplastic growth in species which do not have a mosaic (Weatherley and Gill, 1987a), and by the differences in myosin expression of new fibers produced when muscle regenerates in zebrafish (no mosaic) and sea bream (which does show mosaic hyperplastic growth) (Rowlerson et al., 1997). However, many problems remain, and some of these are outlined below.

Classical satellite cells (i.e., cells with heterochromatic nuclei and very little cytoplasm, and which lie under the basal lamina of mature muscle fibers) have been found in the muscle of several fish species. Those found in the eel, sea bass, and carp were suggested to be a source of new fibers (Willemse and van den Berg, 1978; Romanello et al., 1987; Veggetti et al., 1990; Alfei et al., 1994). However, there are also fish (such as mullet) which show mosaic hyperplastic growth (Carpenè and Veggetti, 1981) but no obvious satellite cell population (Veggetti, 1991). Furthermore, during the initial stages of mosaic hyperplasia, there may still be no basal lamina around the mature fibers (this develops quite late: Veggetti et al., 1990; Stoiber and Sänger, 1996), and scattered, morphologically undifferentiated cells seen in the larval muscle have been proposed as the source of new fibers during subsequent mosaic hyperplastic growth (Johnston et al., 1995; Stoiber and Sänger, 1996). This hypothesis is supported by results of immunostaining for

5BrdU incorporation and PCNA expression, which show that nuclear proliferation is active throughout muscle in fish at and just prior to the onset of mosaic hyperplasia (Rowlerson et al., 1995; Johnston et al., 1998; Veggetti et al., 1999).

In juvenile carp, a quantitative analysis of muscle growth indicated that, although satellite cells were present, some morphologically undifferentiated myogenic cells must also be present, because the satellite cells were predominantly post-mitotic and too few to account for the number of nuclei needed for the observed hyperplastic and hypertrophic growth in these fish (Koumans et al., 1990, 1991, 1993a,b; Koumans and Akster, 1995). Although it is tempting to attribute the source of new fibers to satellite cells formed earlier, and the nuclei needed for hypertrophic growth to proliferation of the morphologically undifferentiated myogenic cells, the true situation is probably much more complicated. It is not known if the satellite cells are directly derived from the other myogenic population or from a separate lineage, nor if these two categories of myogenic cell make distinct contibutions to muscle growth mechanisms (see also Johnston, 1999). There is evidence from a study of myogenesis in larval herring that in small new fibers all nuclei had undergone recent divisions, which means they did not originate from post-mitotic "founder" cells produced much earlier (Johnston et al., 1998), but the situation may be different at later ages.

Clearly, we need to know a lot more about the origin of the new fibers. Identification of the proposed morphologically undifferentiated "myogenic" cell population is important, but must remain very tentative until appropriate biochemical markers are available. These will include myogenic regulatory factors (MRFs) from the MyoD family, some of which have now been found in fish (Rescan et al., 1995; Kobiyama et al., 1998; Johnston et al., 1999; Watabe, 1999). High levels of expression of MRFs (Kobiyama et al., 1998) and fibroblast growth factor-6 (Rescan, 1998) have been observed to parallel hyperplastic growth in juvenile carp and trout. Characteristics of satellite cells isolated from fish muscle (and especially their behavior *in vitro*) are discussed in more detail in Chapter 4.

C. Relative Contributions of Hyperplasia and Hypertrophy to Post-Embryonic Muscle Growth

1. LARVAL GROWTH

Hypertrophic and hyperplastic growth occur with different time courses over the larval period, and have been described for several species (e.g., Veggetti et al., 1990; Rowlerson et al., 1995; Alami-Durante et al., 1997; Johnston et al., 1998). The variable length in the pause in hyperplastic growth which occurs immediately following hatching in many species was mentioned in Section III.B.2.

Rearing embryonic and/or larval stages at different temperatures has been shown in several species to have opposite effects on hypertrophic and hyperplastic growth. In herring, higher temperatures result in greater hyperplasia but reduced hypertrophy (Vieira and Johnston, 1992; Johnston et al., 1995). This effect can be explained on theoretical grounds on the basis that temperature has a greater effect on DNA replication (where the rate-limiting step is enzymatic) than on protein synthesis (where the rate-limiting step is diffusion of ribosomal components into the cytoplasm) (van der Have and de Jong, 1996).

However, in several other species an increase in hypertrophic growth and a decrease in hyperplastic growth occurred in response to an increase in rearing temperature (Stickland et al., 1988; Usher et al., 1994; Hanel et al., 1996; Galloway et al., 1998). In the case of the salmon and trout embryos, at least part of the temperature effect could be attributed to reduced oxygen availability leading to reduced hyperplastic growth (Matschak et al., 1997, 1998). Also, for a given size, larvae which are growing slowly due to dietary restriction show reduced hyperplastic growth (Alami-Durante et al., 1997; Galloway et al., 1999a), whereas hypertrophic growth is less affected (Galloway et al., 1999a). Competition for limited metabolic resources could be a common mechanism here; reduced calorie intake in growing mammals is known to preferentially reduce nuclear replication and hyperplasia (Cheek and Hill, 1970), perhaps because DNA replication is energetically more expensive than protein synthesis. However, effects on hyperplasia and hypertrophy are not always opposite; in plaice an increased rearing temperature increased both hyperplasia and hypertrophy (Brooks and Johnston, 1993). Effects of rearing temperature are discussed in more detail in Chapter 6.

2. Juvenile Growth

The relative contributions of hyperplasia and hypertrophy to muscle growth throughout the juvenile period to adult size have been studied in many species, and it is found that hypertrophic growth persists long after hyperplastic growth has ceased (e.g., Weatherley et al., 1980a, 1988; Stickland, 1983; Kundu and Mansuri, 1990; Veggetti et al., 1990, 1993; Kiessling et al., 1991; Koumans et al., 1993a,b; Meyer-Rochow et al., 1993; Rowlerson et al., 1995, and others mentioned below). Fast-growing fish generally show greater hyperplasia than slow-growing fish of the same age (Weatherley et al., 1979; Weatherley and Gill, 1987a; Higgins and Thorpe, 1990; Kiessling et al., 1991; Meyer-Rochow et al., 1993; Valente et al., 1999), but a large part of this effect is related to the size reached at the time of sampling (Kiessling et al., 1991). If size is taken into account, alterations in growth rate during the juvenile period generally have parallel effects on hypertrophy and hyperplasia, but there are some exceptions to this. In small fish red muscle fiber diameter was more affected by growth rate and ration level than body size and a very slow growth rate was associated with a preferential reduction of white muscle fiber hypertrophy, whereas, in large fish the dependence on body size weakened and other factors also contributed to the balance between hypertrophic and hyperplastic growth (Kiessling et al., 1991). Different effects of growth rate on the relative contribution of hyperplasia and hypertrophy in fast-white compared to slow-red muscle have also been observed by Valente *et al.*, 1999.

IV. FACTORS AFFECTING MUSCLE GROWTH

A. Diet: Ration, Composition, Feeding Behavior, and Social Hierarchy Effects

1. JUVENILE PERIOD

As might be expected, there is a huge literature referring to the effects of diet upon somatic growth in fish, much of it referring to juvenile (post-metamorphic) stages (e.g., Lovell, 1989; Watanabe, 1992; Houlihan *et al.*, 1993). Biochemical studies have identified increased net protein synthesis in muscle linked to a larger dietary intake (reviewed by Houlihan *et al.*, 1993), and low protein turnover as a strategy for efficient growth (Carter *et al.*, 1998). Atrophy of white muscle fibers occurs under conditions of severe nutritional restriction (Johnston, 1981; Moon, 1983; Maddock and Burton, 1994), and reduced ration size obviously reduces muscle growth (reviewed by Weatherley and Gill, 1987a; see also Kiessling *et al.*, 1991).

During juvenile life reduced rations tend to affect both hypertrophy and hyperplasia in equal measure (Weatherley and Gill, 1987a: p.154; Kiessling *et al.*, 1991), although in small fish a very reduced ration level giving a very slow growth rate was associated with a preferential reduction of white muscle fiber hypertrophy (Kiessling *et al.*, 1991)

Optimal diet compositions have been established for a small number of aquaculture species at various life stages (see e.g., Jobling *et al.*, 1993; Lovell, 1989; Murai, 1992), and these tend to be used as the starting point for the empirical development of suitable diets for new species now being introduced into commercial aquaculture (e.g., Cardenete *et al.*, 1997). The source of protein, and especially its amino acid composition, has a particularly important effect on fish growth (Murai, 1992). Muscle-specific effects of diet composition are seen with pigments and fatty acid contents (reflected in intramuscular fat, affecting flesh quality—see Chapter 7) and in the response to vitamin E/selenium deficiency (leads to muscle degeneration: Lovell, 1989; Lopez-Albors *et al.*, 1995; Ferguson, 1989), but there is very little work specifically linking diet composition to muscle growth mechanisms. A study by Fauconneau *et al.* (1997) did not find any specific effect of dietary lipid composition on fast-white muscle growth, but a comparison of farmed and wild sea bream revealed differences both in fatty acid and trace element contents of muscle (attributable to differences in diet) and in the myosin light chain content of red muscle (Carpenè *et al.*, 1998). This latter effect was tentatively attributed to differences in hypertrophic growth of red muscle, but it is not clear if that also resulted from the different diet or from other aspects of the lifestyle, such as exercise.

Investigations of the effects of diet on growth may be complicated by a number of factors affecting feeding behavior, such as temperature, photoperiod, social interactions, etc. (Cutts *et al.*, 1998; Gélineau *et al.*, 1996; Hossain *et al.*, 1998; Jobling and Koskela, 1996; Ryer and Olla, 1996; Metcalfe *et al.*, 1995, Azzaydi *et al.*, 1999; McCarthy *et al.*, 1999). In juvenile salmon, social dominance is thought to underlie their separation into upper modal (fast-growing) and lower modal (slow-growing) groups of fish in early autumn of their first year (Metcalfe *et al.*, 1989), and the consequent effects on muscle growth (hyperplastic growth of fast-white muscle is reduced in the lower modal group; Higgins and Thorpe, 1990). The introduction of methods enabling the dietary intake of individual fish to be identified (McCarthy *et al.*, 1993) means that effects of diet on muscle growth mechanisms (hypertrophy and hyperplasia) could now be studied directly. Interpretation of the *means* by which any such effects are brought about would involve knowledge of the hormonal status, which is strongly associated with nutritional status (LeBail and Boeuf, 1997; MacKenzie *et al.*, 1998); some aspects of this topic are addressed in Chapter 8.

2. LARVAL LIFE

The transition from endogenous to exogenous feeding which follows hatching is a critical stage in larval development, and is reflected in the overall degree of muscle development at that time (Koumans and Akster, 1995). It is significant that the resumption of stratified hyperplastic growth which occurs some days after hatching in several species (Section III.B.2) follows shortly after the transition to exogenous feeding. Of course, in species with large eggs where first feeding is relatively delayed the association with hyperplastic phases is different; in salmon and brown trout, the mosaic hyperplastic phase has begun by the time of first feeding (Johnston and McLay, 1997; Killeen, 1999).

Larval diets have a number of special requirements, principally the need for live food of carefully graded sizes according to larval age (usually starting with rotifers and then moving on to brine shrimp); much research effort has gone into improving the nutritional value of live food, and is now being directed toward the development of suitable formula feeds for larvae (Day, 1998). The effects of suboptimal diets on muscle growth in larvae tend to be assessed in only terms of somatic growth and survival, but effects on muscle have been studied in the carp and cod, and it was found that hyperplastic growth was preferentially reduced (Alami-Durante *et al.*, 1997; Galloway *et al.*, 1999a).

B. Hormonal Manipulations

These mainly involve growth hormone (GH), although thyroid hormone and sex steriods can also act as growth promoters (Higgs *et al.*, 1975, 1977; Weatherley and Gill, 1987a; Sumpter, 1992; see also Chapter 8). Administration of exogenous GH has been repeatedly demonstrated to improve growth rates, especially in fish which are growing suboptimally (Higgs et al., 1975; Weatherley and Gill, 1982, 1987b; Agellon et al., 1988; Down et al., 1988; Fauconneau et al., 1997). Similarly, administration of the growth hormone mediator, IGF-I (insulin-like growth factor–I), also stimulates growth (McCormick et al., 1992). Because handling may cause stress which inhibits growth (Schulte et al., 1989; Pickering et al., 1991), and would in any case be uneconomic commercially, some research effort has been directed into finding ways of administering exogenous GH in ways which do not require fish to be handled individually (e.g., in a "protected" form in the diet: Moriyama et al., 1993; Tsai et al., 1994). However, attention has more recently been directed toward the production of transgenic lines which show increased expression of GH or the IGFs which mediate many of its growth-promoting effects.

A brief summary of the growth characteristics of some GH-transgenic fish is given by Gong and Hew, 1995. Insertion of an additional IGF or GH gene into the genome does not necessarily result in increased growth or in healthy fish (Chen et al., 1995), although improved growth has been reported by several authors (Zhang et al., 1990; Chen et al., 1995; Devlin et al., 1995). If the gene is to be usefully expressed it must be accompanied by a promoter which functions correctly in the host genome (without having negative side effects). The situation is further complicated by the fact that the GH levels do not necessarily correlate with growth rates (Sumpter et al., 1991; Sumpter, 1992), and IGFs are also nutritionally regulated and do not always mirror GH levels (Thissen et al., 1994; Duan and Plisetskaya, 1993). Furthermore, commercial exploitation of transgenic fish is likely to experience vigorous consumer and public resistance which will not be easily resolved (at least in Europe), so for the present these fish are of use principally in studies of the mechanisms of muscle growth and factors regulating them.

Until recently, relatively little attention was paid to the effects of IGFs and GH specifically on cellular hypertrophy and hyperplasia in fish muscle. However, Weatherley and Gill (1982) and Fauconneau et al. (1997) have shown that administration of GH causes an increase in fast-white muscle hyperplasia in juvenile rainbow trout. Furthermore, whereas the administration of GH increased both hyperplasia and hypertrophy in fish which, though small, did have a mosaic white muscle, hyperplastic growth could not be induced in fish which lacked a post-larval mosaic white muscle (Weatherley and Gill, 1987a,b). This suggests that GH maximizes the intrinsic growth potential (hyperplastic as well as hypertrophic) of mosaic fast-white muscle, but cannot induce hyperplasia *de novo*.

In sea bream there is a relatively good correlation between GH and IGF-I levels and growth rates (Funkenstein et al., 1989; Pérez-Sánchez et al., 1994; Funkenstein and Cohen, 1996; Funkenstein et al., 1997), so this species is a good candidate for the investigation of GH and IGF expression during development.

IGFs are known to promote proliferation and differentiation of myoblasts (Stewart and Rotwein, 1996), and IGF-I has been shown to be expressed, together with its receptor, in lateral muscle of young sea bream larvae, especially in the apical regions where the early phase of hyperplastic growth is still under way (Funkenstein et al., 1997; Perrot et al., 1999). There is pressure to improve growth rates of sea bream, because in this species (like the sea bass, also farmed in the Mediterranean area) growth slows down markedly in the juvenile "on-growing" period before fish reach marketable size (leading to low profitability; Harache and Paquotte, 1996). It is hoped that an understanding of the regulation of GH and IGF expression in sea bream may permit identification of their role(s) in regulating muscle growth, and perhaps provide the basis for identifying strains with improved growth potential. The action and regulation of GH and the IGFs in growth in fish at molecular and cellular levels are also the object of intense research effort, and are discussed in Chapters 3 and 8.

In addition to its role in promoting early larval survival (Ayson and Lam, 1993), and in metamorphosis (especially in flatfish), thyroid hormone has been found to affect contractile protein expression during muscle development in fish (Yamano et al., 1991; Chanoine et al., 1992; Martinez et al., 1995). However, its effects specifically on hyperplastic and hypertrophic growth of muscle fibers have apparently not received any attention. The mechanisms of action of thyroid hormone on muscle at biochemical level, and its interaction with other hormones, are considered in Chapter 8.

C. Genetic Effects: Strains, Sex, Triploids, and Hybrids

1. STRAINS

Strain differences within species have been noted in several species, often regarding characteristics which appear to be appropriate for the particular environmental conditions experienced by these fish, such as growth rates, developmental landmarks, and metabolic efficiency at different temperatures (e.g., Svåsand et al., 1996; Hunt von Herbing et al., 1996; Hunt von Herbing and Boutilier, 1996; Sumpter, 1992). Differences affecting muscle growth in naturally occurring strains and even between "families" include variations in myotomal fiber number (Greer-Walker et al., 1972; Johnston and McLay, 1997), in the responses of muscle growth in larval herring to altered temperatures (Johnston et al., 1998) and in relative contributions of hypertrophy and hyperplasia to muscle growth in lacustrine and riverine southern smelt (Meyer-Rochow and Ingram, 1993). Strain differences in hyperplastic growth of white muscle have also been observed in juvenile farmed rainbow trout (Fauconneau et al., 1997; Valente et al., 1998, 1999).

The natural variations in growth rate which occur within any population can be exploited by selective breeding, and a very successful selection for a fast growth rate has been achieved for salmonids in commercial aquaculture. For example, over the period of 1983–1993 the time required for farmed salmon growing under optimal conditions to reach commercial weight decreased by about 40%—a very substantial improvement. Recent work suggests that faster growth potential may already be signaled by a higher metabolic rate in embryos at hatching (Metcalfe et al., 1995) and individual growth rates identified in small juveniles held singly persist over several weeks suggesting that they are intrinsic characteristics (Wang et al., 1998).

Even if genetic improvement results in faster growth, this may well be due to factors which do not affect muscle growth in any specific way, but simply improve "fitness" for aquaculture conditions. It is therefore interesting to know if, in the faster growing fish, muscle growth mechanisms simply reflect a more advanced developmental stage (e.g., in salmon, Higgins and Thorpe, 1990) or show a specific effect on hyperplasia. In some strains of fast-growing rainbow trout at least, there seems to be a sustained increase in hyperplastic growth in fast-white muscle (Valente et al., 1998, 1999).

2. SEX

In some fish species a marked sexual dimorphism in body size appears as juveniles reach sexual maturity. In tilapia, where males grow faster, the introduction of "supermale" breeders giving rise to almost 95% male progeny has been a commercial success (Roderick, 1998). In hake, this dimorphism is reflected in reduced hyperplastic and hypertrophic muscle growth in the smaller male (Calvo, 1989). Techniques also exist for the production of all-female fish (using sex-reversed neomales for breeding), and have been very successful for species such as trout where males show commercially undesirable growth characteristics (Bromage, 1992; Purdom, 1993). In the hermaphrodite sea bream *Sparus aurata*, transformation from male to female is accompanied by a considerable increase in body weight and a significant hypertrophy of fast-white muscle fibers (Rowlerson et al., 1995).

3. TRIPLOIDY

There has been considerable interest in triploid strains since the observation that triploid individuals could occur in some fish stocks, and that as triploidy is associated with sterility, this could avoid production losses associated with sexual maturation in some farmed species (e.g., Thorgaard and Gall, 1979; Thorgaard, 1986). Various methods for producing triploids have been described (Solar et al., 1984; Bromage, 1992; Purdom, 1993; Malison et al., 1993; Colombo et al., 1995; Garrido-Ramos et al., 1996). Gonads in mature fish can account for about 25% of body weight, so an optimistic view of the value of a sterile triploid strain is that the metabolic effort which would otherwise be used for gonadal development can be directed instead to muscle growth. In practice, however, faster growth is not guaranteed, and although some triploids show improved growth rates others show little better or even worse growth than in diploids (Solar *et al.,* 1984; Mol *et al.,* 1994; Carter *et al.,* 1994; Habicht *et al.,*1994; Galbreath and Thorgaard, 1994; Withler *et al.,* 1995, 1998; Qin *et al.,* 1998; Bonnet *et al.,* 1999). The triploids are also at a disadvantage when raised together with diploids (they seem to be less aggressive, competing less well for food: Bromage, 1992; Carter *et al.,* 1994).

Furthermore, triploidy does directly affect muscle growth, but the overall effect seems to be that the additional number of chromosomes per nucleus seems to result in larger but fewer muscle fibers (Suresh and Sheehan 1998a,b; Johnston *et al.,* 1999), and a reduction in the number of myogenic cells in growing muscle (Greenlee *et al.,* 1995b; Johnston *et al.,* 1999) which may limit growth potential (see Section III.B). However, triploid strains do offer other advantages, which can tip the commercial balance in their favor; for example, the impact of escaped farmed fish on the natural environment is obviously lessened if they are sterile, and in salmon farming sterile triploids are of value because they do not show the seasonal growth and losses due to early sexual maturation which occurs in normal diploids.

4. Hybrids

Hybrid fish (diploid and triploid) can also be produced artificially between related species (Purdom, 1993), and there is currently some interest in using this approach to obtain fish with particular combinations of commercially valuable characteristics such as disease resistance, preferred temperature range, sterility, etc. Results so far have been variable; growth characteristics of hybrids vary from clearly inferior to at least comparable to those of the parental species (Blanc *et al.,* 1992; Habicht *et al.,* 1994; Galbreath and Thorgaard, 1994).

An important commercial consideration is that accelerated growth needs to be maintained through to commercial size. In the case of the sparid hybrid Pantex, a cross between female *Pagrus major* and male *Dentex dentex,* an initially very high growth rate is not maintained beyond the first year, so the time taken to reach commercial size is not significantly shorter than for other farmed sparid species (Poli *et al.,* 1999). Unfortunately, there is currently no information about hyperplastic and hypertrophic growth of muscle fibers in these species.

D. Season

Many fish show seasonal changes in growth rate, and a major factor in this is water temperature (see below). Photoperiod can also affect growth and developmental changes such as smoltification in salmon, but probably exerts any effect on muscle only indirectly (Mäkinen and Ruohonen, 1992; Sumpter, 1992; Smith *et al.,* 1993; Coggan, 1997; Dickhoff *et al.,* 1997; Sánchez-Vázquez and Tabata, 1998). It is well known that increasing temperature favors growth up to an optimal value, above which net growth falls (because of the cost of the increasing metabolic rate) before reaching the survival limit (reviewed by Jobling, 1997; Larsson and Berglund, 1998). The effects of increasing temperature on muscle growth at different life stages have also been described for several species. These effects include acclimation responses such as phenotype transformations of existing muscle fibers (reviewed by Johnston, 1993b, 1994; see also Chapter 6), effects on protein growth and growth efficiency (McCarthy et al., 1994), and changes in muscle growth dynamics.

During the early (i.e., embryonic and larval) stages, changes in rearing temperature can affect the number and/or size of muscle fibers and other myogenic cells produced, although the exact effect can vary between species and between years (see Section III.C; Stickland et al., 1988; Calvo and Johnston, 1992; Vieira and Johnston, 1992; Brooks and Johnston, 1993; Johnston, 1993b; Usher et al., 1994; Nathanailides et al., 1995a, 1996; Hanel et al., 1996; Johnston and McLay, 1997; Matschak et al., 1997, 1998; Johnston et al., 1998; Galloway et al., 1998; Galloway et al., 1999a; Johnston, 1999). Effects on proliferation competent myogenic cells present in the muscle are of particular interest, because they could influence the long term growth potential (see Johnston, 1999 and Section IV.F).

An increase in the ambient temperature can also affect juvenile growth, stimulating fiber hyperplasia and therefore resulting in the appearance of new fibers throughout the lateral muscle (Ennion et al., 1995; Nathanailides et al., 1995b, 1996). The seasonal appearance of new fibers in mullet muscle reported by Carpenè and Veggetti (1981) was most likely promoted by the increase in water temperature during the summer; it was prevented by keeping the fish at winter temperatures throughout the summer period and occurred earlier in fish brought to summer temperatures in early spring (Carpenè et al., 1983). A seasonal effect on hyperplasia (presumably mediated by changes in feeding) was also shown by Higgins and Thorpe (1990). There was a large reduction in the proportion of smallest diameter fibers in lower modal group salmon during the winter months of their first year, and in the winter months of upper modal group fish in their second year. Variations in the proportions of small diameter fibers in fast-white muscle of rainbow trout, which might be related to season, have also been reported by Kiessling et al. (1991).

E. Exercise

Training regimes may produce hypertrophy of slow-red and/or fast-white muscle fibers in some species, although significant effects on hyperplasia are not usually observed (reveiwed by Davison, 1989, 1997; see also Totland et al., 1987; Sänger, 1992; Young and Cech, 1992 and Chapter 7). The outcome is variable depending on the species, swimming habits (stayers, sprinters), and the type of exercise imposed. For example, there is some evidence to suggest that hypertrophy of white muscle fibers occurs only if the training regime actually results in an increase in burst swimming (Totland et al., 1987; Hinterleitner et al., 1992). Endurance training tends to result in an increase in oxidative enzymes, and in the relative size of the red muscle layer, but does not affect white muscle fiber diameters or hyperplasia (Davie et al., 1986). From an aquaculture viewpoint, there is evidence that regular light exercise promotes better growth, although in salmonids this is at least partly the result of reduced agonistic behavior between fish and improved food intake (e.g., Christiansen and Jobling, 1990, East and Magnan, 1987; Totland et al., 1987) rather than by direct effect on muscle growth mechanisms. Higher levels of exercise may have negative effects in some species (Davison, 1997), even causing myoseptal damage (Christiansen et al., 1992).

F. "Growth History" Effects

Can growth at one life stage affect later growth potential? This is currently a topic of some interest in mammals, where there is good evidence that, for example, nutritional restriction during early gestation can have substantial negative effects on later muscle growth (Stickland and Dwyer, 1996; Rehfeldt et al., 1993). In fish, egg (and therefore yolk) size influences larval size and survival (Baynes and Howell, 1996; Marteinsdottir and Steinarsson, 1998) and experiments which have tested the effects of temperature on early growth in fish indicate that this may affect the number of myogenic cells in the muscle (Stickland et al., 1988; Vieira and Johnston, 1992; Usher et al., 1994) and in some cases affect later hyperplastic and hypertrophic growth (Nathanailides et al., 1995a; Johnston et al., 1998). There is also some evidence for growth history effects on somatic growth in juvenile fish (Mäkinen and Ruohonen, 1992), and if rearing conditions could be found which potentiate hyperplastic muscle growth in this later phase it would be of immediate practical interest because this brings the fish to commercial size.

V. TRENDS FOR FUTURE RESEARCH ON MUSCLE GROWTH IN FISH, ESPECIALLY AQUACULTURE SPECIES

The (potential) benefits to be derived from genetic improvement (strain selection, hybrids, transgenics) were mentioned in Sections IV.B and IV.C, but the most successful method so far (strain selection) has been largely restricted to salmonids. Other aquaculture species would undoubtedly benefit from this approach (Knibb et al., 1996). There is also a need for information relating growth mechanisms (specifically hypertrophy and hyperplasia) in *muscle* to other factors (e.g., nutrition) influencing overall growth performance.

More attention is also likely to be directed toward growth history effects (Section IV.F), and especially the influence of larval growth on later juvenile growth.

The possibility of using larval growth conditions to manipulate the behavior and/or number of myogenic cells which will give rise to the mosaic hyperplastic growth phase is obviously very attractive, and poses very interesting biological questions. Although both hyperplastic and hypertrophic growth depend on the availability of myogenic cells, hypertrophy has a functional maximum in most fish of about 200 μm for fiber diameter, which limits its overall contribution to growth. The hyperplastic process contributes rather more, especially during juvenile stages, but wanes as the fish increases in size, and we do not yet know if this is because myogenic cells are intrinsically limited in the number of mitoses they can complete, because of a decline in some physiological stimulus for their proliferation or because of a progressive insensitivity to previously effective stimuli (cf. Quinn et al., 1990; Schultz and McCormick, 1993; see also Chapter 4).

In higher vertebrates, conditions such as "double muscling" and other genetically determined faster growth rates are attributed to a prolonged proliferative phase during embryonic and fetal myogenesis (Penney et al., 1993; Remignon et al., 1995; Picard et al., 1995), and a higher proliferative activity of satellite cells (Merly et al., 1998). Double-muscling in cattle, and a greatly increased hypertrophic and hyperplastic growth of muscle in mice, have been shown to be due to mutations in the gene for myostatin (GDF-8), a negative regulator of muscle growth (McPherron and Lee, 1997; McPherron et al., 1997). Clearly, a better understanding of the factors which control the proliferation of the myogenic cells in fish muscle would be of immense benefit in the search for improved rearing conditions, and strains, which would reach commercial size more rapidly.

REFERENCES

Agellon, L. B., Emery, C. J., Jones, J. M., Davies, S. L., Dingle, A. D., and Chen, T. T. (1988). Promotion of rapid growth of rainbow trout (*Salmo gairdneri*) by a recombinant fish growth hormone. *Can. J. Fish. Aquat. Sci.* **45,** 146–151.

Akster, H. A., Koumans, J. T. M., Cuelenaere, J., and Osse, J. W. M. (1995). Uptake of tritiated thymidine in muscle of juvenile carp. *J. Fish Biol.* **47,** 165–167.

Alami-Durante, H., Fauconneau, B., Rouel, M., Escaffre, A. M., and Bergot, P. (1997). Growth and multiplication of white skeletal muscle fibres in carp larvae in relation to somatic growth rate. *J. Fish Biol.* **50,** 1285–1302.

Alfei, L., Maggi, F., Parvopassu, F., Bertoncello, G., and de Vita, R. (1989). Postlarval muscle growth in fish: a DNA flow cytometric and morphometric analysis. *Bas. Appl. Histochem.* **33,** 147–158.

Alfei, L., Colombari, P. T., Cavallo, D., Eleuteri, P. and De Vita, R. (1993). Use of 5′-bromodeoxyuridine immunohistochemistry to examine proliferative activity of fish tissues. *Eur. J. Histochem.* **37,** 183–189.

Alfei, L., Onali, A., Spanò, L., Colombari, P. T., Altavista, P. L., and De Vita, R. (1994). PCNA/cyclin expression and BrdU uptake define proliferating myosatellite cells during hyperplastic muscle growth of fish (*Cyprinus carpio* L). *Eur. J. Histochem.* **38,** 151–162.

Amthor, H., Christ, B., and Patel, K. (1999). A molecular mechanism enabling continuous embryonic muscle growth—a balance between proliferation and differentiation. *Development* **126**, 1041–1053.

Arndt, S. K. A., Benfey, T. J., and Cunjak, R. A. (1994). A comparison of RNA concentrations and ornithine decarboxylase activity in Atlantic salmon (*Salmo salar*) muscle tissue, with respect to specific growth rates and diel variations. *Fish Physiol. Biochem.* **13**, 463–471.

Ayala, M. D., López-Albors, O., Gil, F., Ramírez-Zarzosa, G, Abellán, E., and Moreno, F. (1999). Red muscle development of gilthead sea bream *Sparus aurata* (L.): structural and ultrastructural morphometry. *Anat. Histol. Embryol.* **28**, 17–21.

Ayson, F. G., and Lam, T. J. (1993). Thyroxine injection of female rabbit fish (*Siganus guttatus*) broodstock: changes in thyroid hormone levels in plasma, eggs, and yolk-sac larvae, and its effects on larval growth and survival. *Aquaculture* **109**, 83–93.

Azzaydi, M., Martínez, F. J., Zamora, S., Sánchez-Vázquez, F. J., and Madrid, J. A. (1999). Effect of meal size modulation on growth performance and feeding rhythms in European sea bass (*Dicentrarchus labrax,* L.). *Aquaculture* **170**, 253–266.

Battram, J. C., and Johnston, I. A. (1991). Muscle growth in the Antarctic teleost, *Notothenia neglecta* (Nybelin). *Antarct. Sci.* **3**, 29–33.

Baynes, S. M., and Howell, B. R. (1996). The influence of egg size and incubation temperature on the condition of Solea solea (L) larvae at hatching and first feeding. *J. Exp. Mar. Biol. Ecol.* **199**, 59–77.

Benfey, T. J., Saunders, R. L., Knox, D. E., and Harmon, P. R. (1994). Muscle ornithine decarboxylase activity as an indication of recent growth of pre-smolt Atlantic salmon, *Salmo salar. Aquaculture* **121**, 125–135.

Blanc, J. M., Poisson H., and Vallée, F. (1992). Survival, growth and sexual maturation of the triploid hybrid between rainbow trout and arctic charr. *Aquat. Living Resour.* **5**, 15–21.

Bobe, J., Andre, S., and Fauconneau, B. (2000). Embryonic muscle development in rainbow trout (*Onchrhynchus mykiss*): a scanning microscopy and immunohistological study. *J. Exp. Zool.* **286**, 379–389.

Bonnet, S., Haffray, P., Blanc, J. M., Vallée, F., Vauchez, C., Faure, A., and Fauconneau, B. (1999). Genetic variation in growth parameters until commercial size in diploid and triploid freshwater rainbow trout (*Oncorhynchus mykiss*) and seawater brown trout (*Salmo trutta*). *Aquaculture* **173**, 359–375.

Bromage, N. (1992). Propagation and stock improvement. *In* "Intensive Fish Farming" (Shepherd, C. J., and Bromage, N. R., eds.), pp. 103–153. Blackwell Scientific Publications, Oxford.

Brooks, S., and Johnston, I. A. (1993). Influence of development and rearing temperature on the distribution, ultrastructure and myosin sub-unit composition of myotomal muscle-fibre types in the plaice *Pleuronectes platessa. Mar. Biol.* **117**, 501–513.

Brooks, S., Vieira, V. L. A., Johnston, I. A., and Macheru, P. (1995). Muscle development in larvae of a fast-growing tropical freshwater fish, the curimatã-pacú. *J. Fish Biol.* **47**, 1026–1037.

Calvo, J. (1989). Sexual differences in the increase of white muscle fibres in Argentine hake, *Merluccius hubbsi,* from the San Matias Gulf (Argentina). *J. Fish Biol.* **35**, 207–214.

Calvo, J., and Johnston, I. A. (1992). Influence of rearing temperature on the distribution of muscle fibre types in the turbot *Scophthalmus maximus* at metamorphosis. *J. Mar. Biol. Ecol.* **161**, 45–55.

Cardasis, C. A., and Cooper, G. W. (1975). An analysis of nuclear numbers in individual muscle fibers during differentiation and growth: a satellite cell—muscle fiber growth unit. *J. Exp. Zool.* **191**, 347–358.

Cardenete, G., Abellan, E., Skalli, A., and Massuti, S. (1997). Feeding *Dentex dentex* with dry diets: growth response and diet utilisation. *In* "Feeding Tomorrow's Fish," Cahiers Options Méditerranéens Vol. 22 (Tacon, A., and Basurco, B., eds.), pp. 141–152. CIHEAM, Zaragoza.

Carpenè, E., and Veggetti, A. (1981). Increase in muscle fibres in the lateralis muscle (white portion) of *Mugilidae* (Pisces, Teleostei). *Experientia* **37**, 191–193.

Carpenè, E., Scapolo, P. A., Mascarello, F., Veggetti, A., and Rowlerson, A. (1983). The influence of temperature on the annual hyperplastic cycle in white muscle of Mugilidae. *Atti S. I. S. Vet.* **37**, 117–118.

Carpenè, E., Martin, B., and Dalla Libera, L. (1998). Biochemical differences in lateral muscle of wild and farmed gilthead sea bream (*Sparus aurata* L.). *Fish Physiol. Biochem.* **19**, 229–238.

Carter, C. G., McCarthy, I. D., Houlihan, D. F., Johnstone, R., Walsingham, M. V., and Mitchell. A. I. (1994). Food consumption, feeding behaviour, and growth of triploid and diploid Atlantic salmon *Salmo salar* L. parr. *Can. J. Zool.* **72**, 609–617.

Carter, C. G., Houlihan, D. F., and Owen, S. F. (1998). Protein synthesis, nitrogen excretion and long-term growth of juvenile *Pleuronectes flesus*. *J. Fish Biol.* **53**, 272–284.

Chanoine, C., Guyot-Lenfant, M., Elattari, A., Saadi, A., and Gallien, C. L. (1992). White muscle differentiation in the eel (*Anguilla anguilla* L.): changes in the myosin isoforms pattern and ATPase profile during post-metamorphic development. *Differentiation* **49**, 69–75.

Cheek, D. B., and Hill, D. E. (1970). Muscle and liver cell growth: role of hormones and nutritional factors. *Fed. Proc.* **29**, 1503–1509.

Chen, T. T., Lu, J. K., Shamblott, M. J., Cheng, C. M., Lin, C. M., Burns, J. C., Reimschuessel, R., Chatakondi, N., and Dunham, R. A. (1995). Transgenic fish: ideal models for basic research and biotechnological applications. *Zool. Stud.* **34**, 215–234.

Christiansen, J. S., and Jobling, M. (1990). The behaviour and the relationship between food intake and growth of juvenile Arctic charr, *Salvelinus alpinus* L., subjected to sustained exercise. *Can. J. Zool.* **68**, 2185–2191.

Christiansen, J. S., Martinez, I., Jobling, M., and Amin, A. B. (1992). Rapid somatic growth and muscle damage in a salmonid fish. *Bas. Appl. Myol.* **2**, 235–239.

Coggan, R. (1997). Growth: ration relationships in the antarctic fish *Notothenia corriceps* Richardson maintained under different conditions of temperature and photoperiod. *J. Exp. Mar. Biol. Ecol.* **210**, 23–35.

Colombo, L., Barbaro, A., Libertini, A., Benedetti, P., Francescon, A., and Lombardo, I. (1995). Artificial fertilisation and induction of triploidy and meioginogenesis in the European sea bass, *Dicentrarchus labrax* L. *J. Appl. Ichthyol.* **11**, 118–125.

Cutts, C. J., Metcalfe, N. B., and Taylor, A. C. (1998). Aggression and growth depression in juvenile Atlantic salmon: the consequences of individual variation in standard metabolic rate. *J. Fish Biol.* **52**, 1028–1037.

Davie, P. S., Wells, R. M. G., and Tetens, V. (1986). Effects of sustained swimming on rainbow trout muscle structure, blood oxygen transport, and lactate dehydrogenase isozymes: evidence for increased aerobic capacity of white muscle. *J. Exp. Zool.* **237**, 159–171.

Davison, W. (1989). Training and its effects on teleost fish. *Comp. Biochem. Physiol.* **94A**, 1–10.

Davison, W. (1997). The effects of exercise training on telost fish, a review of recent literature. *Comp. Biochem. Physiol.* **117A**, 67–75.

Day, O. (1998). Formulated feeds for marine fish larvae. *Fish Farmer* **21**, 14–15.

Delalande, J. M., and Rescan, P. Y. (1999). Differential expression of two nonallelic MyoD genes in developing and adult myotomal musculature of the trout (*Oncorhynchus mykiss*). *Dev. Genes Evol.* **209**, 432–437.

Devlin, R. H., Yesaki, T. Y., Donaldson, E. M., Du, S. J., and Hew, C. L. (1995). Production of germ-line transgenic Pacific salmonids with dramatically increased growth performance. *Can. J. Fish. Aquat. Sci.* **52**, 1376–1384.

Dickhoff, W. W., Beckman, B. R., Larsen, D. A., Duan, C., and Moriyama, S. (1997). The role of growth in endocrine regulation of salmon smoltification. *Fish Physiol. Biochem.* **17**, 231–236.

Down, N. E., Donaldson, E. M., Dye, H. M., Langley, K., and Souza, L. M. (1988). Recombinant

bovine somatotropin more than doubles the growth rate of Coho salmon (*Oncorhynchus kisutch*) acclimated to sea water and ambient winter conditions. *Aquaculture* **68**, 141–155.
Duan, C., and Plisetskaya, E. M. (1993). Nutritional regulation of insulin-like growth factor-I mRNA expression in salmon tissues. *J. Endocrinol.* **139**, 243–252.
East, P., and Magnan, P. (1987). The effect of locomotor activity on the growth of brook charr, *Salvelinus fontinalis* Mitchill. *Can. J. Zool.* **65**, 843–846.
Egginton, S., and Johnston, I. A. (1982). Muscle fibre differentiation and vascularisation in the juvenile European eel (*Anguilla anguilla* L.). *Cell Tissue Res.* **222**, 563–577.
El-Fiky, N., and Wieser, W. (1988). Life styles and patterns of development of gills and muscles in larval cyprinids (*Cyprinidae;* Teleostei). *J. Fish Biol.* **33**, 135–145.
Ennion, S., Gauvry, L., Butterworth, P., and Goldspink, G. (1995). Small-diameter white myotomal muscle fibres associated with growth hyperplasia in the carp (*Cyprinus carpio*) express a distinct myosin heavy chain gene. *J. Exp. Biol.* **198**, 1603–1611.
Ennion, S., Wilkes, D., Gauvry, L., Alami-Durante, H., Goldspink, G. (1999). Identification and expression analysis of two developmentally regulated myosin heavy chain gene transcripts in carp (*Cyprinus carpio*). *J. Exp. Biol.* **202**, 1081–1090.
FAO (1998). Aquaculture production statistics 1987–1996. *Fish. Circ.* **815**, revision 10.
Fauconneau, B., André, S., Chmaitilly, J., LeBail, P. Y., Krieg, F., and Kaushik, S. J. (1997). Control of skeletal muscle fibres and adipose cells size in the flesh of rainbow trout. *J. Fish Biol.* **50**, 296–314.
Ferguson, H. W. (1989). Musculoskeletal system. *In* "Systemic Pathology of Fish," pp. 215–229. Iowa State University Press.
Funkenstein, B., and Cohen, I. (1996). Ontogeny of growth hormone protein and mRNA in the gilthead sea bream *Sparus aurata. Growth Regulat.* **6**, 16–21.
Funkenstein, B., Silbergeld, A., Cavari, B., and Laron, Z. (1989). Growth hormone increases plasma levels of insulin-like growth factor (IGF-I) in a teleost, the gilthead seabream (*Sparus aurata*). *J. Endocrinol.* **120**, R19–R21.
Funkenstein, B., Almuly, R., and Chan, S. J. (1997). Localisation of IGF-I and IGF-I receptor mRNA in *Sparus aurata* larvae. *Gen. Comp. Endocrinol.* **107**, 291–303.
Galbreath, P. F., and Thorgaard, G. H. (1994). Viability and fresh-water performance of atlantic salmon (*Salmo salar*) x brown trout (*Salmo trutta*) triploid hybrids. *Can. J. Fish Aquat. Sci.* **51**, 16–24.
Galloway, T. F., Kjørsvik, E., and Kryvi, H. (1998). Effect of temperature on viability and axial muscle development in embryos and yolk sac larvae of the Northeast Arctic cod (*Gadus morhua*). *Mar. Biol.* **132**, 559–567.
Galloway, T. F., Kjørsvik, E., and Kryvi, H. (1999a). Muscle growth and development in Atlantic cod larvae (*Gadus morhua* L.) related to different somatic growth rates. *J. Exp. Biol.* **202**, 2111–2120.
Galloway, T. F., Kjørsvik, E., and Kryvi, H. (1999b). Muscle growth in yolk-sac larvae of the Atlantic halibut as influenced by temperature in the egg and yolk-sac stage. *J. Fish Biol.* **55 (Suppl. A)**, 26–43.
Garrido-Ramos, M., de la Herrán, R., Lozano, R., Cárdenas, S., Rejón, C. R., and Rejón, M. R. (1996). Induction of triploidy in offspring of gilthead seabream (*Sparus aurata*) by means of heat shock. *J. Appl. Ichthyol.* **12**, 53–55.
Gauvry, L., and Fauconneau, B. (1996). Cloning of a trout fast skeletal myosin heavy chain expressed both in embryo and adult muscles and in myotubes neoformed *in vitro. Comp. Biochem. Physiol.* **115B**, 183–190.
Gélineau, A., Mambrini, M., Leatherland, J. F., and Boujard, T. (1996). Effect of feeding time on hepatic nucleic acid, plasma T_3, T_4, and GH concentrations in rainbow trout. *Physiol. Behav.* **59**, 1061–1067.

Gibson, S., and Johnston, I. A. (1995). Temperature and development in larvae of the turbot *Scophthalmus maximus. Mar. Biol.* **124,** 17–25.

Gong, Z. and Hew, C. L. (1995). Transgenic fish in aquaculture and developmental biology. *Curr. Top. Dev. Biol.* **30,** 177–214.

Grant, G. C. (1996). RNA-DNA ratios in white muscle tissue biopsies reflect recent growth rates of adult brown trout. *J. Fish Biol.* **48,** 1223–1230.

Greenlee, A. R., Dodson, M. V., Yablonka-Reuveni, Z., Kersten, C. A., and Cloud, Y. G. (1995a). *In vitro* differentiation of myoblasts from skeletal muscle of rainbow trout. *J. Fish Biol.* **46,** 731–747.

Greenlee, A. R., Kersten, C. A., and Cloud, Y. G. (1995b). Effects of triploidy on rainbow trout myogenesis *in vitro. J. Fish Biol.* **46,** 381–388.

Greer-Walker, M., Burd, A. C., and Pull, G. A. (1972). The total numbers of white skeletal muscle fibres in cross section as a character for stock separation in North sea herring (*Clupea harengus* L.). *J. Cons. Int. Explor. Mer.* **34,** 238–243.

Habicht, C., Seeb, J. E., Gates, R. B., Brock, I. R., and Olito, C. A. (1994). Triploid coho salmon outperform diploid and triploid hybrids between coho salmon and chinook salmon during their first year. *Can. J. Fish. Aquat. Sci.* **51,** 31–37.

Hanel, R., Karjalainen, J., and Wieser, W. (1996). Growth of swimming muscles and its metabolic cost in larvae of whitefish at different temperatures. *J. Fish Biol.* **48,** 937–951.

Harache, Y., and Paquotte, P. (1996). The development of marine fish farming in Europe: a parallel with salmon culture. *In* "Seabass and Seabream Culture: Problems and Prospects" (Chatain, B., Saroglia, M., Sweetman, J., and Lavens, P., eds.), Annex B, pp. 369–378. European Aquaculture Society Handbook.

Higgins, P. J. (1990). The histochemistry of muscle in juvenile atlantic salmon, *Salmo salar* L. *J. Fish Biol.* **37,** 521–529.

Higgins, P. J., and Thorpe, J. E. (1990). Hyperplasia and hypertrophy in the growth of skeletal muscle in juvenile Atlantic salmon, *Salmo salar* L. *J. Fish Biol.* **37,** 505–519.

Higgs, D. A., Donaldson, E. M., Dye, H. M., and McBride, J. R. (1975). A preliminary investigation of the effect of bovine growth hormone on growth and muscle composition of coho salmon (*Oncorhyncus kisutch*). *Gen. Comp. Endocrinol.* **27,** 240–253.

Higgs, D. A., Fagerlund, U. H. M., McBride, J. R., Dye, H. M., and Donaldson, E. M. (1977). Influence of combinations of bovine growth hormone, 17α-methyltestosterone, and L-thyroxine on growth of yearling coho salmon (*Oncorhynchus kisutch*). *Can. J. Zool.* **55,** 1048–1056.

Hinterleitner, S., Huber, M., Lackner, R., and Wieser, W. (1992). Systemic and enzymatic responses to endurance training in two cyprinid species with different life styles (Teleostei: Cyprinidae). *Can. J. Fish. Aquat. Sci.* **49,** 110–115.

Hossain, M. A. R., Beveridge, M. C. M., and Haylor, G. S. (1998). The effects of density, light and shelter on the growth and survival of African catfish (*Clarias gariepinus* Burchell, 1822) fingerlings. *Aquaculture* **160,** 251–258.

Houlihan, D. F., Mathers, E. M., and Foster, A. (1993). Biochemical correlates of growth rate in fish. *In* "Fish Ecophysiology" (Rankin, J. C., and Jensen, F. B., eds.), pp. 45–71. Chapman & Hall, London.

Hunt von Herbing, I., and Boutilier, R. G. (1996). Activity and metabolism of larval Atlantic cod (*Gadus morhua*) from Scotian Shelf and Newfoundland source populations. *Mar. Biol.* **124,** 607–617.

Hunt von Herbing, I., Boutilier, R. G., Miyake, T., and Hall, B. K. (1996). Effects of temperature on morphological landmarks critical to growth and survival in larval Atlantic cod (*Gadus morhua*). *Mar. Biol.* **124,** 593–606.

Jobling, M. (1997). Temperature and growth: modulation of growth rate via temperature change. *In*

"Global Warming: Implications for Freshwater and Marine Fish" (Wood, C. M., and McDonald, D. G., eds.), pp. 225–252. C.U.P.

Jobling, M., and Koskela, J. (1996). Interindividual variations in feeding and growth in rainbow trout during restricted feeding and in a subsequent period of compensatory growth. *J. Fish Biol.* **49**, 658–667.

Jobling, M., Jørgensen, E. H., Arnesen, A. M., and Ringø, E. (1993). Feeding, growth and environmental requirements of Arctic charr: a review of aquaculture potential. *Aquaculture Int.* **1**, 20–46.

Johnston, I. A. (1981). Quantitative analysis of muscle breakdown during starvation in the marine flatfish *Pleuronectes platessa*. *Cell Tissue Res.* **214**, 369–386.

Johnston, I. A. (1982). Capillarisation, oxygen diffusion distances and mitochondrial content of carp muscles following acclimation to summer and winter temperatures. *Cell Tissue Res.* **222**, 325–337.

Johnston, I. A. (1993a). Temperature influences muscle differentiation and the relative timing of organogenesis in herring (*Clupea harengus*) larvae. *Mar. Biol.* **116**, 363–379.

Johnston, I. A. (1993b). Phenotypic plasticity of fish muscle to temperature change. In "Fish Ecophysiology" (Rankin, J. C., and Jensen, F. B., eds.), pp. 322–340. Chapman & Hall, London.

Johnston, I. A. (1994). Development and plasticity of fish muscle with growth. *Bas. Appl. Myol.* **4**, 353–368.

Johnston, I. A. (1999). Muscle development and growth: potential implications for flesh quality in fish. *Aquaculture* **177**, 99–115.

Johnston, I. A., and Horne, Z. (1994). Immunocytological investigations of muscle differentiation in the Atlantic herring (*Clupea harengus:* Teleostei). *J. Mar. Biol. Ass. U.K.* **74**, 79–91.

Johnston, I. A., and McLay, H. A. (1997). Temperature and family effects on muscle cellularity at hatch and first feeding in Atlantic salmon (*Salmo salar* L.). *Can. J. Zool.* **75**, 64–74.

Johnston, I. A., Vieira, V. L. A., and Abercromby, M. (1995). Temperature and myogenesis in embryos of the Atlantic herring *Clupea harengus*. *J. Exp. Biol.* **198**, 1389–1403.

Johnston, I. A., Cole, N. J., Abercromby, M., and Vieira, V. L. A. (1998). Embryonic temperature modulates muscle growth characteristics in larval and juvenile herring. *J. Exp. Biol.* **201**, 623–646.

Johnston, I. A., Strugnell, G., McCracken, M. C., and Johnstone, R. (1999). Muscle growth and development in normal-sex ratio and all-female diploid and triploid Atlantic salmon. *J. Exp. Biol.* **202**, 1991–2016.

Kiessling, A., Storebakken, T., Åsgård, T., and Kiessling, K.-H. (1991). Changes in the structure and function of the epaxial muscle of rainbow trout (*Oncorhynchus mykiss*) in relation to ration and age. I. Growth dynamics. *Aquaculture* **93**, 335–356.

Kiessling, A., Larsson, L., Kiessling, K.-H., Lutes, P. B., Storebakken, T., and Hung, S. S. S. (1995). Spawning induces a shift in energy metabolism from glucose to lipid in rainbow trout white muscle. *Fish Physiol. Biochem.* **14**, 439–448.

Kilarski, W. (1990). Histochemical characterization of myotomal muscle in the roach, *Rutilis rutilis* (L.). *J. Fish Biol.* **36**, 353–362.

Killeen, J. R. (1999). The effects of temperature on development and growth of muscle in the trout (*Salmo trutta* L.). Thesis, University of St. Andrews (U.K.).

Killeen, J. R., McLay, H. A., and Johnston, I. A. (1999). Temperature and neuromuscular development in embryos of the trout (*Salmo trutta* L.). *Comp. Biochem. Physiol.* **122A**, 53–64.

Knibb, W., Gorshkova, G., and Gorshkov, S. (1996). Potential gains through genetic improvement: selection and transgenesis. In "Seabass and Seabream Culture: Problems and Perspectives" (Chatain, B., Saroglia, M., Sweetman, J., and Lavens, P., eds.), pp. 176–188. European Aquaculture Society Handbook.

Kobiyama, A., Nihei, Y., Hirayama, Y., Kikuchi, K., Suetake, H., Johnston, I. A., and Watabe, S. (1998). Molecular cloning and developmental expression patterns of the MyoD and MEF2 families of muscle transcription factors in the carp. *J. Exp. Biol.* **201,** 2801–2813.

Koumans, J. T. M., and Akster, H. A. (1995). Myogenic cells in development and growth of fish. *Comp. Biochem. Physiol.* **110A,** 3–20.

Koumans, J. T. M., Akster, H. A., Dulos, G. J., and Osse, J. W. M. (1990). Myosatellite cells of *Cyprinus carpio* (Teleostei) *in vitro:* isolation, recognition and differentiation. *Cell Tissue Res.* **261,** 173–181.

Koumans, J. T. M., Akster, H. A., Booms, G. H. R., Lemmens, C. J. J., and Osse, J. W. M. (1991). Numbers of myosatellite cells in white axial muscle of growing fish: *Cyprinus carpio* L. (Teleostei). *Am. J. Anat.* **192,** 418–424.

Koumans, J. T. M., Akster, H. A., Booms, G. H. R., and Osse, J. W. M. (1993a). Growth of carp (*Cyprinus carpio*) white axial muscle; hyperplasia and hypertrophy in relation to the myonucleus/sarcoplasm ratio and the occurence of different subclasses of myogenic cells. *J. Fish Biol.* **43,** 69–80.

Koumans, J. T. M., Akster, H. A., Booms, R. G. H., and Osse J. W. M. (1993b). Influence of fish size on proliferation and differentiation of cultured myosatellite cells of white axial muscle of carp (*Cyprinus carpio* L.). *Differentiation* **53,** 1–6.

Kundu, R., and Mansuri, A. P. (1990). Growth dynamics of myotomal muscle fibres in a carangid, *Carianx melabaricus* (Cuv. and Val.). *J. Fish Biol.* **36,** 21–27.

Larsson, S., and Berglund, I. (1998). Growth and food consumtion of 0+ Arctic charr fed pelleted or natural food at six different temperatures. *J. Fish Biol.* **52,** 230–242.

LeBail, P.-Y., and Boeuf, G. (1997). What hormones may regulate food intake in fish? *Aquat. Living Resour.* **10,** 371–379.

López-Albors, O., Gil, F., Ramírez-Zarzosa, G., Latorre, R., García-Alcázar, A., Abellán, E., Blanco, A., Vázquez, J. M., and Moreno, F. (1995). Early muscle injuries in a standard reared stock of sea bass *Dicentrarchus labrax* (L.). *Aquaculture* **138,** 69–76.

Lone, K. P., and Ince, B. W. (1983). Cellular growth responses of rainbow trout (*Salmo gairdneri*) fed different levels of dietary protein, and an anabolic steriod ethylestrenol. *Gen. Comp. Endocrinol.* **49,** 32–49.

Lovell, T. (1989). The nutrients. *In* "Nutrition and Feeding of Fish" pp. 11–71. Van Nostrand Reinhold, New York.

Luquet, P., and Durand, G. (1970). Evolution de la teneur en acides nucléiques de la musculature épaxiale au cours de la croissance chez la truite arc-en-ciel (*Salmo gairdneri*): roles respectifs de la multiplication et du grandissement cellulaires. *Ann. Biol. Anim. Biochem. Biophys.* **10,** 481–493.

McCarthy, I. D., Houlihan, D. F., Carter, C. G., and Moutou, C. (1993). Variation in individual food consumption rates of fish and its implications for the study of fish nutrition and physiology. *Proc. Nutr. Soc.* **52,** 427–436.

McCarthy, I. D., Houlihan, D. F., and Carter, C. G. (1994). Individual variation in protein turnover and growth efficiency in rainbow trout, *Oncorhynchus mykiss* (Walbaum). *Proc. R. Soc. Lond. B.* **257,** 141–147.

McCarthy, I. D., Moksness, E., Pavlov, D. A., and Houlihan, D. F. (1999). Effects of water temperature on protein synthesis and protein growth in juvenile Atlantic wolffish (*Anarhichas lupus*). *Can. J. Fish. Aquat. Sci.* **56,** 231–241.

McCormick, S. D., Kelley, K. M., Young, G., Nishioka, R. S., and Bern, H. A. (1992). Stimulation of coho salmon growth by insulin-like growth factor-I. *Gen. Comp. Endocrinol.* **86,** 398–406.

MacKenzie, D. S., VanPutte, C. M., Leiner, K. A. (1998). Nutrient regulation of endocrine function in fish. *Aquaculture* **161,** 3–25.

McLaughlin, R. L., Ferguson, M. M., and Noakes, D. L. G. (1995). Concentrations of nucleic-acids

and protein as indexes of nutritional-status for recently emerged brook trout (*Salvelinus fontinalis*). *Can. J. Fish. Aquat. Sci.* **52,** 848–854.
McPherron, A. C., and Lee, S.-J. (1997). Double-muscling in cattle due to mutations in the myostatin gene. *Proc. Natl. Acad. Sci. U.S.A.* **94,** 12457–12461.
McPherron, A. C., Lawler, A. M., and Lee, S.-J. (1997). Regulation of skeletal muscle mass in mice by a new TGF-β superfamily member. *Nature* **387,** 83–90.
Maddock, D. M., and Burton, M. P. M. (1994). Some effects of starvation on the lipid and skeletal muscle layers of the winter flounder, *Pleuronectes americanus*. *Can. J. Zool.* **72,** 1672–1679.
Mäkinen, T., and Ruohonen, K. (1992). The effect of "growth history" on the actual growth of Finnish rainbow trout (*Oncorhynchus mykiss* Walbaum) stock. *J. Appl. Ichthyol.* **8,** 51–61.
Malison, J. A., Procarione, L. S., Held, J. A., Kayes, T. B., and Amundson, C. H. (1993). The influence of triploidy and heat and hydrostatic pressure shocks on the growth and reproductive development of juvenile yellow perch (*Perca flavescens*). *Aquaculture* **116,** 121–133.
Marteinsdottir, G., and Steinarsson, A. (1998). Maternal influence on the size and viability of Iceland cod *Gadus morhua* eggs and larvae. *J. Fish Biol.* **52,** 1241–1258.
Martinez, I., Christiansen, J. S., Ofstad, R., and Olsen, R. L. (1991). Comparison of myosin isoenzymes present in skeletal and cardiac muscles of the Arctic charr *Salvelinus alpinus* (L). Sequential expression of different MHC during development of the fast white skeletal muscle. *Eur. J. Biochem.* **195,** 743–753.
Martinez, I., Dreyer, B., Agersborg, A., Leroux, A., and Boeuf, G. (1995). Effects of T3 and rearing temperature on growth and skeletal myosin heavy chain isoform transition during early development in the salmonid *Salvelinus alpinus* (L). *Comp. Biochem. Physiol.* **112B,** 717–725.
Mascarello, F., Rowlerson, A., Radaelli, G., Scapolo, P. A., and Veggetti, A. (1995). Differentiation and growth of muscle in the fish *Sparus aurata* (L): I. Myosin expression and organisation of fibre types in lateral muscle from hatching to adult. *J. Musc. Res. Cell Motil.* **16,** 213–222.
Mathers, E. M., Houlihan, D. F., McCarthy, I. D., and Burren, L. J. (1993). Rates of growth and protein synthesis correlated with nucleic acid content in fry of rainbow trout, *Oncorhynchus mykiss*: effects of age and temperature. *J. Fish Biol.* **43,** 245–263.
Matschak, T. W., and Stickland, N. C. (1995). The growth of Atlantic salmon (*Salmo salar* L.) myosatellite cells in culture at two different temperatures. *Experientia* **51,** 260–266.
Matschak, T. W., Stickland, N. C., Mason, P. S., and Crook, A. R. (1997). Oxygen availability and temperature affect embryonic muscle development in Atlantic salmon (*Salmo salar* L). *Differentiation* **61,** 229–235.
Matschak, T. W., Hopcroft, T., Mason, P. S., Crook, A. R., and Stickland, N. C. (1998). Temperature and oxygen tension influence the development of muscle cellularity in embryonic rainbow trout. *J. Fish Biol.* **53,** 581–590.
Matsuoka, M. (1998). Development of the lateral muscle in the Japanese sardine *Sardinops melanostictus*. *Fish. Sci.* **64,** 83–88.
Merly, F., Magras-Resch, C., Rouaud, T., Fontaine-Perus, J., and Gardahaut, M. F. (1998). Comparative analysis of satellite cell properties in heavy- and light-weight strains of turkey. *J. Musc. Res. Cell Motil.* **19,** 257–270.
Metcalfe, N. B., Huntingford, F. A., Graham, W. D., and Thorpe. J. E. (1989). Early social status and the development of life history strategies in Atlantic salmon. *Proc. R. Soc. Lond. B* **236,** 7–19.
Metcalfe, N. B., Taylor, A. C., and Thorpe, J. E. (1995). Metabolic rate, social status and life-history strategies in Atlantic salmon. *Anim. Behav.* **49,** 431–436.
Meyer-Rochow, V. B., and Ingram, J. R. (1993). Red-white muscle distribution and fibre growth dynamics: a comparison between lacustrine and riverine populations of the Southern smelt *Retropinna retropinna* Richardson. *Proc. R. Soc. Lond. B* **25,** 85–92.
Miglavs, I., and Jobling, M. (1989). Effects of feeding regime on food consumption, growth rates and

tissue nucleic acids in juvenile Arctic charr, *Salvelinus alpinus,* with particular respect to compensatory growth. *J. Fish Biol.* **34,** 947–957.

Mol, K., Byamungu, N., Cuisset, B., Yaron, Z., Ofir, M., Mélard, C., Castelli, M., and Kühn, E. R. (1994). Hormonal profile of growing male and female diploids and triploids of the blue tilapia, *Oreochromis aureus,* reared in intensive culture. *Fish Physiol. Biochem.* **13,** 209–218.

Moon, T. W. (1983). Changes in tissue ion contents and ultrastructure of food-deprived immature American eels, *Anguilla rostrata* (Lesueur). *Can. J. Zool.* **61,** 812–821.

Moriyama, S., Yamamoto, H., Sugimoto, S., Abe, T., Hirano, T., and Kawauchi, H. (1993). Oral administration of recombinant salmon growth hormone to rainbow trout, *Oncorhynchus mykiss. Aquaculture* **112,** 99–106.

Moss, F. P., and Leblond, C. P. (1971). Satellite cells as the source of nuclei in muscles of growing rats. *Anat. Rec.* **170,** 421–436.

Murai, T. (1992). Protein nutrition of rainbow trout. *Aquaculture* **100,** 191–207.

Nathanailides, C., Lopez-Albors, O., and Stickland, N. C. (1995a). Influence of prehatch temperature on the development of muscle cellularity in posthatch Atlantic salmon (*Salmo salar*). *Can. J. Fish Aquat. Sci.* **52,** 675—680.

Nathanailides, C., Lopez-Albors, O., and Stickland, N. C. (1995b). Temperature- and developmentally-induced variation in the histochemical profile of myofibrillar ATPase activity in carp. *J. Fish Biol.* **47,** 631–640.

Nathanailides, C., Lopez-Albors, O., Abellan, E., Vazquer, J. M., Tyler, D. D., Rowlerson, A., and Stickland, N. C. (1996). Muscle cellularity in relation to somatic growth in the European sea bass *Dicentrarchus labrax* (L.). *Aquaculture Res.* **27,** 885–889.

O'Connell, C. P. (1981). Development of organ systems in the Northern anchovy, *Engraulis mordax,* and other teleosts. *Am. Zool.* **21,** 429–446.

Patruno, M., Radaelli, G., Mascarello, F., and Candia Carnevali, M. D. (1998). Muscle growth in response to changing demands of functions in the teleost *Sparus aurata* (L.) during development from hatching to juvenile. *Anat. Embryol.* **198,** 487–504.

Pelletier, D., Guderley, H., and Dutil, J. D. (1993). Does the aerobic capacity of fish muscle change with growth rates? *Fish Physiol. Biochem.* **12,** 83–93.

Pelletier, D., Blier, P. U., Lambert, Y., and Dutil, J.-D. (1995). Deviation from the general relationship between RNA concentration and growth rate in fish. *J. Fish Biol.* **47,** 920–922.

Penney, R. K., Prentis, P. F., Marshall, P. A., and Goldspink, G. (1983). Differentiation of muscle and the determination of ultimate tissue size. *Cell Tissue Res.* **228,** 375–388.

Pérez-Sánchez, J., Marti-Palanca, H., and Le Bail, P. Y. (1994). Seasonal changes in circulating growth hormone (GH), hepatic GH-binding and plasma insulin-like growth factor-I immunoreactivity in a marine fish, gilthead sea bream, *Sparus aurata. Fish Physiol. Biochem.* **13,** 199–208.

Perrot, V., Moiseeva, E. B., Gozes, Y., Chan, S. J., Ingleton, P., and Funkenstein, B. (1999). Ontogeny of the insulin-like growth factor system (IGF-I, IGF-II, and IGF-1R) in Gilthead seabream (*Sparus aurata*): expression and cellular localization. *Gen. Comp. Endocrinol.* **116,** 445–460.

Picard, B., Gagnière, H., Robelin, J., and Geay, Y. (1995). Comparison of the fetal development of muscle in normal and double-muscled cattle. *J. Muscle Res. Cell Motil.* **16,** 629–639.

Picard, B., Lefaucheur, L., Fauconneau, B., Remignon, H., Cherel, Y., and Barrey, E. (1998). Production and utilisation of monoclonal antibodies against myosin heavy chain isoforms in different species. *Prod. Anim.* **11,** 145–147.

Pickering, A. D., Pottinger, T. G., Sumpter, J. P., Carragher, J. F., and Le Bail, P. Y. (1991). Effects of acute and chronic stress on the levels of circulating growth hormone in the rainbow trout, *Oncorhynchus mykiss. Gen. Comp. Endocrinol.* **83,** 86–93.

Poli, B. M., Parisi, G., Lupi, P., Mecatti, M., Bonelli, A., Zampacavallo, G., Gualtieri, M., and Mascini, M. (1999). Qualitative traits and shelf life in *Pagrus major* x *Dentex dentex* hybrids stored at refrigerated condition and comparison with European gilthead seabream (*Sparus au-*

rata). In "XXXIII International Symposium on New Species for Mediterranean Aquaculture" (Enne, G., and Greppi, G. F., eds.), pp. 329–336. Elsevier, Amsterdam.

Powell, R. L., Dodson, N. W., and Cloud, J. G. (1989). Cultivation and differentiation of satellite cells from skeletal muscle of the rainbow trout *Salmo gairdneri*. *J. Exp. Zool.* **250**, 333–338.

Purdom, C. E. (1993). "Genetics and Fish Breeding" 1st ed., Chapman & Hall, London.

Qin, J. G., Fast, A. W., and Ako, H. (1998). Growout performance of diploid and triploid Chinese catfish *Clarias fuscus*. *Aquaculture* **166**, 247–258.

Quinn, L. S., Ong, L. D., and Roeder, R. A. (1990). Paracrine control of myoblast proliferation and differentiation by fibroblasts. *Dev. Biol.* **140**, 8–19.

Radaelli, G., Rowlerson, A., Mascarello, F., and Veggetti, A. (1999). Methods for investigating muscle growth in fish: examples and potential applications. *In* "XXXIII International Symposium on New Species for Mediterranean Aquaculture" (Enne, G., and Greppi, G. F., eds.), pp. 353–359. Elsevier, Amsterdam.

Ramírez-Zarzosa, G., Gil, F., Latorre, R., Ortega, A., García-Alcaráz, A., Abellán, E., Vázquez, J. M., López-Albors, O., Arencibia, A., and Moreno, F. (1995). The larval development of lateral musculature in the gilthead sea bream *Sparus aurata* and sea bass *Dicentrarchus labrax*. *Cell Tissue Res.* **280**, 217–224.

Rehfeldt, C., Fiedler, I., Weikard, R., Kanitz, E., and Ender, K. (1993). It is possible to increase skeletal muscle fibre number *in utero*. *Biosci. Rep.* **13**, 213–220.

Remignon, H., Gardahaut, M. F., Marche, G., and Ricard, F. H. (1995). Selection for rapid growth increases the number and the size of muscle fibres without changing their typing in chickens. *J. Muscle Res. Cell Motil.* **16**, 95–102.

Rescan, P.-Y., Gauvry, L., and Paboeuf, G. (1995). A gene with homology to myogenin is expressed in developing myotomal musculature of the rainbow trout and in vitro during the conversion of myosatellite cells to myotubes. *FEBS Lett.* **362**, 89–92.

Rescan, P.-Y. (1997). Identification in a fish species of two Id (inhibitor of DNA binding/differentiation)-related helix-loop-helix factors expressed in the slow oxidative muscle fibers. *Eur. J. Biochem.* **247**, 870–876.

Rescan, P.-Y. (1998). Identification of a fibroblast growth factor-6 (FGF6) gene in a non-mammalian vertebrate: continuous expression of FGF6 accompanies muscle fiber hyperplasia. *Biochem. Biophys. Acta–Gene Structure Expression* **1443**, 305–314.

Roderick, E. (1998). Latest research benefits versatile tilapia. *Fish Farmer* **21**, 33–34.

Romanello, M. G., Scapolo, P. A., Luprano, S., and Mascarello, F. (1987). Post-larval growth in the lateral white muscle of the eel, *Anguilla anguilla*. *J. Fish Biol.* **30**, 161–172.

Rowlerson, A., Mascarello, F., Radaelli, G., and Veggetti, A. (1995). Differentiation and growth of muscle in the fish *Sparus aurata* (L): II. Hyperplastic and hypertrophic growth of lateral muscle from hatching to adult. *J. Muscle Res. Cell Motil.* **16**, 223–236.

Rowlerson, A., Radaelli, G., Mascarello, F., and Veggetti, A. (1997). Regeneration of skeletal muscle in two teleost fish: *Sparus aurata* and *Brachydanio rerio*. *Cell Tissue Res.* **289**, 311–322.

Ryer, C. H., and Olla, B. L. (1996). Growth depensation and aggression in laboratory-raised coho salmon the effect of food distribution and ration size. *J. Fish Biol.* **48**, 686–694.

Sánchez-Vázquez, F. J., and Tabata, M. (1998). Circadian rhythmus of demand-feeding and locomotor activity in rainbow trout. *J. Fish Biol.* **52**, 255–267.

Sänger, A. M. (1992). Effects of training on axial muscle of two cyprinid species: *Chondrostoma nasus* (L:) and *Leuciscus cephalus* (L.). *J. Fish Biol.* **40**, 637–646.

Sänger, A. M. (1993). Limits to the acclimation of fish muscle. *Rev. Fish Biol. Fish.* **3**, 1–15.

Scapolo, P.-A., Veggetti, A., Mascarello, F., and Romanello, M. G. (1988). Developmental transitions of myosin isoforms and organisation of the lateral muscle in the teleost *Dicentrarchus labrax* (L.). *Anat. Embryol.* **178**, 287–295.

Sin, F. Y. T. (1997). Transgenic fish. *Rev. Fish Biol. Fish.* **7**, 417–441.

Schulte, P. M., Down, N. E., Donaldson, E. M., and Souza, L. M. (1989). Experimental administration of recombinant bovine growth hormone to juvenile Rainbow trout (*Salmo gairdneri*) by injection or by immersion. *Aquaculture* **76,** 145–156.
Schultz, E., and McCormick, K. M. (1993). Cell biology of the satellite cell. *In* "Molecular and Cell Biology of Muscular Dystrophy" (Partridge, T., ed.), pp. 190–209. Chapman & Hall, London.
Smith, I. P., Metcalfe, N. B., Huntingford, F. A., and Kadri, S. (1993). Daily and seasonal patterns in the feeding behaviour of Atlantic salmon (*Salmo salar* L.) in a sea cage. *Aquaculture* **117,** 165–178.
Solar, I. I., Donaldson, E. M., and Hunter, G. A. (1984). Induction of triploidy in rainbow trout (*Salmo gairdneri* Richardson) by heat shock, and investigation of early growth. *Aquaculture* **42,** 57–67.
Stewart, C. E. H., and Rotwein, P. (1996). Growth, differentiation, and survival: multiple physiological functions for insulin-like growth factors. *Physiol. Rev.* **76,** 1005–1026.
Stickland, N. C. (1983). Growth and development of muscle fibres in the rainbow trout (*Salmo gairdneri*). *J. Anat.* **137,** 323–333.
Stickland, N. C., and Dwyer, C. M. (1996). *In* "Molecular Physiology of Growth." (Loughna, P. T., and Pell, J. M., eds.), pp. 135–150. Cambridge University Press.
Stickland, N. C., White, R. N., Mescall, P. E., Crook, A. R., and Thorpe, J. E. (1988). The effect of temperature on myogenesis in embryonic development of the Atlantic salmon (*Salmo salar* L.). *Anat. Embryol.* **178,** 253–257.
Stoiber, W., and Sänger, A. M. (1996). An electron microscopic investigation into the possible source of new muscle fibres in teleost fish. *Anat. Embryol.* **194,** 569–579 .
Stoiber, W., Haslett, J. R., Goldschmidt, A., and Sänger, A. (1998). Patterns of superficial fibre formation in the European pearlfish (*Rutilus frisii meidingeri*) provide a general template for slow muscle development in teleost fish. *Anat. Embryol.* **197,** 485–496.
Stoiber, W., Haslett, J. R., and Sänger, A. M. (1999). Myogenic patterns in teleosts: what does the present evidence really suggest? *J. Fish Biol.* **55 (Suppl. A),** 84–99.
Sumpter, J. P. (1992). Control of growth of rainbow trout (*Oncorhynchus mykiss*). *Aquaculture* **100,** 299–320.
Sumpter, J. P., Le Bail, P. Y., Pickering, A. D., Pottinger, T. G., and Carragher, J. F. (1991). The effect of starvation on growth and plasma growth hormone concentrations of rainbow trout (*Oncorhynchus mykiss*). *Gen. Comp. Endocrinol.* **83,** 94–102.
Suresh, A. V., and Sheehan, R. J. (1998a). Muscle fibre growth dynamics in diploid and triploid rainbow trout. *J. Fish Biol.* **52,** 570–587.
Suresh, A. V., and Sheehan, R. J. (1998b). Biochemical and morphological correlates of growth in diploid and triploid rainbow trout. *J. Fish Biol.* **52,** 588–599.
Sutrave, P., Kelly, A. M., and Hughes, S. H. (1990). *ski* can cause selective growth of skeletal muscle in transgenic mice. *Genes Dev.* **4,** 1462–1472.
Svåsand, T., Jørstad, K. E., Otterå, H., and Kjesbu, O. S. (1996). Differences in growth performance between Arcto-Norvegian and Norwegian coastal cod reared under identical conditions. *J. Fish Biol.* **49,** 108–119.
Thorgaard, G. H. (1986). Ploidy manipulation and performance. *Aquaculture* **57,** 57–64.
Thorgaard, G. H., and Gall, G. A. (1979). Adult triploids in a rainbow trout family. *Genetics* **93,** 961–973.
Thissen, J.-P., Ketelslegers, J.-M., and Underwood, L. (1994). Nutritional regulation of the insulin-like growth factors. *Endocr. Rev.* **15,** 80–101.
Totland, G. K., Kryvi, H., Jødestol, K. A., Christiansen, E. N., Tangerås, A., and Slinde, E. (1987). Growth and composition of the swimming muscle of adult Atlantic salmon (*Salmo salar* L.) during long-term sustained swimming. *Aquaculture* **66,** 299–313.
Tsai, H. J., Kuo, J. C., Lou, S. W., and Kuo, T. T. (1994). Growth enhancement of juvenile striped mullet by feeding recombinant yeasts containing fish growth hormone. *Prog. Fish Cult.* **56,** 7–12.

Usher, M. L., Stickland, N. C., and Thorpe, J. E. (1994). Muscle development in Atlantic salmon (*Salmo salar*) embryos and the effect of temperature on muscle cellularity. *J. Fish Biol.* **44**, 953–964.

Valente, L. M. P., Gomes, E. F. S., and Fauconneau, B. (1998). Biochemical growth characterisation of fast and slow-growing rainbow trout strains: effect of cell proliferation and size. *Fish Physiol. Biochem.* **18**, 213–224.

Valente, L. M. P., Rocha, E., Gomes, E. F. S., Silva, M. W., Olivereira, M. H., Monteiro, R. A. F., and Fauconneau, B. (1999). Growth dynamics of white and red muscle fibres in fast- and slow-growing strains of rainbow trout. *J. Fish Biol.*, **55**, 675–691.

van der Have, T. M., and de Jong, G. (1996). Adult size in ectotherms: temperature effects on growth and differentiation. *J. Theor. Biol.* **183**, 329–340.

van Raamsdonk, W., van't Veer, L., Veeken, K., Heyting, C., and Pool, C. W. (1982). Differentiation of muscle fiber types in the Teleost *Brachydanio rerio,* the zebrafish. Posthatching development. *Anat. Embryol.* **164**, 51–62.

van Raamsdonk, W., Mos, W., Smit-Onel, M. J., van der Laarse, W. J., and Fehres, R. (1983). The development of the spinal motor column in relation to the myotomal muscle fibres in the Zebrafish (*Brachydanio rerio*). I. Posthatching development *Anat. Embryol.* **167**, 125–139.

Veggetti, A. (1991). Differenziamento ed accrescimento del muscolo laterale di teleostei oggetto di acquicoltura. *Atti S. I. S. Vet.* **45**, 67–73.

Veggetti, A., Mascarello, F., Scapolo, P. A., and Rowlerson, A. (1990). Hyperplastic and hypertrophic growth of lateral muscle in *Dicentrarchus labrax* (L.). An ultrastructural and morphometric study. *Anat. Embryol.* **182**, 1–10.

Veggetti, A., Mascarello, F., Scapolo, P. A., Rowlerson, A., and Candia Carnevali, M. D. (1993). Muscle growth and myosin isoform transitions during development of a small teleost fish, *Poecilia reticulata* (Peters) (Atheriniformes, Poeciliidae): a histochemical, immunohistochemical, ultrastructural and morphometric study. *Anat. Embryol.* **187**, 353–361.

Veggetti, A., Rowlerson, A., Radaelli, G., Arrighi, S., and Domeneghini, C. (1999). Posthatching development of the gut and lateral muscle in the sole, *Solea solea* (L). *J. Fish Biol.* **55 (Suppl. A)**, 44–65.

Vieira, V. L. A., and Johnston, I. A. (1992). Influence of temperature on muscle-fibre development in larvae of the herring *Clupea harengus. Mar. Biol.* **112**, 333–341.

Vieira, V. L. A., and Johnston, I. A. (1999). Temperature and neuromuscular development in the tambaqui. *J. Fish Biol.* **55 (Suppl. A)**, 66–83.

Wang, N., Hayward, R. S., and Noltie, D. B. (1998). Variation in food consumption, growth, and growth efficiency among juvenile hybrid sunfish held individually. *Aquaculture* **167**, 43–52.

Watabe, S. (1999). Myogenic regulatory factors and muscle differentiation during ontogeny in fish. *J. Fish Biol.* **55 (Suppl. A)**, 1–18.

Watanabe, T. (1992). Nutrition and growth. *In* "Intensive Fish Farming" (Shepherd, C. J., and Bromage, N. R., eds.), pp. 154–197. Blackwell Scientific Publications, Oxford.

Waterman, R. E. (1969). Development of the lateral musculature in the teleost *Brachydanio rerio:* A fine structural study. *Am. J. Anat.* **125**, 457–494.

Weatherley, A. H., and Gill, H. S. (1982). Influence of bovine growth hormone on the growth dynamics of mosaic muscle in relation to somatic growth of rainbow trout, *Salmo gairdneri* Richardson. *J. Fish Biol.* **20**, 165–172.

Weatherley, A. H., and Gill, H. S. (1984). Growth dynamics of white myotomal muscle fibres in the bluntnose minnow, *Pimephales notatus* Rafinesque, and comparison with rainbow trout, *Salmo gairdneri* Richardson. *J. Fish Biol.* **25**, 13–24.

Weatherley, A. H., and Gill, H. S. (1985). Dynamics of increase in muscle fibres in fishes in relation to size and growth. *Experientia* **41**, 353–354.

Weatherley, A. H., and Gill, H. S. (1987a). "The Biology of Fish Growth" Academic Press, London.

Weatherley, A. H., and Gill, H. S. (1987b). Growth increases produced by bovine growth hormone in grass pickerel, *Esox americanus vermiculatus* (LeSueur) and the underlying dynamics of muscle fiber growth. *Aquaculture* **65**, 55–66.

Weatherley, A. H., Gill, H. S., and Rogers, S. C. (1979). Growth dynamics of muscle fibres, dry weight and condition in relation to somatic growth rate in yearling rainbow trout (*Salmo gairdneri*). *Can. J. Zool.* **57**, 2385–2392.

Weatherley, A. H., Gill, H. S., and Rogers, S. C. (1980a). The relationship between mosaic muscle fibres and size in rainbow trout (*Salmo gairdneri*). *J. Fish Biol.* **17**, 603–610.

Weatherley, A. H., Gill, H. S., and Rogers, S. C. (1980b). Growth dynamics of mosaic muscle fibres in fingerling rainbow trout (*Salmo gairdneri*) in relation to somatic growth rate. *Can. J. Zool.* **58**, 1535–1541.

Weatherley, A. H., Gill, H. S., and Lobo, A. F. (1988). Recruitment and maximal diameter of axial muscle fibres in teleosts and their relationship to somatic growth and ultimate size. *J. Fish Biol.* **33**, 851–859.

Willemse, J. J., and van den Berg, P. G. (1978). Growth of striated muscle fibres in the M. Lateralis of the European eel *Anguilla anguilla* (L.) (Pisces Teleostei). *J. Anat.* **125**, 447–460.

Withler, R. E., Beacham, T. D., Solar, I. I., and Donaldson, E. M. (1995). Fresh-water growth, smolting, and marine survival and growth of diploid and triploid coho salmon (*Oncorhynchus-Kisutch*). *Aquaculture* **136**, 91–107.

Withler, R. E., Clarke, W. C., Blackburn, J., and Baker, I. (1998). Effect of triploidy on growth and survival of pre-smolt and post-smolt coho salmon (*Oncorhynchus kisutch*). *Aquaculture* **168**, 413–422.

Yamano, K., Miwa, S., Obinata, T., and Inui, Y. (1991). Thyroid hormone regulates developmental changes in muscle during flounder metamorphosis. *Gen. Comp. Endocrinol.* **81**, 464–472.

Young, P. S., and Cech, J. J. (1992). Improved growth, swimming performance, and muscular development in exercise-conditioned young-of-the-year striped bass (*Morone saxatilis*). *Can. J. Fish. Aquat. Sci.* **50**, 703–707.

Zhang, P., Hayat, M., Joyce, C., Gonzales-Villaseñor, L. I., Lin, C. M., Dunham, R. A., Chen, T. T., and Powers, D. A. (1990). Gene transfer, expression and inheritance of pRSV-rainbow trout-GH cDNA in the common carp, *Cyprinus carpio* (Linnaeus). *Mol. Reprod. Dev.* **25**, 3–13.

Zimmerman, A. M., and Lowery, M. S. (1999). Hyperplastic development and hypertrophic growth of muscle fibers in the white seabass (*Atractoscion nobilis*). *J. Exp. Zool.* **284**, 299–308.

6

GENETIC AND ENVIRONMENTAL DETERMINANTS OF MUSCLE GROWTH PATTERNS

IAN A. JOHNSTON

I. Introduction
II. Myogenesis in Salmonids
   A. Embryonic Phase
   B. Germinal Zone Phase
   C. Juvenile and Adult Growth
   D. The Role of Satellite Cells in Fiber Recruitment and Hypertrophy
   E. Molecular Markers of Muscle Satellite Cells
III. Methods for Quantifying Growth Patterns in Muscle
IV. Genetic Variation in Muscle Growth Patterns
   A. Variation Between Populations
   B. Ploidy Manipulation
V. Environmental Influences on Muscle Fiber Growth Patterns
   A. Temperature
   B. Influence of Oxygen Tension
   C. Implications for Fisheries Management and Conservation
VI. Exercise as a Stimulus for Muscle Growth
VII. Mechanisms Underlying Differences in Muscle Growth Patterns
   A. Number of Muscle Satellite Cells
   B. The Regulation of Satellite Cell Behavior
   C. Regulation of Muscle Fiber Hypertrophy
VIII. Implications for Flesh Quality in Farmed Fish
   A. Muscle Cellularity and Texture
   B. Color Visualization of the Fillet
   C. Commercial Implications
   References

I. INTRODUCTION

Myotomal muscle is the single most abundant tissue in fish, comprising 65% of the body mass in salmonid species. Slow and fast muscle fiber types can be distinguished in the myotomes of fish embryos (van Raamsdonk *et al.*, 1974;

Devoto *et al.,* 1996; Blagden *et al.,* 1997). Additional muscle fiber types are formed to replace the embryonic muscle fibers during the free-swimming larval stages (Matsuoka and Iwai, 1984; Brooks and Johnston, 1993; Martinez and Christiansen, 1994; Mascarello *et al.,* 1995). Slow sustainable swimming speeds are powered by the superficial red muscle layer on either side of the body (Bone, 1966; Johnston *et al.,* 1977). The red muscle is composed of slow twitch fibers, which contain a high volume density of mitochondria (20–50%) and are well supplied with capillaries (Bone, 1978; Johnston, 1981). As swimming speed increases, faster contracting pink muscle fibers become active (Johnston *et al.,* 1977). In common carp (*Cyprinus carpio*), the pink fibers have a high capacity for anaerobic glycogenolysis and a moderate aerobic capacity (Johnston *et al.,* 1977). They express fast muscle myosin light chains (Johnston *et al.,* 1977) and have distinct isoforms of myosin heavy chains from those found in red and white muscle fibers (Scapolo and Rowlerson, 1987). The white muscle, which comprises more than 90% of the myotome in most species, is reserved for high speed cruising and fast-starts associated with predation and escape behavior. It is composed of fast twitch fibers (Altringham and Johnston, 1988) that are tightly packed with myofibrils (Nag, 1972). Contraction of the fast muscle is dependent on anaerobic metabolic pathways (Driedzic and Hochachka, 1978). The anatomical arrangement of the red and white muscle fibers in the Atlantic salmon (*Salmo salar*) is illustrated in Fig. 1. Original references to studies on the structure and physiology of different muscle fiber types in fish can be found in several recent reviews (Johnston and Altringham, 1991; Altringham and Ellerby, 1999).

Post-embryonic growth of the muscle tissue involves an increase in the number and diameters of fibers and a contemporary remodeling of the associated connective tissue, nerve, and blood supply. Muscle fibers are differentiated structures and the nuclei required for new fiber recruitment and hypertrophy are derived from an undifferentiated stem cell population of mononuclear cells. During the late embryo, larval, and fry stages distinct germinal zones of new fiber production can often be distinguished (Veggetti *et al.,* 1990; Rowlerson *et al.,* 1995; Johnston and McLay, 1997; Galloway *et al.,* 1999). The final phase of growth involves the activation of muscle satellite cells (Koumans *et al.,* 1991; Johnston *et al.,* 1995). These cells are either located beneath the basal lamina of muscle fibers or are interspersed between the muscle fibers, particularly in the early life stages (Veggetti *et al.,* 1990; Johnston, 1993). The relative importance of these three phases of muscle growth varies between fiber types and with the growth rate and ultimate body size of the species (Weatherley *et al.,* 1988). Zebrafish, (*Danio rerio*) reach a body length of only 5 cm. The number of red muscle fibers continues to increase with body length whereas the number of white muscle fibers reaches a plateau at 1.6 cm and subsequent growth is entirely by fiber hypertrophy (van Raamsdonk *et al.,* 1983). In contrast, fiber recruitment continues in the white muscle of Atlantic cod (*Gadus morhua* L.) in excess of 100 cm fork length (Greer-Walker, 1970).

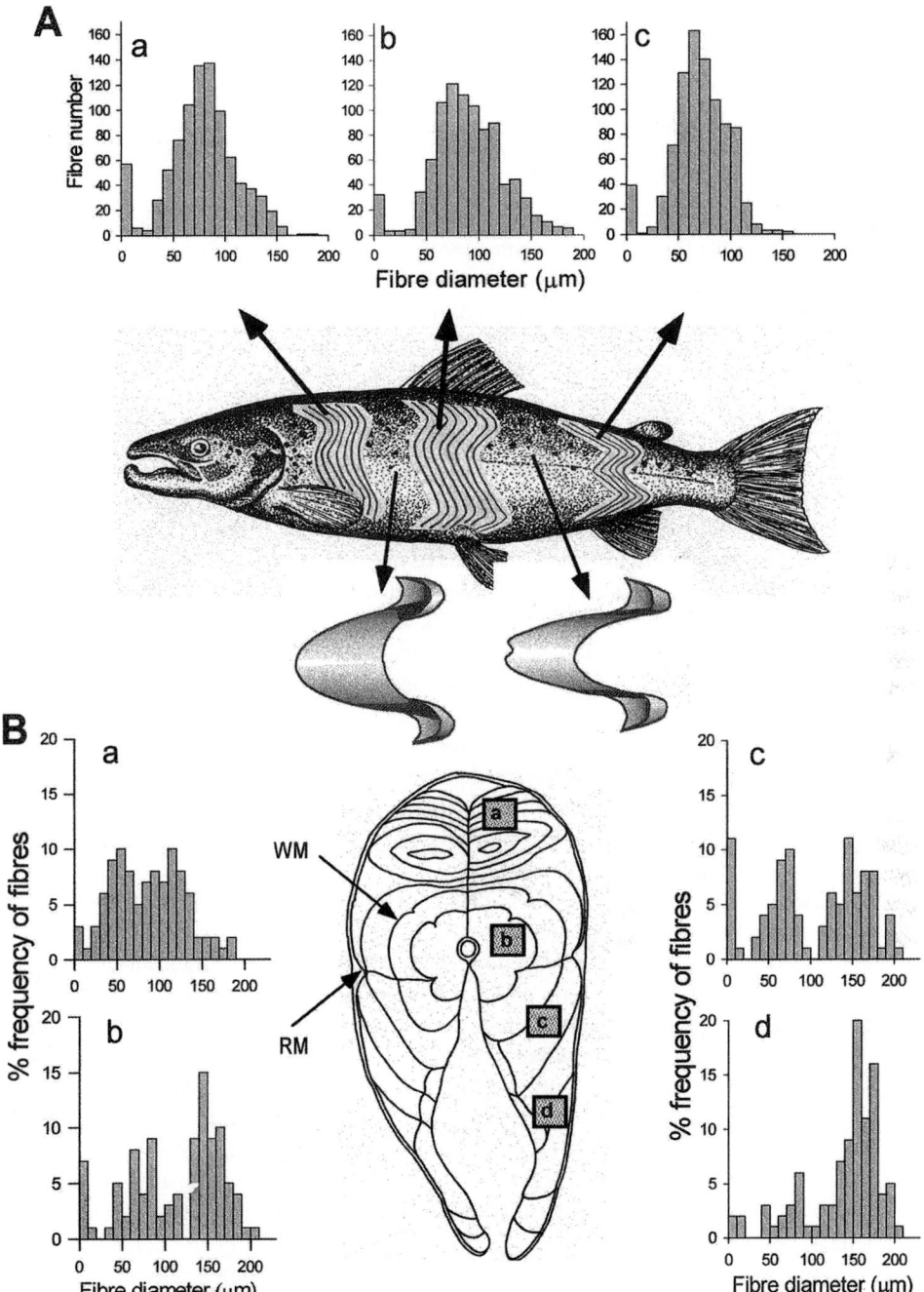

Fig. 1. The anatomical arrangement and size distribution of myotomal muscle fibers in the Atlantic salmon (*Salmo salar* L.). (A) A fish with the skin removed to reveal the segmentally arranged myotomes. Individual muscle fibers insert into connective tissue sheets called myosepta. The shapes of the rostral and caudal myosepta are illustrated below the fish. The distribution of white muscle fiber diameter at three points along the body (a–c) is illustrated for a 2-kg fish (900 measurements were made per site). (B) Illustrates a cross-section through the trunk at the level of the first dorsal fin ray showing the arrangement of red (RM) and white (WM) muscle fibers. The distribution of white fiber diameter in four regions of the cross-section (a–d) is illustrated.

Muscle fiber recruitment continues for longer in female than male Argentine hake, *Merluccius hubbsi,* correlating with their significantly greater maximum length; 90 vs. 70 cm (Calvo, 1989).

Reproductive isolation promotes genetic differentiation in fish populations resulting in phenotypic variation in morphology, physiology, behavior, and life history characteristics. Phenotypic variation in growth characteristics between populations can also be the result of environmental factors, such as water chemistry, temperature, day-length, and food availability, acting at some particular stage of ontogeny. The complex nature of growth makes it particularly difficult to disentangle genetic from environmental causes of phenotypic variation in populations. There is evidence that the number and behavior of the muscle stem cell population is under both genetic and environmental control. Several studies have demonstrated variations in muscle growth patterns between populations of the same species. For example, Atlantic herring (*Clupea harengus*) from the Bank stock in the North Sea are smaller and have only half the number of white fibers per trunk cross-section at sexual maturity as fish that spawn in the nearby Blackwater estuary (Greer-Walker *et al.,* 1972). In the common New Zealand smelt *Retropina retropina,* the recruitment of white muscle fibers has been shown to occur at shorter body lengths in the dwarf lacustrine form than in the larger riverine form (Meyer-Rochow and Ingram, 1993). There is evidence that the landlocked populations in North Island lakes are the result of introductions from the lower Waikato river by early settlers and are, therefore, recently descended from the diadromous population (Jolly, 1967). In this case, differences in muscle growth characteristics between populations may be largely environmental in origin (Meyer-Rochow and Ingram, 1993).

The Atlantic salmon (*S. salar* L.) is an anadromous species that returns to its native stream to spawn. It occurs over a wide latitudinal range and there is strong evidence for genetic differences in populations both between and within river systems (Verspoor, 1997). Genetic and environmental factors combine to influence life history factors including the time spent in freshwater and the age and size at sexual maturity, all of which affect growth rate (Thorpe, 1989), and have the potential to modify muscle growth patterns. For example, muscle fiber recruitment in hatchery-reared parr is significantly greater in fish destined to smolt after one than two years in freshwater (Higgins and Thorpe, 1990). The decision to smolt in a particular year varies between families (Thorpe, 1987), but is strongly influenced by prior feeding opportunity and can be modified by temperature and daylength (Metcalfe *et al.,* 1988; Thorpe, 1989).

This chapter reviews what is known about genetic and environmental determinates of muscle growth patterns in fish with particular reference to salmonids. The molecular and cellular mechanisms underlying the plasticity of muscle fiber recruitment are discussed as are ecological and practical implications of the research for fisheries and aquaculture.

II. MYOGENESIS IN SALMONIDS

A. Embryonic Phase

Muscle growth has been described throughout the life cycle in rainbow trout (*Oncorhynchus mykiss;* Stickland, 1983; Kiessling *et al.,* 1991) and Atlantic salmon (*S. salar;* Higgins and Thorpe, 1990; Johnston *et al.,* 1999; 2000a). Both species have large eggs and develop over a protracted period at relatively low temperatures. The embryos hatch at an advanced stage of development compared with zebrafish and most marine fish larvae (Blaxter, 1988). There have been several descriptive studies of the embryonic phase of muscle development (Nag and Nursall, 1972; Proctor *et al.,* 1980; Stoiber and Sänger, 1996). The rather limited information available suggests that the embryonic phase of myogenesis in salmonids is broadly similar to that described in zebrafish.

Members of the MyoD family of myogenic regulatory factors (MRFs) play a key role in the commitment and differentiation of mesoderm cells to a muscle lineage (see Watabe (2001), Chapter 2, this volume). MRFs are basic helix-loop-helix transcription (bHLH) factors which form dimers with other bHLH factors and bind to a consensus sequence, termed E-box (CANNTG), present in the promoter regions of many muscle-specific genes (Weintraub *et al.,* 1994). The primary MRFs, MyoD and Myf-5, are required for myogenic determination, whereas the secondary MRFs, myogenin and myf-6, are downstream transcription factors involved in differentiation (Rudnicki and Jaenisch, 1995). Two non-allelic MyoD encoding genes (TmyoD and TmyoD2) have been identified in the rainbow trout that were probably duplicated during the tetraploidization of the salmonid genome (Rescan and Gauvry, 1996). The temporal and spatial expression patterns of the two MyoD encoding genes differ during embryonic development. TmyoD expression is first detected at the mid-gastrula stage in two parallel cords of cells on either side of the elongating embryonic shield (Delalande and Rescan, 1999). Prior to somite formation adaxial cells can be identified in rainbow trout (Delalande and Rescan, 1999) similar to those described in zebrafish (see Devoto *et al.,* 1996 and Currie and Ingram, 2001, Chapter 1, this volume). It has been shown that in zebrafish the adaxial cells commit to a slow muscle fate under the influence of the glycoprotein, sonic hedgehog, and migrate through the somite to form the superficial layer of embryonic fibers (Devoto *et al.,* 1996; Blagden *et al.,* 1997). In contrast, the lateral pre-somitic cells have been shown to remain in approximately the same position and are transformed into the embryonic fast muscle fibers in zebrafish (Devoto *et al.,* 1996) and the European Pearlfish (*Rutilus frisii meidingeri;* Stoiber *et al.,* 1998). In rainbow trout, TmyoD, but not myogenin, is expressed in the adaxial cells prior to somite formation (Delalande and Rescan, 1999).

Myotube formation begins in the most rostral somites at about 9-somite stage in brown trout (*Salmo trutta*; Killeen et al., 1999a) and at around the 15-somite stage in Atlantic salmon (Johnston et al., 1999). Two classes of myotube can be distinguished morphologically. Mononuclear cells, corresponding to the muscle pioneers, are present on either side of the notochord. Multinucleated myotubes are also formed in the lateral mesoderm from the fusion of two to five myoblasts. The expression of TmyoD was found to be restricted to the adaxial cells under the 30-somite stage. Expression spread to the lateral myotubes as the myotomes began to acquire their characteristic chevron shape (Delalande and Rescan, 1999). TmyoD2 was not expressed until around the 15-somite stage when it was localized in all the cells of the posterior compartment of the somite. However, by the time the myotomes had acquired their chevron shape TmyoD2 transcripts progressively disappeared from the inner part of the somite until they were only present in the most superficial layers of muscle fibers (Delalande and Rescan, 1999). In adult rainbow trout, TmyoD is expressed equally in red and white muscle fibers. In contrast, TmyoD2 and myogenin are preferentially expressed in the slow muscle (Delalande and Rescan, 1999).

Actin and myosin filaments and the initial stages of myofibril assembly were first observed in the rostral somites at about the 30-somite stage in both Atlantic salmon (Johnston et al., 1999) and brown trout (Killeen et al., 1999a). In brown trout, the first primary motor neurones emerge from the ventral region of the spinal cord in 30-somite stage embryos (Killeen et al., 1999a). By about the 40- to 45-somite stage in both species the first 10 or more somites showed intense staining for acetylcholinesterase (AchE) at the myosepta (Killeen et al., 1999a; Johnston et al., 1999). At this point spontaneous rhythmic contractions are first observed in the anterior part of the trunk. In brown trout, the rostral progression of myofibril assembly and innervation with respect to somite stage was relatively independent of temperature (Killeen et al., 1999a). However, in Atlantic herring (*C. harengus*) myofibril synthesis and the acquisition of AchE activity at the myosepta was progressively delayed with respect to somite stage as the rearing temperature was reduced from 15 to 5°C (Johnston et al., 1997), and these effects persist until the early larval stages (Johnston et al., 1998). There is considerable interspecific variation in the degree of muscle differentiation at the stage the embryos become free swimming. Larvae of the tambaqui, *Colossomoa macropomum,* a fast-growing fish from the Amazon region, hatch at a primitive stage of development prior to the formation of eyes and jaws and before the majority of somites have developed motor axons (Vieira and Johnston, 1996; 1999). Only the superficial layer of fibers contain organized myofibrils. In this case, numerous gap-like junctions are observed between the muscle fibers and somite boundaries suggesting a role for electrical coupling in the generation of swimming movements (Vieira and Johnston, 1999).

Salmon and trout spend the yolk-sac stage buried in the gravel and become free swimming around the time of first feed. Nag and Nursall (1972) found that muscle fibers with the ultrastructural characteristics of adult red fibers were only apparent in free-swimming larvae. At hatch in Atlantic salmon, a single layer of embryonic slow fibers is present around the entire lateral surface of the myotome. This layer thickens to two to three fibers in the region of the lateral line nerve at the major horizontal septum. These superficial muscle fibers initially express fast myosin light chain (MLC3), as do the inner white muscle fibers (Johnston and McLay, 1997). During the yolk-sac stage adult slow (red) muscle fibers are added externally to this superficial muscle layer and the expression of MLC3 is gradually switched off starting at the level of the major horizontal septum (Johnston and McLay, 1997).

B. Germinal Zone Phase

A second phase of myogenesis begins in the late embryo. Discrete germinal zones of myoblasts are observed at the lateral apices of the myotomes in Atlantic salmon that are the sites of the new fiber production (Johnston and McLay, 1997). This results in a gradation of muscle fiber diameter, which increases away from the germinal zones toward the central part of the myotome. Similar germinal zone phases of myogenesis have been described in a wide range of species (see Rowlerson and Veggetti, this volume) and this is probably the predominant means of post-embryonic fiber recruitment in species that only attain a small ultimate body size such as zebrafish (van Raamsdonk *et al.,* 1983). It has been suggested on the basis of an ultrastructural study that recruitment from the germinal zones is initiated by the attachment of presumptive myogenic cells that originate from and proliferate in the adjacent mesenchymal tissue lining (Stoiber and Sänger, 1996).

C. Juvenile and Adult Growth

In species that reach a large ultimate size muscle fiber recruitment continues into the adult stages of the life cycle (Weatherley *et al.,* 1988). This final phase of recruitment involves a class of persistent myoblasts analogous to the satellite or myosatellite cells first described in the mouse (Mauro, 1961). In adult fish, satellite cells are observed between the sarcolemma and basal membrane of muscle fibers (Koumans and Akster, 1995). However, in the larval and early juvenile stages of several species including sea bass (*Dicentrarchus labrax* (L.); Veggetti *et al.,* 1990), Atlantic herring (*C. harengus* L.; Johnston, 1993) and Atlantic salmon (Johnston *et al.,* 2000b), myogenic precursors are also observed interspersed between the muscle fibers. Presumptive satellite cells are numerous in the white muscle of late embryo stages in the rainbow trout (Stoiber and Sänger, 1996).

The pattern of muscle growth in rainbow trout has been studied over the whole life cycle for fish reared entirely in freshwater (Stickland, 1983) or transferred from freshwater to seawater 12 months post-hatch (Kiessling et al., 1991). A comparison of the results in these two studies highlights the importance of adequate statistical sampling of both numbers of fish and different regions of the myotome. The ratio of red to white muscle cross-sectional area remained at about 1:25 between 2 and 71 cm fork length (Stickland, 1983). The length of individual muscle fibers was linearly related to body length, increasing from less than 1 mm in 5-cm fish to about 8 mm in 45-cm trout (Kiessling et al., 1991). Stickland (1983) investigated myogenesis in relatively caudal myotomes (0.66 body lengths from the snout) in a total of 17 trout. One hundred white muscle fibers were sampled from each of three regions (superficial, middle, and deep) at the level of the median horizontal septum. In a more extensive study, Kiessling et al. (1991) studied 450 individual fish, sampling muscle from dorsal, lateral, and ventral regions of the myotomes anterior to the first dorsal fin ray. Kiessling et al. (1991) found that the smallest size class of white muscle fibers (<10 μm) was present even in 70-cm fish weighing around 2.5 kg. Furthermore, the proportion of smaller size classes of white fiber was much higher in the dorsal than the lateral regions of the myotome. This indicates that muscle fiber recruitment continued throughout the period studied. In contrast, Stickland (1983) found that the number of white muscle fibers increased rapidly in the lateral part of the myotome until fish were 40 cm long, and then more slowly, until it reached a plateau in 65-cm fish. Frequency histograms of muscle fiber diameter revealed that the smallest size class of white fibers (0–10 μm) was absent in fish bigger than 52 cm, such that subsequent growth in muscle girth was entirely by fiber hypertrophy. Indeed there were no white fibers smaller than 90 μm diameter in 71-cm fish (Stickland, 1983). However, the results of Kiessling and co-workers suggest that fibers in the lateral region of the myotome may not be representative of the whole trunk cross-section. Thus fiber recruitment probably continues to longer body lengths in rainbow trout than reported by Stickland (1983), although differences in muscle growth patterns due to genetic background, rearing conditions, or some other factor cannot be ruled out.

The number of white muscle fibers per trunk cross-section in Atlantic salmon increased from around 5000 at hatch to 10,000 at first feeding, 150,000 in 150,000 in seawater-adapted smolts, 650,000 in 1 sea-winter fish, and more than 1,000,000 in 2 sea-winter fish (Johnston 1999, Johnston et al., 1999). Thus the number of fibers increases 30-fold in the freshwater stages of the life cycle and a further 6–8 times during 2 years of seawater growth. The maximum diameter of white muscle fibers is around 240 μm and is not achieved until around 6 months after seawater adaptation (Johnston et al., 2000a).

The continuous recruitment of muscle fibers produces a range of diameters. The smaller diameter fibers in rainbow trout tend to stain more intensely for mitochondrial enzymes than the larger diameter fibers, leading early workers to think that the deep muscle contained a mosaic of red and white muscle fibers (Boddeke et al., 1959). However, Johnston et al. (1975) found that both small and large diameter fibers in rainbow trout had similar staining characteristics for myosin

ATPase activity (mATPase), and suggested that the small diameter fibers were growth stages of the white muscle fibers. In gilthead sea bream (*Sparus auratus;* Mascarello et al., 1995; Ramírez-Zarzosa et al., 1998) and sea bass (*D. labrax;* Ramiírez-Zarzosa et al., 1998), the mATPase of the small diameter white fibers is more alkaline stable than the larger diameter fibers. Recently, a cDNA clone of myosin heavy chain (MHC) has been isolated from the white muscle of common carp that is associated with fiber recruitment. The RNA transcripts of this MHC gene only hybridized to small diameter fibers consistent with the transient expression of a developmental-stage-specific myosin isoform (Ennion et al., 1995). The white muscle in the yolk-sac larvae of plaice (*Pleuronectes platessa* L.; Brooks and Johnston, 1993); and Atlantic herring (*C. harengus* L.; Johnston et al., 1997) express an embryonic isoform of the phosphorylated myosin light chain 2 (MLC2). The embryonic MLC2 is progressively replaced by the adult MLC2 isoform during the larval stages. Adult white muscle contains a trace of the embryonic MLC2 isoform. It has been suggested that the embryonic myosin light chain isoform is expressed transiently in small diameter fibers and may be a potential marker for recruitment (Brooks and Johnston, 1993; Johnston et al., 1997). Seasonal cycles of muscle fiber recruitment have been observed in the mullet (*Mugil cephalus*) with a higher proportion of small diameter fibers in the spring (Carpenè and Veggetti, 1981).

Muscle growth is also accompanied by a dramatic rise in the number of myonuclei. In common carp, the percentage of nuclei associated with muscle fibers (satellite cells plus myonuclei) was found to be independent of fish length and was 54% for white muscle and 32% for red muscle (Koumans et al., 1991). Koumans et al. (1991) suggested that the abundant capillary endothelial nuclei probably accounted for the higher percentage of non-muscle nuclei in red muscle. The percentage of satellite cells in the white muscle of common carp declined from 5% to less than 1% as standard length increased from 5–20 cm, respectively. The corresponding percentage of satellite cells to total myonuclei in red muscle were 11 and 3%, respectively (Koumans et al., 1991). Although the percentage of satellite cells declined with growth, their total number remained relatively constant over this size range (Koumans et al., 1994).

D. The Role of Satellite Cells
   in Muscle Fiber Recruitment and Hypertrophy

It is generally accepted that the nuclei for muscle growth are derived from the satellite cells. At the ultrastructural level a wide diversity of nuclear morphology was observed in the white muscle of Atlantic salmon fry (Fig. 2A–C). Spindle-shaped mononuclear cells were relatively abundant at first feeding and occurred between muscle fibers (Fig. 2A). Multinucleated myotubes were also common on the surface of muscle fibers. Myotubes are formed from the fusion of a number of mononuclear myoblasts (Fig. 2B, C). In some cases mononuclear cells were

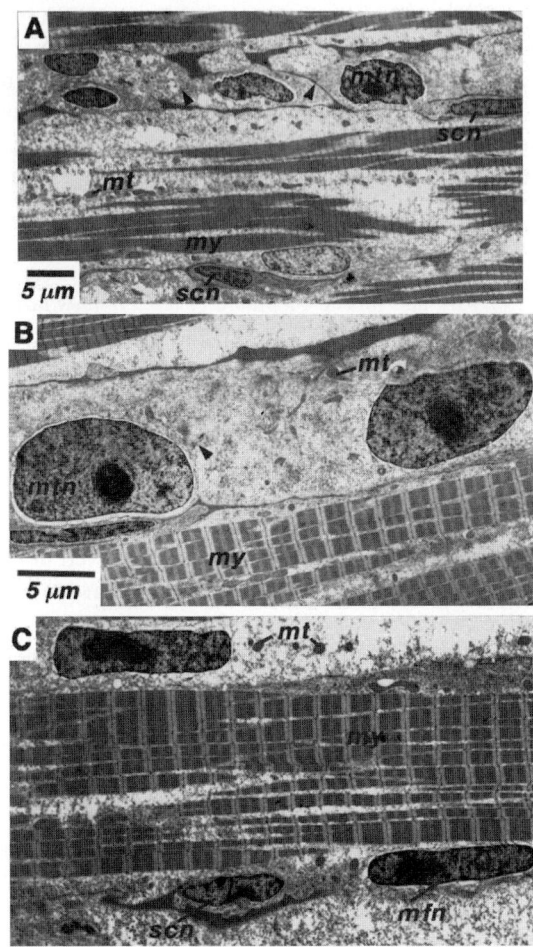

Fig. 2. Electron micrographs of sagittal sections of white muscle from Atlantic salmon (*Salmo salar* L.) fry at first feeding. (A) The formation of a myotube on the surface of a white muscle fiber. The arrowheads indicate the cell membrane of a myoblast prior to fusion with the myotube of the left-hand side of the micrograph. Note the wide range of nuclear morphology. (B) A mature myotube in which only a few remnants of the myoblast cell membranes are present (arrowhead). At this stage there are no myofibrils and only small mitochondria (mt), which may represent an early stage of their biogenesis. (C) A segment of a white muscle fiber showing two elliptical myonuclei (mfn). Also illustrated is a mononuclear cell (scn) that appears to be fusing with the muscle fiber membrane. Other abbreviations: mtn, myotube nucleus; mt, mitochondrion; scn, satellite cell nucleus; my, myofibril; mfn, myofiber nucleus. From Johnston *et al.* (2000b).

observed that appeared to be fusing with maturing fibers containing differentiated myonuclei (Fig. 2C). Johnston *et al.*, (1998) incubated Atlantic herring larvae in 5′-bromo-2-deoxyuridine to label dividing nuclei. The presence of labeled nuclei in myotubes indicated that the mononuclear cells from which they were derived had recently divided.

The study of muscle satellite cells in fish is much less advanced than it is for mammals (see Fauconneau and Paboeuf, 2001, Chapter 4, this volume), and the cellular mechanisms underlying fiber recruitment and hypertrophy are largely unknown. Current ideas are therefore largely based on evidence from tissue culture studies with mammalian myocytes. Mammalian satellite cells can occur in a quiescent and an activated state. Quiescent satellite cells are thought to correspond to a muscle stem cell population (Schultz, 1996). The activated satellite cell undergoes an asymmetric division to produce another stem cell and a daughter cell that divides a limited number of further times prior to differentiating (Quinn et al., 1988; Cornelison and Wold, 1997). The length of the proliferation phase and hence the number of myonuclei produced from each stem cell division is under the control of antagonistic and highly complex signaling pathways. In fish, a major uncertainty is whether there are separate muscle stem cell populations for fiber recruitment and fiber hypertrophy. Apoptosis (programmed cell death) is a normal part of myogenesis during the development of mammals (Dominov et al., 1998) and following exercise and certain pathological conditions in adults (Podhorska-Okolow et al., 1998). Apoptosis may also have a role in fish myogenesis although this has yet to be investigated. A possible model for the role of satellite cells in post-embryonic muscle growth in fish is presented in Fig. 3. In the absence of experimental data all such schemes must remain highly speculative.

E. Molecular Markers of Muscle Satellite Cells

Satellite cells identified by their location and morphology in electron micrographs presumably correspond to a mixture of muscle stem cells and their division products committed to differentiation. Recently, a number of molecular markers have been identified that are specific for muscle satellite cells in mammals, including the c-met tyrosine kinase receptor (Cornelison and Wold, 1997). The ligand for the c-met tyrosine kinase receptor is hepatocyte growth factor/scatter factor (HGF/SF), which stimulates mouse satellite cells to enter the cell cycle (Tatsumi et al., 1998). In mammals, c-met stains both muscle stem cells and their division products committed to differentiation (Cornelison and Wold, 1997). All cells destined to differentiate express one or more members of the MyoD family of transcription factors, MyoD, Myf-5, myogenin, and Myf-6, that are involved in the initiation and maintenance of the differentiation program (Cornelison and Wold, 1997; see Watabe, 2001, Chapter 2, this volume). Thus dual immunohistochemical labeling with c-met and MRFs as primary antibodies can potentially distinguish between muscle stem cells and cells committed to differentiation. Proliferating cells can be identified by the expression of Proliferating Cell Nuclear Antigen (PCNA), an accessory protein of DNA polymerase (see Rowlerson and Veggetti, 2001, Chapter 5, this volume). At first feeding in Atlantic salmon, 24.5%

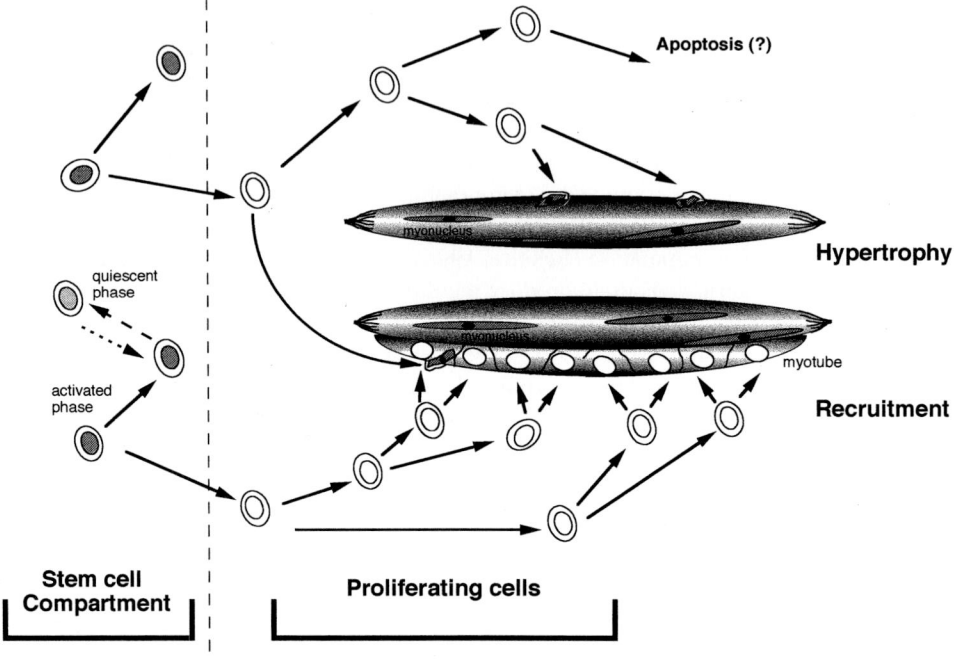

Fig. 3. A speculative model for the involvement of satellite cells in fiber recruitment and hypertrophy in fish muscle. The possibility of separate muscle stem cell populations for the provision of nuclei for recruitment and hypertrophy is illustrated. Equally plausible is a single muscle stem cell population with the fate of proliferating cells being decided by local signaling pathways. The muscle stem cells are believed to exist in both a quiescent (Go) phase of the cell cycle and in an activated state. Activated muscle stem cells probably undergo an asymmetric division to produce another stem cell and a cell committed to differentiation after a limited number of further divisions. Cells committed to differentiation express myogenic regulatory factors and have one of two fates: they either fuse together on the surface of an existing muscle fiber to form multinucleated myotubes or else they are absorbed into muscle fibers as they hypertrophy. The myonuclei have a distinct morphology from the mononuclear cells and are incapable of division.

of the total nuclei in the white muscle were observed in mononuclear cells (Johnston et al., 2000b). A similar proportion of the muscle nuclei (22.5%) stained with an antibody against PCNA (Johnston et al., 2000b). However, only about 80% of the PCNA-positive nuclei were present in cells that stained for c-met (Johnston et al., 2000b). It seems likely that these cells correspond to the muscle satellite population and the remaining mononuclear cell types correspond to fibroblasts, other connective tissue cells, and capillary endothelial cells. The great majority of the c-met-positive cells also stained for one or more MRF: 72.4% for MyoD, 76.7% for Myf-5, 62.1% for myogenin, and 48.7% for Myf-6, indicating that they corresponded to the division products of proliferating stem cells (Fig. 4).

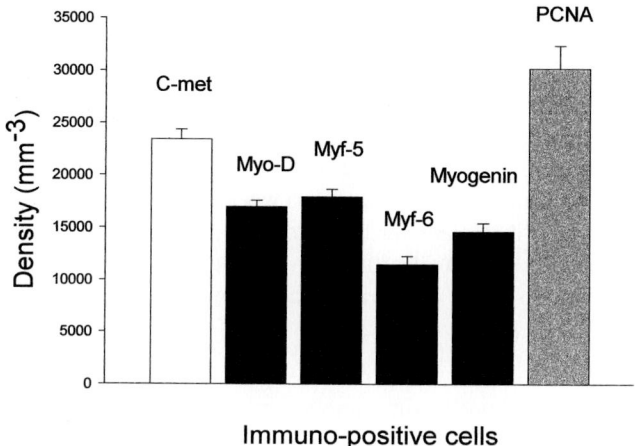

Fig. 4. Immunohistochemistry of mononuclear cells from the white muscle of Atlantic salmon (*Salmo salar*) fry at first feed. The density of cells immunopositive for c-met, MyoD, Myf-5, Myf-6, myogenin, and proliferating cell nuclear antigen (PCNA) is shown. The bars represent average values and standard error bars are shown, n = 10 fish per stain. From Johnston *et al.* (2000b).

Although Myf-6 has not yet been cloned and sequenced in fish, the cross-reaction of salmon satellite cells with anti-mouse Myf-6 suggests that a homologous transcription factor is present. Other markers for satellite cells have been identified in mammalian muscle including the winged helix transcription factor, myocyte nuclear factor-β (Yang *et al.*, 1997), and extracellular signal regulated kinases 1 and 2 (Yablonka-Reuvini *et al.*, 1999). The expression patterns and utility of these factors as molecular markers of muscle satellite cells in fish has yet to be examined.

III. METHODS FOR QUANTIFYING GROWTH PATTERNS IN MUSCLE

The myomeres are composed of a series of overlapping cones and therefore a transverse section of the trunk will section several myotomes at different levels. The number and diameter of muscle fibers varies along the length of the body and in different regions of the cross-section (Fig. 1). There may be discrete zones of new fiber production producing gradients of fiber diameter, particularly during juvenile stages. For this reason fiber number is best counted in one half of the cross-section and not just a particular segment. Counting all the fibers is impractical for very large fish and so fiber number has to be estimated. In this case it is important to obtain an accurate measure of the total cross-sectional area of the muscle and cut sufficient tissue blocks to obtain a representative sample of all regions. Different muscle types can be identified using histochemical (Johnston et al., 1974) or immunocytochemical techniques (Rowlerson et al., 1995). In many published studies, the cross-sectional areas or diameters of 50–200 muscle fibers per fish have been measured and the results plotted as a frequency distribution histogram, often with a limited attempt at statistical analysis. However, nonparametric smoothing techniques have distinct advantages over methods that are dependent on a histogram of the realizations when dealing with a continuous variable such as muscle fiber diameter. Johnston et al. (1999) used a smooth kernel function to obtain an estimated probability density function of muscle fiber diameters. The kernel estimate used was of the following form:

$$\hat{f}(y) = \sum_{i=1}^{n} w(y - y_i; h_i)$$

where f is the estimated probability function, y_i is the ith observation from the list of n, h is the smoothing parameter, and w is the kernel function (Silverman, 1986; Bowman and Azzalini, 1997). Around 800–1000 measurements of muscle fiber diameter per cross-section is recommended. The average diameter can either be calculated using a computer algorithm or expressed as the diameter of the equivalent circle from a measurement of the cross-sectional area. Fig. 5 shows a comparison between a histogram and a smooth probability function calculated for muscle fibers from Atlantic salmon. Further statistical analysis should take into account the hierarchical structure of the data, which results in two sources of variation—variation among fish and variation within fish (Johnston et al., 1999). One strategy is to sample randomly n fish with replacement and then to sample

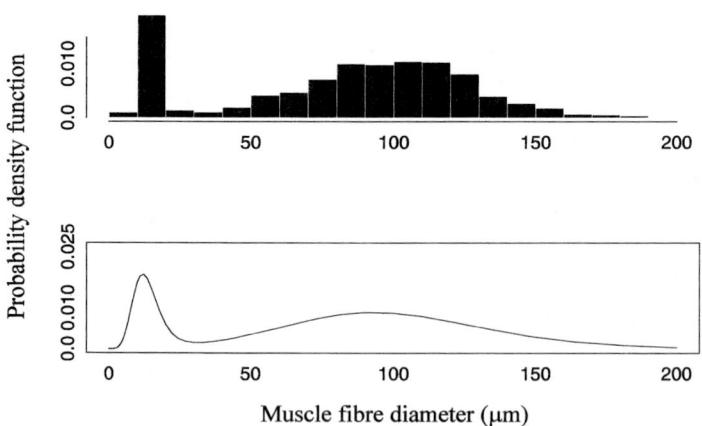

Fig. 5. Methods for investigating the distribution of fiber diameter in fish muscle. The figure compares a histogram (top) and the corresponding smooth probability density function (bottom) calculated for 800 muscle fiber diameters from a triploid normal sex ratio (Atlantic salmon (*Salmo salar* L.). From Johnston et al. (1999).

randomly with or without replacement, m fibers within a fish. For samples of 10 or more fish sampling fibers without replacement is preferred (Davison and Hinkley, 1997). For smaller samples of fish a smooth bootstrap sample can be used as described by Silverman (1986) and Davison and Hinkley (1997). Estimates of the variability of the average probability density function for a group of fish provide valuable information on the structure of the distribution, e.g., whether it is unimodal or bimodal, etc., and allows specific hypotheses to be tested statistically (see Johnston et al., 1999). Finally, many papers report measurements of the average muscle fiber diameter. This is not always a useful parameter since for much of the life cycle average fiber diameter is a complex function of both fiber recruitment and hypertrophy which tends to decrease and increased the mean value, respectively.

IV. GENETIC VARIATION IN MUSCLE GROWTH PATTERNS

A. Variation Between Populations

Wild Atlantic salmon tend to home to their natal streams to spawn thus promoting reproductive isolation and genetic differentiation between populations (Verspoor, 1997). There is significant phenotypic diversity of salmon between and within river systems with respect to body size and age at maturity (Thorpe and Mitchell, 1981), morphology (Riddell and Leggett, 1981), and physiological characteristics including digestion and growth rates (Taylor, 1981; Nicieza et al., 1994). In some cases, population differences in characters have been shown to be heritable in breeding experiments (Riddell and Leggett, 1981; Nicieza et al., 1994; Thorpe et al., 1983).

There is some evidence for genetic differences in muscle growth patterns between populations of salmonids, particularly with respect to the number and size distributions of muscle fibers. Valente et al. (1998) determined the concentrations of RNA, DNA, and protein in the skeletal muscle of a fast- (Cornec) and slow- (Mirwart) growing strain of rainbow trout (*O. mykiss*). The higher DNA:protein ratio observed in the Cornec strain was interpreted in terms of a smaller average cell size, but a larger number of cells per unit weight compared to the Mirwart strain. In a subsequent paper, the percentage of fibers less than 25 μm diameter was shown to be greater in the fish from the fast- than the slow-growing strain (Valente et al., 1999).

Atlantic salmon populations show major differences in their age at sexual maturity and the timing of their spawning migration (Taylor, 1981). Johnston et al., (2000a) compared the pattern of muscle growth in the seawater stages of a predominantly early maturing (strain X) and late maturing (strain Y) population. Around 90% of strain X matured after their first sea-winter compared with 10%

for strain Y. Each population had been reared for 5 generations by a major Scottish salmon farmer and comprised more than 500 fish representing 6 different families per strain. The fish were PIT-tagged and introduced together in April into two adjacent 5 × 5 × 5 m sea cages and reared under semi-commercial conditions on the West Coast of Scotland. Measurements of the number and size distribution of white muscle fibers were made from frozen sections prepared from 5–12 blocks so as to sample all regions of the myotomal cross-section. The distribution of muscle fiber diameters in different age-classes was investigated using non-parametric smoothing and bootstrapping techniques.

After the first 6 months in seawater (April–October 1997) the body mass of the fish increased at a significantly faster rate in strain X than in strain Y (Fig. 6A). From January 1998 onward, the cross-sectional area of white muscle at the level of the first dorsal fin ray was greater in strain X than strain Y for a given fork length, indicating a more stocky body shape (Fig. 6B). The number of white muscle fibers per trunk cross-section was around 150,000 in June 1997 and 250,000 in July 1997 for both populations (Fig. 7A). However, after this initial period, the relative contributions of fiber recruitment and hypertrophy to muscle growth were quite distinct in the two populations. Between July and August 1997, fish from strain Y recruited twice as many white fibers for each square millimeter increase in muscle cross-sectional area than fish from strain X (Fig. 7B). As a result, in August the number of white muscle fibers per trunk cross-section was 26% higher in strain Y than strain X (Fig. 7A). However, the number of white muscle fibers in strain X had caught up with strain Y by January with 545,000 fibers per trunk cross-section in each population (Fig. 7A). In both populations, the recruitment of new fibers declined in the period from August 1997 to January 1998 (Fig. 7B) as water temperature and day-length fell. Similar seasonal declines in muscle fiber recruitment have been reported for the freshwater stages of the life cycle (Higgins and Thorpe, 1990). For strain X the number of fibers recruited between January and March 1998 (~ 30 mm^{-2} muscle) was only about 10% of that observed between June and July 1997 shortly after transfer to seawater. In strain Y, there was no significant increase in the number of white fibers between January and September 1998, such that fiber number remained relatively constant between 42 and 63 cm fork length (Johnston et al., 2000a) and growth was entirely by fiber hypertrophy (Fig. 7B). In contrast, the number of fibers recruited in the predominantly early maturing population (strain X) increased again in the spring and early summer, reaching around five times the winter level (~ 150 mm^{-2} muscle; Fig. 7B), equivalent to 718,000 per trunk cross-section by June (Fig. 7A). Thus, the increase in muscle girth that was observed between January and the final harvest sample was the result of a combination of fiber recruitment and hypertrophy in the early maturing population but entirely due to fiber hypertrophy in the late maturing population. These results point to an association between fast muscle growth and fiber recruitment as has been observed for juvenile rainbow

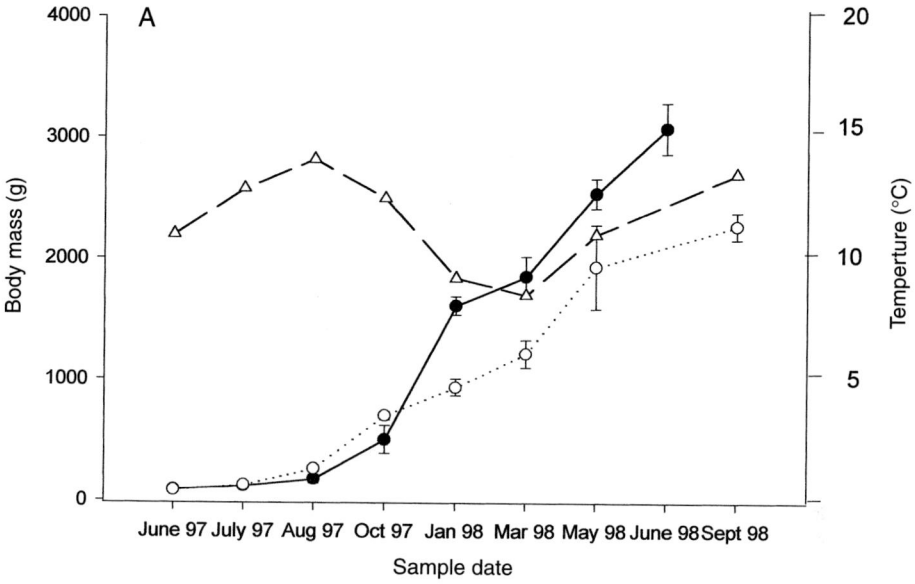
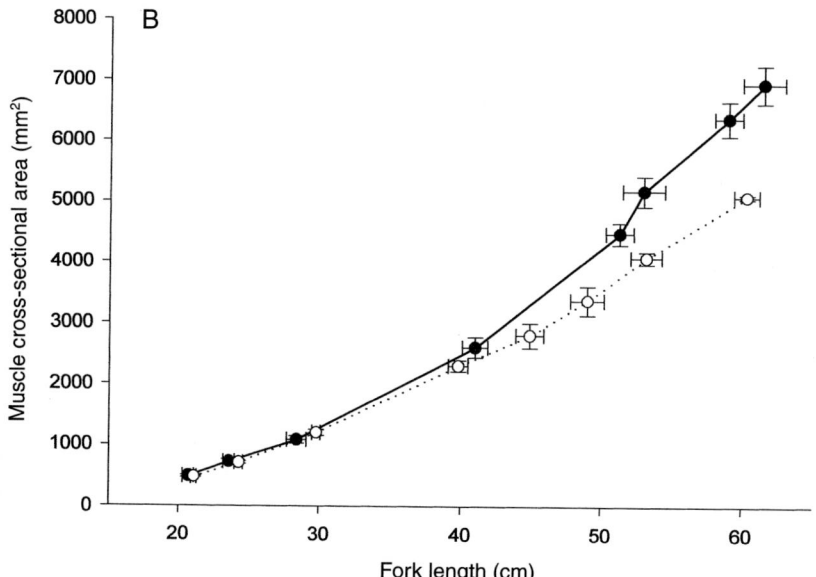

Fig. 6. Seawater growth in a predominantly early (closed circles; strain X) and late (open circles; strain Y) maturing populations of Atlantic salmon (*Salmo salar* L.). The fish were transferred to sea in April 1997 and harvested in June 1998 (strain X) and September 1998 (strain Y). The triangles show the average seawater temperature during the trial. (A) illustrates the growth in body mass and (B) the relationship between the total cross-sectional area of white muscle, at the level of the first dorsal fin ray, and fork length. The symbols represent mean ± SE for 6–18 fish per sample. From Johnston *et al.* (2000a).

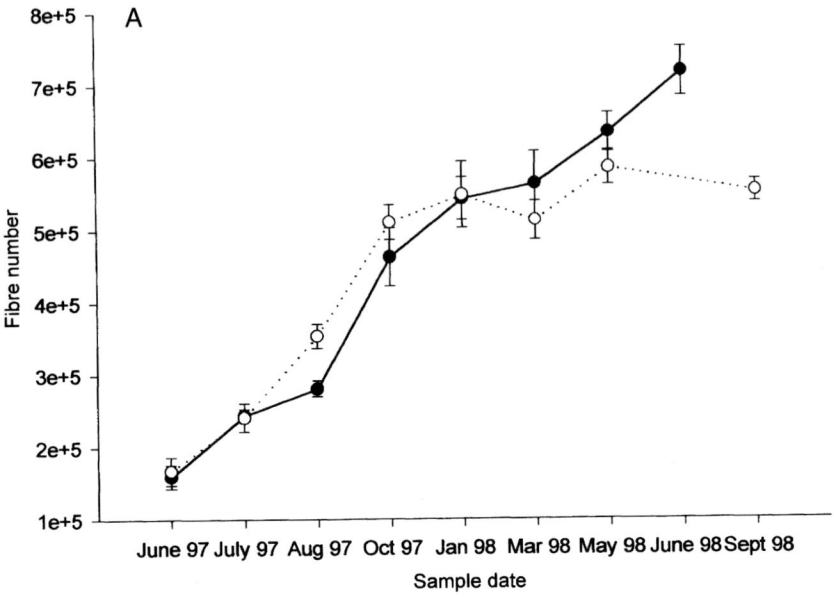

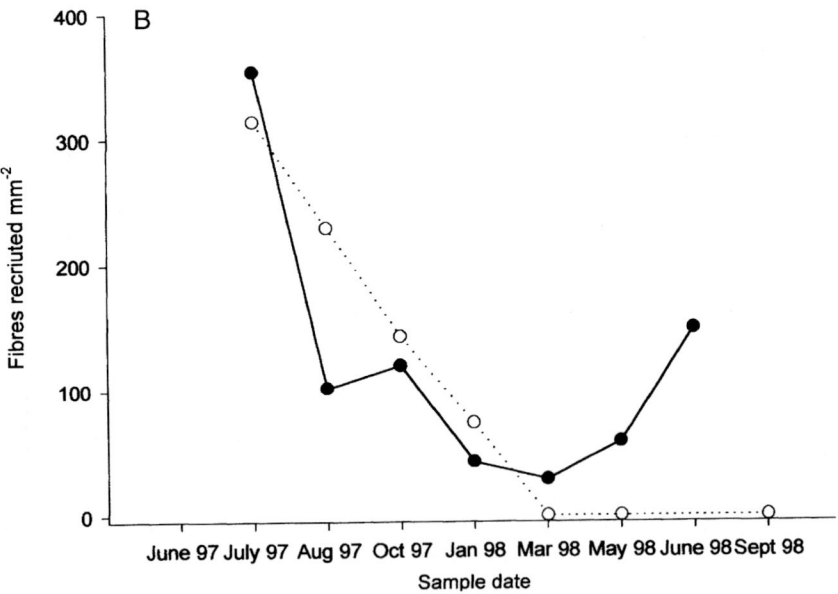

Fig. 7. Muscle fiber recruitment in a predominantly early (closed circles; strain X) and late (open circles; strain Y) maturing population of Atlantic salmon (*Salmo salar* L.). The fish were transferred to sea in April 1997 and harvested in June 1998 (strain X) and September 1998 (strain Y). (A) illustrates the growth in the number of white muscle fibers at the level of the first dorsal fin ray and (B) the number of fibers recruited for each 1 mm² increment in muscle cross-sectional area since the previous sample. Thus the points for July 1997 represent the growth between June and July etc. The symbols represent in (A) mean ± SE for 6–18 fish per sample. From Johnston *et al.* (2000a).

trout (Weatherley et al., 1979) and larvae of Atlantic cod (*G. morhua*; Galloway et al., 1999) and common carp (Alami-Durante et al., 1997) fed different diets and ration levels.

Johnston et al., (2000a) found that the majority of Atlantic salmon sampled in the first 5 months after seawater transfer had a unimodal distribution of white muscle fiber diameter. Muscle fiber distributions became bimodal in the autumn following seawater transfer. In all subsequent samples the right-hand peak of the distribution was shifted toward higher diameters in strain X compared to strain Y, indicating a greater rate of fiber hypertrophy in the early maturing population. For example, after 13 months in seawater the 95th percentile of white fiber diameter calculated from the smooth distributions was at 215 μm in strain X and 171 μm in strain Y (Fig 8). In March 1998 and subsequent samples, the left-hand peak of relatively small fibers was shifted to higher diameters in strain Y relative to strain X, reflecting the cessation of new fiber recruitment in strain Y. Immature fish within strain X had a lower density of small diameter fibers and the right-tail of the distribution was shifted to higher diameters relative to fish that had begun to sexually mature. Thus greater muscle fiber hypertrophy was observed in immature than maturing fish of the same strain. It was concluded that the superior growth performance of the early maturing population was associated with a longer period of fiber recruitment and greater fiber hypertrophy than in the late maturing population, although these differences were not directly related to sexual maturation (Johnston et al., 2000a). These results indicate complex genetic differences in the program of muscle fiber recruitment with ontogeny and quite possibly different responses to environmental cues such as temperature and day-length.

B. Ploidy Manipulation

In general, there are fewer but larger cells in the organs and tissues of triploid than diploid fish (Small and Benfey, 1987). The effects of triploidy on growth performance in salmonids is variable and may be influenced by induction method, husbandry practices, and the stage of the life cycle being compared (Thorgaard, 1986). In the case of Atlantic salmon, growth performance is often as good or better in triploids than diploids when they are reared separately (Carter et al., 1994). Triploid fish potentially have environmental benefits in aquaculture since individuals that escape from sea cages cannot breed with their wild counterparts.

Greenlee et al. (1995) reported that the yield of mononuclear cells per gram of muscle tissue was significantly higher in primary cell cultures derived from diploid than triploid rainbow trout. This is consistent with the general finding of reduced cell numbers per unit volume in triploids. If triploid treatment alters the number of muscle satellite cells this would be expected to have profound consequences for the provision of nuclei for muscle fiber recruitment and hypertrophy. Suresh and Sheehan (1998) estimated that there were 10% fewer white fibers per unit cross-sectional area in triploid than diploid rainbow trout. Surprisingly,

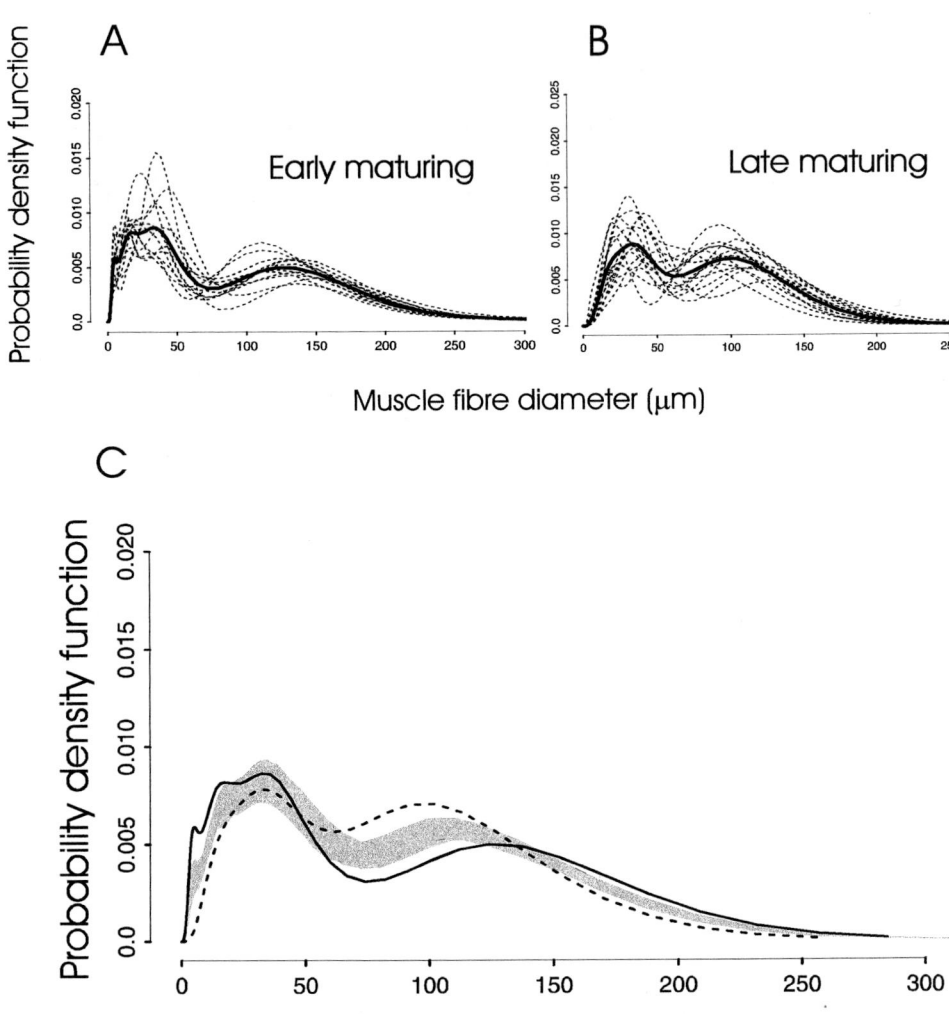

Fig. 8. Distribution of muscle fiber diameter in the white muscle of early (A) and late (B) maturing populations of Atlantic salmon (*Salmo salar* L.). Measurements were made from a cross-section at the level of the first dorsal fin ray. Fish were sampled at the end of the growth trial in June (early maturing strain) and September (late maturing strain). The dotted lines represent the smooth probability density function (pdf) of muscle fiber diameter in individual fish and the solid line the average pdf per group. (C) The average probability density function is shown for the early maturing strain (solid line) and the late maturing strain (dotted line). The shaded area enclosed 100 bootstrap estimates of the total population, combined strains. Areas where the solid and dotted lines lie outside the shaded area provide evidence for differences in the density distributions between the populations. From Johnston *et al.* (2000a).

6. GENETIC AND ENVIRONMENTAL DETERMINANTS

however, in fish greater than 3 cm fork length the histograms of the frequency distribution of fiber diameters were similar in diploid and triploid fish.

Recently, we investigated the patterns of muscle growth in populations of normal-sex-ratio (NSR) and all-female (AF) diploid and triploid Atlantic salmon (*S. salar* L.); (Johnston *et al.*, 1999). All-female fish do not show the deteriorative changes associated with sexual maturation and hence have potential advantages in aquaculture. AF fish were produced by fertilizing normal eggs with the milt from genetic females (XX) that had been treated with male hormones to produce functional males. In Atlantic salmon smolts, c-met expressing cells (satellite cells) corresponded to 17.5% of the total muscle nuclei and were 24% more abundant in diploid than in triploid fish. About 85% of the c-met positive cells also stained for the myogenic regulatory factor myf-6, indicating they were committed to muscle differentiation (Johnston *et al.*, 1999). The embryonic phase of myogenesis was relatively little affected by ploidy status. In late embryos, 46-day post-fertilization at 6°C, when around 2800 fibers per cross-section had been formed, there were just 7% more fibers in diploids than triploids in NSR fish and there were no significant differences in fiber number with ploidy status in the AF population (Johnston *et al.*, 1999). However, at first feeding there were around one third more white muscle fibers in diploids than triploids in the NSR population and this difference was maintained during seawater and freshwater growth (Fig. 9A). At 800 days post-hatch, the average number of white muscle fibers calculated from the fitted regressions was 540,000 for NSR diploids, 363,600 for NSR triploids, 525,400 for AF diploids, and 369,900 for AF triploids (Johnston *et al.*, 1999). Diploids had a higher body mass for a given fork length in both NSR and AF fish. Triploid fish tended to have larger diameter fibers than their diploid counterparts (Fig. 9B). For example, 839 days post-hatch in the NSR population the fiber density function was bimodal and the right-hand peak was shifted to higher diameters in triploid than diploid individuals (Fig. 9B). Thus the lower density of satellite cells in triploid than diploid salmon was associated with a reduced importance of fiber recruitment to the growth in muscle girth, but a compensatory increase in fiber hypertrophy. Presumably the lower density of nuclei in triploids was compensated for by an increase in nuclear size such that the nuclear to cytoplasmic ratio as relatively unaffected, and this may be a critical factor for fiber hypertrophy.

V. ENVIRONMENTAL INFLUENCES ON MUSCLE FIBER GROWTH PATTERNS

A. Temperature

The rate of development and growth of fish embryos increases as water temperature is increased until the upper lethal temperature of the species is approached. Stickland *et al.* (1988) made the important observation that not all

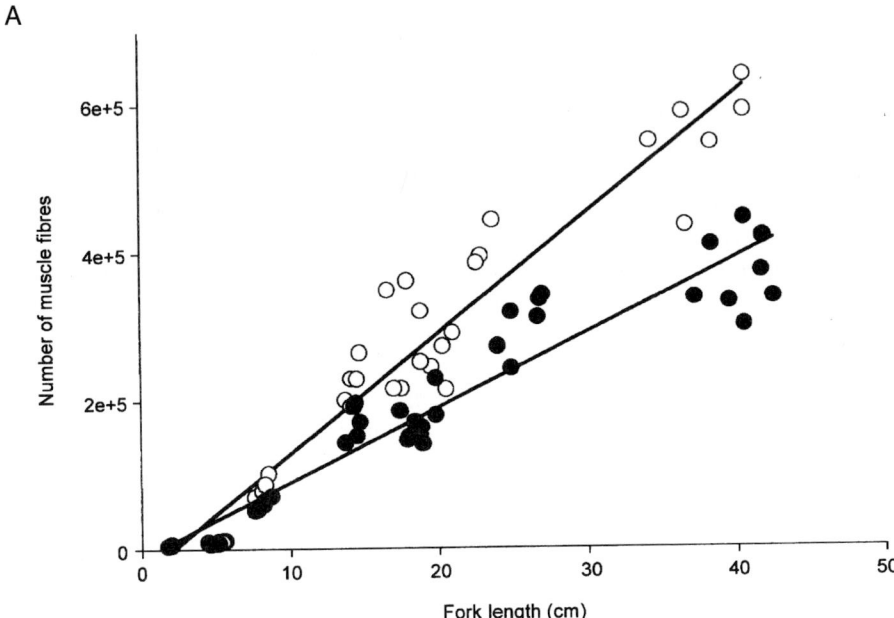

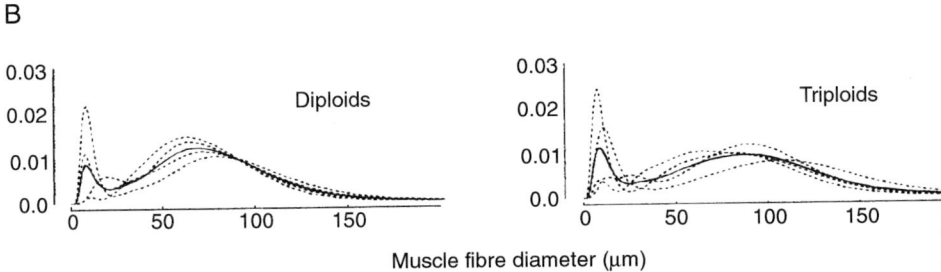

Fig. 9. Influence of ploidy on muscle growth in diploid and triploid Atlantic salmon (*Salmo salar* L.). (A) illustrates the relationship between the number of white muscle fibers at the level of the pelvic fin insertions and fork length for diploid (open circles) and triploid (closed circles) normal sex ratio Atlantic salmon (*Salmo salar* L.). The lines were fitted with a first order polynomial. (B) illustrates the smooth probability function of white muscle fiber diameter for individual fish (dotted lines) and each group (solid line) 839 days post-hatch. From Johnston et al. (1999).

processes associated with myogenesis increase to the same extent as temperature rises, resulting in differences in muscle cellularity. In their first published experiment, eggs from a single Atlantic salmon family were incubated at either ambient temperature fluctuating around 1.6°C or in heated water at a constant 10°C (Stickland et al., 1988). The cross-sectional area of the muscle at the level

6. GENETIC AND ENVIRONMENTAL DETERMINANTS

of somites 10–15 was not significantly different between the two groups from embryonic stages 28–33 as defined by Gorodilov (1983). Stage 33 corresponded to hatching in both groups and occurred after approximately 54 days at 10°C and 127 days at ambient temperature. Significantly, the number of muscle fibers increased much more in the ambient group than the heated group in the period just prior to hatching (Fig. 10). As a result of hatching, salmon alevins in the ambient group had significantly more fibers that were of lower average cross-sectional area than those in the heated group (Stickland et al., 1988). Similar findings were reported by Nathanailides et al. (1995) studying the progeny of a single cross of wild salmon from the River Almond, Perthshire, that had been reared at ambient stream temperatures fluctuating around 5°C or at a constant 11°C. In this study, the cross-sectional area of the white muscle was also similar in the two groups at hatch, and the ambient group had some 4500 fibers per cross-section in caudal myotomes compared with only 3200 in the heated group. Nathanailides et al. (1995) also found that the total density of muscle nuclei was higher in the ambient than the heated groups. The same pattern of results was obtained by Usher et al. (1994), although at hatch in this study the total number of fibers per myotomal cross-section in embryos incubated at 5°C was almost double the number at 11°C.

Although the effects of temperature on muscle fiber number in Atlantic salmon

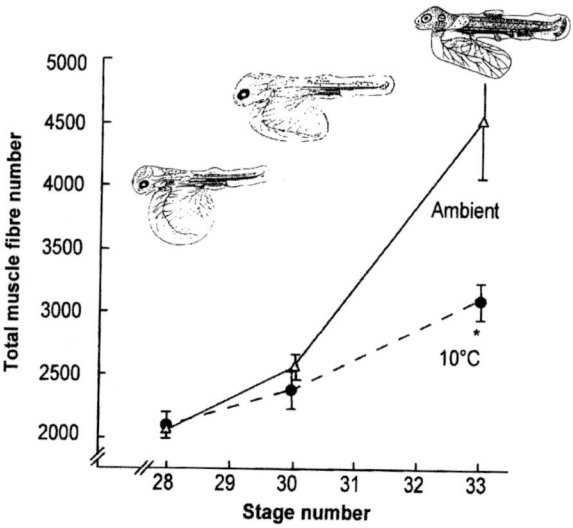

Fig. 10. The influence of egg temperature on the number of white muscle fibers per cross-section in relation to embryonic (Gorodilov) stage in Atlantic salmon. Eggs from a single cross were incubated at either fluctuating ambient river temperature (averaging 1.6°C) or in water heated to 10°C until they hatched at stage 33. The values represent mean ± SE, 5 or 6 fish per group. Redrawn and adapted from Stickland et al. (1988).

embryos is consistent across studies its impact on average fiber size is not. Johnston and McLay (1997) investigated early myogenesis in five families of salmon derived from wild-caught fish caught in the River Shin, Sutherland, Scotland. Eggs were incubated in a hatchery at ambient temperature averaging 4.3°C or in water heated to 8°C. At hatch, embryos incubated at 8°C had a significantly lower total cross-sectional area of both red and white muscles, reduced numbers of fibers and nuclei, and a reduced average cross-sectional area of white muscle fibers (Johnston and McLay, 1997). The differences in muscle fiber number between temperature groups were of the order of 6–12%. There was evidence for significant variation between families and differences in the family-within-temperature variance component, although such effects were relatively small compared with the main temperature and developmental-stage effects. The results from this study were consistent with the general observation that salmon eggs incubated at low temperatures produce larger alevins. Hamor and Garside (1977) reported that alevins hatched at 5°C were larger than those reared at 10°C, but that subsequent yolk efficiency was higher at 10°C. Johnston and McLay (1997) found that for white muscle, fiber number almost doubled, whereas average fiber cross-sectional area increased around three-fold, between hatching and first feeding. However, the differences between temperature groups were no longer significant at first feeding, indicating a growth advantage in the heated groups. These results provide an indication that temperature has different effects on muscle growth at the embryo stage and in the period following hatching when fibers are recruited from distinct germinal zones and from the satellite cell population (Johnston and McLay, 1997). Nathanailides and co-workers (1995) found that Atlantic salmon embryos reared at ambient temperature (1–8°C), had more, smaller diameter, white muscle fibers but a similar total muscle cross-sectional area to a group heated to 11°C. However, 3 weeks after first feeding, there had been more fiber hypertrophy in the ambient group such that total muscle cross-sectional area was greater than in the heated group, but there was no "catch-up" in fiber number (Nathanailides et al., 1995). Differences in experimental stock origin, incubation temperatures, and other factors that influence larval growth and size, such as egg size and quality, probably underlie the different responses observed between studies.

The effects of temperature on embryonic and early larval myogenesis have been investigated for some other fish species including common carp (*C. carpio* L.; Alami-Durante et al., 1997), whitefish (Hanel et al., 1996), rainbow trout (Killeen et al., 2000), plaice (*P. platessa* L.; Brooks and Johnston, 1993), turbot (*Scophthalmus maximus;* Gibson and Johnston, 1995), and Atlantic cod (*G. morhua* L.; Galloway et al., 1998). One-day-old cod larvae hatched at shorter lengths at 1°C and had more, but smaller diameter, white muscle fibers than those at either 5 or 8°C (Galloway et al., 1998). These differences in muscle cellularity were no longer apparent between temperature groups in 5-day-old larvae at the time of first feeding. The volume density of mitochondria was in the range 35–38% for red muscle and 12–14% for white muscle at all temperatures studied (Galloway *et al.*, 1998). However, at first feeding the white muscle fibers had a less mature appearance in 1°C larvae and the volume density of myofibrils was only 49% compared with 64% in larvae reared at 8°C. It was concluded that 1°C was close to, or below, the lower thermal tolerance limit for normal development in the Northeast Arctic cod stock (Galloway *et al.*, 1998). In contrast, egg incubation temperature was found to influence mitochondrial abundance in spring-spawning Atlantic herring (Vieira and Johnston, 1992). For example, in 1-day-old larvae, the fraction of fiber volume occupied by mitochondria in the embryonic slow muscle layer ranged from 46% at 15°C to 38.8% at 5°C. In the embryonic white muscle the volume density of mitochondria was 26.1% at 15°C and 15.9% at 5°C (Vieira and Johnston, 1992).

Hatching is not a precise developmental stage. For example, brown trout embryos reared at 2°C hatched with significantly more fin rays and more advanced body pigmentation than embryos reared at 10°C (Killeen *et al.*, 2000). Killeen *et al.* (1999b) developed a quantitative scoring system for the brown trout based on an assessment of a wide range of developmental characters. The cross-sectional area of white muscle fibers and fiber number was found to be greater for embryos reared at 2°C than 10°C both with respect to developmental score (Fig. 11A) and fork length (Fig. 11B). The effects of temperature on the cross-sectional area of red muscle were similar but the higher incubation temperature depressed fiber number to an even greater extent (Killeen, 1999). No difference was found in the effects of temperature on early myogenesis between the offspring of resident and migratory forms of the brown trout (Killeen, 1999).

In order to put these early developmental effects of temperature in a stronger ecological context, we have recently studied myogenesis in wild-caught fish from the Aberdeenshire Dee, in northeastern Scotland (Fig. 12A) incubated under the temperature regimes of their natal streams (McLay *et al.*, 2000; Johnston *et al.*, 2000b). Tagging and radiotracking studies have shown that spring-running fish return to the colder headwaters of the river whereas autumn-running fish spawn lower down in the catchment (Youngson *et al.*, 1994). In all but a few summer months the upland streams are significantly cooler than tributaries toward the river mouth (Webb and McLay, 1996). In our study, fish were captured in the final stages of their spawning migration to an upland (Baddoch) and a lowland (Sheeoch) tributary. Ova from 11 females were fertilized with the milt from 11 males from each site to produce 11 unique families. The eggs were split into two groups and incubated in a hatchery under the simulated temperature regimes of either the Baddoch and Sheeoch, based on averaged *in situ* recordings made between 1994 and 1996 (Fig. 12B). On average the Baddoch temperature regime was 2.8°C cooler than that of the Sheeoch (McLay *et al.*, 2000). Our first aim was to test the hypothesis that early myogenesis in spring- and autumn-running salmon

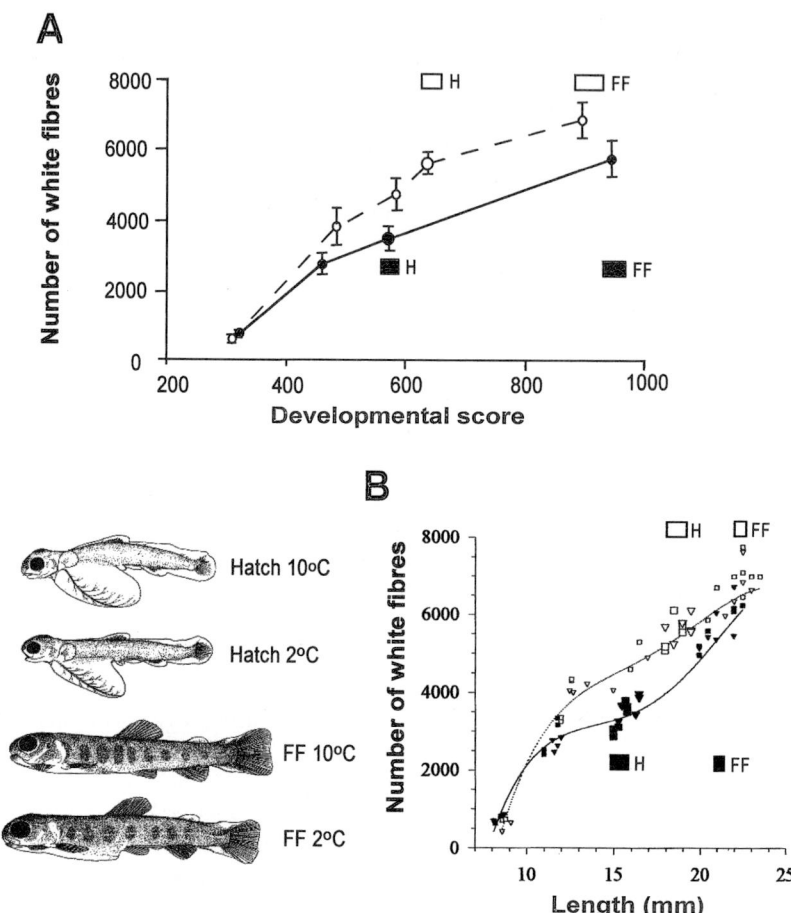

Fig. 11. The influence of temperature on the number of white muscle fibers per trunk cross-section in the brown trout (*Salmo trutta*). (A) fiber number in relation to developmental score (stages at hatch and first feeding are illustrated) and (B) fiber number in relation to body length. Fish were reared in a hatchery at either 2°C (open symbols) or 10°C (closed symbols). The squares and inverted triangles represent the offspring of anadromous and of freshwater resident females, respectively. The range of developmental scores and body lengths that the fish attained at hatch (H) and first feeding (FF) is illustrated. The scoring system is based on the quantitative assessment of numerous developmental features with a maximum score of 1000 following the transition to exogenous feeding. From Killeen (1999b) and Kileen (1999).

showed different responses to temperature, providing evidence for local adaptation between populations. Alevins from the different temperature groups both hatched with 19 caudal fin rays and were probably at a comparable developmental stage. Although at hatch there was around 20% more red muscle at the level of

6. GENETIC AND ENVIRONMENTAL DETERMINANTS

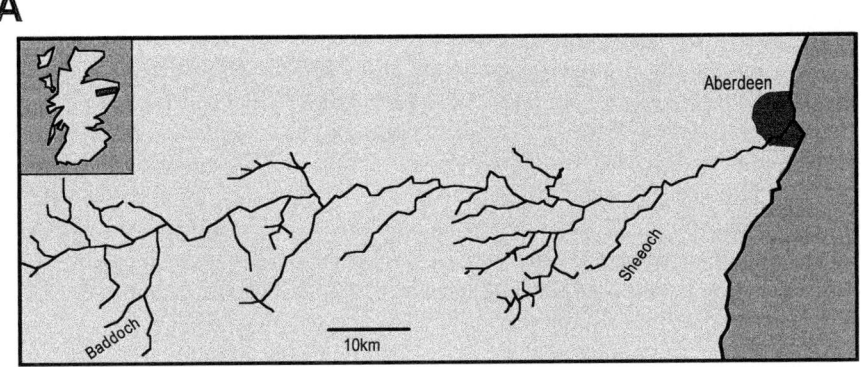

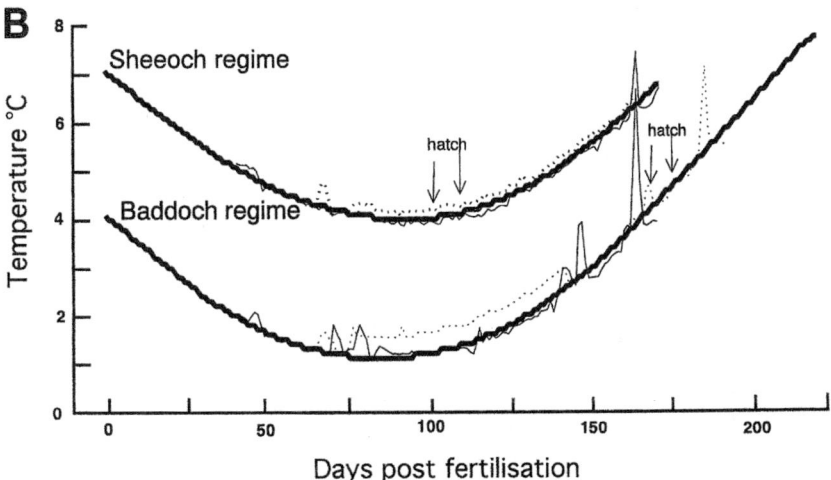

Fig. 12. (A) The River Dee system in Aberdeenshire, Scotland, illustrating the location of the lowland (Sheeoch) and upland (Baddoch) tributaries studied. (B) shows temperature data obtained from thermocouples and data loggers placed in salmon reeds on the Sheeoch and Baddoch in 1994–1995 and 1995–1996. The dotted line shows the smoothed temperature profiles simulated in the hatchery and the dashed line the actual temperatures. From Johnston et al. (2000b).

the adipose fin, this difference had disappeared by first feeding (Johnston et al., 2000b). In contrast, there were major effects on the amount and cellularity of the white muscle in fish incubated under the two thermal regimes, and the response to temperature differed between the populations (Table I).

The cross-sectional area of white muscle was 10.4% greater in Sheeoch fish incubated at the Baddoch than Sheeoch temperature regimes until hatching. For the Baddoch source population, embryos incubated under the cooler Baddoch temperature regime had a 19.6% greater cross-sectional area of white muscle in caudal myotomes than embryos reared at Sheeoch temperatures. Interestingly, by first feeding the average difference in white muscle cross-sectional area had narrowed to 4.5% in the Sheeoch population, but had increased to 30.5% in the Baddoch population (Table I). The differences in muscle cross-sectional area between temperature groups were largely a function of differences in muscle fiber number with only minor, though statistically significant, differences in average fiber diameter (Table I). Thus at first feeding there were significantly more white muscle fibers in embryos reared under the cooler than warmer temperature regimes with the largest difference found for the Baddoch population (Table I). The number of white fibers recruited between hatch and first feeding was 2077 and 1882 per trunk cross-section for the Sheeoch population reared at Baddoch and Sheeoch temperatures, respectively. The corresponding numbers for the Baddoch source population were 2910 and 2069, respectively (Table I). Thus fish of Baddoch origin produced around 50% more fibers during the yolk-sac stage at their native thermal regime than did the other groups.

At hatch, there were 60 and 24% more nuclei in the white muscle of alevins incubated under the warmer than cooler temperature regimes for the Sheeoch and Baddoch populations, respectively, although the differences had disappeared by first feeding (Johnston et al., 2000b). In the Sheeoch population, there were around 20% more c-met expressing mononuclear cells in fish reared under the simulated thermal regime of the Sheeoch than the Baddoch streams (Table II). In contrast, the density of c-met^{+ve} cells was independent of egg incubation temperature in the Baddoch population (Table II). Since the number of satellite cells is thought to remain relatively constant throughout the juvenile and adult stages,

Table I.

Influence of the Tributary of Origin and Early Thermal Experience on the Amount and Cellularity of White Muscle from the Myotomes of Atlantic Salmon (*Salmo salar* L.)

Population		Sheeoch		Baddoch	
Thermal regime		Sheeoch	Baddoch	Sheeoch	Baddoch
Cross-sectional area	H	0.86 ± 0.049	0.95 ± 0.059	0.90 ± 0.048	1.07 ± 0.031
muscle (mm²)	FF	2.74 ± 0.11	2.87 ± 0.14	3.05 ± 0.10	3.98 ± 0.11
Number of fibers	H	4110 ± 102	4752 ± 96	4296 ± 103	4526 ± 79
per myotome	FF	6187 ± 97	6634 ± 105	6336 ± 74	7436 ± 114
Average fiber	H	16.38 ± 0.29	14.85 ± 0.41	16.5 ± 0.23	16.39 ± 0.22
diameter (μm²)	FF	20.95 ± 0.27	20.37 ± 0.38	21.68 ± 0.3	22.97 ± 0.22
Nuclear	H	429884 ± 14445	269014 ± 12478	311070 ± 8921	251880 ± 5405
density (mm^{-3})	FF	167804 ± 2839	164146 ± 6467	153466 ± 3642	141600 ± 1947

Note: Values represent mean ± SE of 40 fish at hatch (H) and 112 fish at first feeding (FF), representing 6 families per population.

From Johnston *et al.* (2000b).

6. GENETIC AND ENVIRONMENTAL DETERMINANTS

Table II.
The Density of Immunopositive Mononuclear Cells (mm^{-3} muscle) in the White Muscle of Atlantic Salmon (*Salmo salar*) of Sheeoch and Baddoch Origin Reared to First Feeding (FF) Under the Simulated Temperature Regime of the Sheeoch (ST) and Baddoch (ST) Tributaries

Antigen	Stage	Sheeoch population		Baddoch population	
		ST	BT	ST	BT
c-met	FF	22,003 ± 740	22,924 ± 1535	29,571 ± 3118	34,103 ± 1668
	FF + 24 weeks	11,157 ± 570	9,108 ± 612	nd[a]	nd
MRFs	FF	24,909 ± 873	20,214 ± 1667	24,223 ± 3579	29,378 ± 2190
	FF + 24 weeks	8,925 ± 544	8,527 ± 289	nd	nd

Note: At first feeding all fish were transferred to constant environmental conditions (12–14°C; 16 hr light: 8 hr dark) for 24 weeks.
[a]nd = not determined.
From Johnston *et al.* (2000).

changes in satellite cell number could potentially provide a mechanism whereby early thermal experience could produce persistent effects on muscle growth patterns. In this case early thermal experience would be expected to affect subsequent muscle fiber recruitment in the Sheeoch but not the Baddoch population.

In order to test this hypothesis fish from the Baddoch and Sheeoch source populations that had been reared under the two thermal regimes to first feeding were subsequently transferred to duplicate tanks and reared at 12–15°C under a constant photoperiodic regime (16 hr light: 8 hr dark) for 40 weeks (Johnston *et al.*, 2000c). After 24 weeks the groups became bimodal in length and only those fish from the upper mode destined to smolt in the first year were studied. The density of satellite cells was determined after 24 weeks in upper mode fish from the Sheeoch. The density of c-met^{+ve} cells per unit volume of muscle was lower than that at first feeding but showed the same pattern with respect to the initial thermal regime (Table II).

As predicted, the pattern of muscle growth varied with egg incubation temperature in fish from the Sheeoch (Fig. 13A), but not the Baddoch source populations (Fig. 13B). Fish from the Sheeoch initially reared under the warmer temperature regime recruited significantly more muscle fibers to reach a given cross-sectional area than fish reared under the cooler Baddoch regime (Fig. 13A). The average of the 5th, 10th, 50th, and 95th percentiles of fiber diameter were correspondingly significantly greater in fish initially reared in the cooler than the warmer regimes (Table III). These results show that early thermal experience has a persistent effect on the number and size distribution of muscle fibers throughout growth in freshwater, but only for the Sheeoch population. A high proportion of small diameter fibers is associated with a fast growth rate (Weatherley *et al.*, 1979; 1988; Higgins and Thorpe, 1990). Since the maximum diameter of fibers is fixed at around 240 μm, the presence of a greater number of muscle fibers could also result in a larger ultimate body size, if maintained throughout the life cycle.

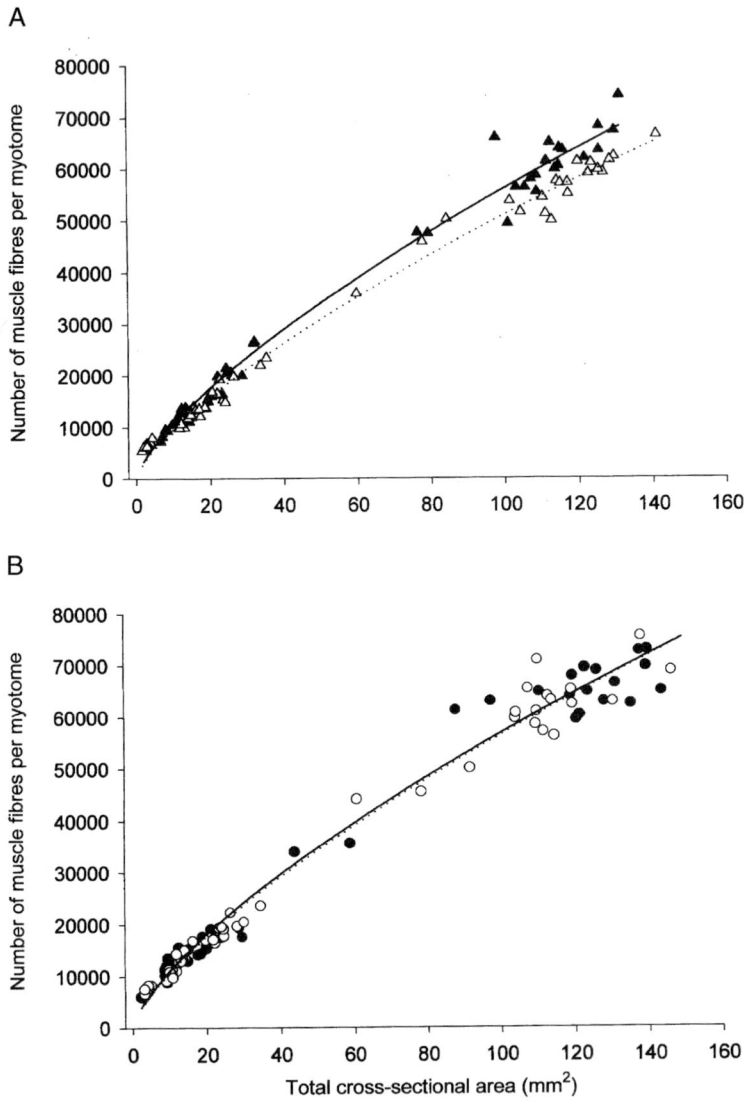

Fig. 13. The relationship between the number of white muscle fibers and the cross-sectional area of white muscle cross-sectional area (mm²) at the level of the adipose fin in the progeny of Atlantic salmon (*Salmo salar* L.) originating from (A) the Sheeoch (triangles) and (B) the Baddoch (circles) tributaries of the River Dee system, Aberdeenshire, Scotland, and grown at constant temperature and photoperiod (see text). Prior to first feeding the fish had been incubated under the simulated natural temperature regimes of the Baddoch (open symbols) and Sheeoch (closed symbols) tributaries (see text for details). Each point represents an individual fish. The data was fitted to a power equation of the form: fiber number = a × muscle cross-sectional area^b. From Johnston *et al.* (2000c).

Table III.
Comparison of the Percentiles for the Mean Probability Density Functions
of Muscle Fiber Diameter in Sheeoch and Baddoch Atlantic Salmon (*Salmo salmo* L.)

	Muscle fiber diameter			
	Sheeoch salmon		Baddoch salmon	
Percentile	Native thermal regime	Baddoch thermal regime	Native thermal regime	Sheeoch thermal regime
5	10.5 ± 0.3	12.2 ± 0.3***	10.4 ± 0.4	10.1 ± 0.3^{NS}
10	14.5 ± 0.4	16.2 ± 0.4**	14.3 ± 0.4	14.0 ± 0.4^{NS}
50	40.2 ± 0.5	42.4 ± 0.4**	40.6 ± 0.5	40.7 ± 0.6^{NS}
95	91.7 ± 0.7	95.7 ± 0.6***	92.6 ± 0.9	92.3 ± 1.0^{NS}
99	115.8 ± 1.3	123.2 ± 1.3***	117.2 ± 1.4	116.4 ± 1.8^{NS}

Note: Eggs were incubated under native and non-native thermal regimes until hatch and the fish transferred to constant conditions at first feeding. Statistical significance between thermal regimes at the $p < 0.01$(**) and $p < 0.001$(***) levels. NS = not significantly different $p > 0.05$: Mann-Whitney Rank Sum-test. Values represent mean ± SE of the percentile values of the probability density estimate of muscle fiber diameter from 16 fish per population/thermal regime. From Johnston *et al.* (2000).

The lack of sensitivity of satellite cell numbers to egg incubation temperature in the Baddoch population may reflect the presence of additional regulatory mechanism(s) that serve to buffer against temperature variation. Our results add to the growing but largely circumstantial evidence for local adaptation between salmon populations. The greater rate of muscle growth in the offspring of Baddoch fish reared at the temperature of their natal stream may confer some fitness advantage in terms of swimming performance and the ability to avoid predators and/or defend territories. Spring-running Atlantic salmon from the Aberdeenshire Dee spawn earlier than fish from lower down in the catchment, and it has been suggested that this helps to compensate for the longer development time at the low temperatures and represents an adaptation to optimize fry emergence to coincide with favorable conditions for growth (Webb and McLay, 1996). However, in order to demonstrate that these responses to early development temperature represent adaptations to local environmental conditions it would be necessary demonstrate some competitive fitness advantage by releasing marked fish into the field.

Other experimental evidence for persistent effects of egg incubation temperature on muscle growth characteristics comes from work on Atlantic herring. Johnston (1993) incubated herring eggs from the spring spawning Clyde population to 5, 8, or 12°C and the number of mononuclear cells was quantified by electron microscopy in 1-day-old larvae. The number of these presumptive satellite cells per square millimeter white muscle cross-sectional area was found to be significantly higher at 8°C (1493 ± 150) than at either 5°C (478 ± 46) or 12°C (924 ± 104; mean ± SE, 5 fish/temperature) (Johnston, 1993). In subsequent experiments, eggs of herring from the same spawning population were incubated at either 5 or 8°C until hatch and maintained at these temperatures until first feeding (Johnston et al., 1998). The fish were then transferred to triplicate tanks and reared for 80 days at ambient temperature and fed to satiation. Fish initially reared at 8°C had a greater cross-sectional area of white muscle, a higher number of fibers, and a larger average fiber diameter, than fish initially reared at 5°C (Johnston et al., 1998), mirroring the previously reported differences in satellite cell density (Johnston, 1993). Atlantic herring populations spawn in almost every season of the year at sea temperatures ranging from 2–17°C. Differences in body mass and the number of white muscle fibers at sexual maturity are well documented for different herring stocks (Greer-Walker et al., 1972), although allozyme (Smith and Jamieson, 1986) and mitochondrial DNA (Jørstad et al., 1991) studies indicate relatively little genetic structuring of stocks except for some isolated Scandinavian fjordic populations. Indeed a lack of correspondence between genetic and morphological variability has been reported for different Atlantic herring populations (Ryman et al., 1984). The effects of egg incubation temperature on muscle stem cell numbers provide one plausible mechanism for producing relatively large phenotypic effects on growth characteristics. Development temperature is also known to influence a range of meristic characteristics including the number of vertebrae in a whole range of fish species (Tåning, 1952).

B. Influence of Oxygen Tension

In Atlantic salmon, there is evidence that the effects of temperature on embryonic myogenesis are influenced by the presence of the egg capsule (Matschak et al., 1995, 1997). It has been suggested that the primary effects of temperature are related to changes in oxygen diffusion across the chorion and perivitelline fluid layer (Matschak et al., 1995, 1997). To test this idea salmon embryos at Gorodilov stage 27 were incubated at either 5 or 10°C in water at 100% air saturation, 50% air saturation, or at 150% air saturation at constant CO_2 levels (Matschak et al., 1997). Other groups of embryos were removed from their egg capsules and incubated in saline solution under the same combinations of temperature and oxygen levels. All the embryos were sampled at stage 31, which is equivalent to hatch. Fiber number was found to be independent of temperature in the dechorinated embryos whereas at 10°C in the intact embryos, fiber number was lower at 10 than at 5°C (Matschak et al., 1997). Fiber number was also reduced at 5°C in low oxygen water in the intact eggs, consistent with a role for oxygen availability in influencing the formation of embryonic muscle fibers. However, an increase in oxygen to 150% saturation at 10°C in the intact eggs failed to produce an increase in fiber number and reducing the oxygen level to the dechorinated embryos was also without effect (Matschak et al., 1997). In contrast, to their previous studies, but in agreement with the findings of Johnston and McLay (1997), white muscle fibers from intact embryos reared at 5°C had a larger average cross-sectional area than those at 10°C. The average fiber cross-sectional area in intact and dechorinated groups was reduced in fish reared in low oxygen, but 150% oxygen saturation had no effect on fiber size in intact embryos (Matschak et al., 1997). Similar results were obtained with rainbow trout that also have large yolky eggs (Matschak et al., 1998). Although somewhat contradictory, these findings provide some experimental support for the idea that temperature effects on embryonic myogenesis are secondary to those of oxygen availability.

C. Implications for Fisheries Management and Conservation

In re-stocking programs, the eggs from wild caught fish are often reared in hatcheries at elevated temperatures compared to their native streams in order to shorten the development period. The findings that egg temperature can modify muscle growth characteristics and that the responses can vary between populations from different environments has important consequences for fisheries management and conservation. It suggests that caution should be exercised both in moving fish between different locations and in rearing embryos and fry at artificially high temperatures. Both practices may represent inefficient stocking strategies since the resulting fish may have reduced fitness relative to the natural populations.

VI. EXERCISE AS A STIMULUS FOR MUSCLE GROWTH

Forced exercise is known to be a powerful stimulus for the hypertrophy of both red and white muscle fibers in fish (Greer-Walker and Pull, 1973; Johnston and Moon, 1980a,b; Hinterleitner et al., 1992; Sänger, 1992). Totland et al. (1987) raised Atlantic salmon with a starting weight of 2 kg in a 20 m raceway for 8 months on a commercial scale. The mean speed of the exercised group was 20 cm s^{-1}, equivalent to an average swimming speed of 0.40–0.45 body lengths s^{-1} over the duration of the experiment, and a maximum speed of 0.8 body lengths s^{-1}. The reference groups were kept in standard sea cages at the same stocking density as fish in the raceway. At the end of the experiment the weight gain was 38% greater in the exercised group largely due to an increase in the mass of the swimming muscle; the mass of the internal organs was similar in both groups. The average cross-sectional area of white muscle fibers was 17% greater in the exercised than reference group largely due to a higher percentage of very large diameter fibers in the fish forced to swim. Exercise training increased glycogen storage levels in both red and white muscle fibers of the salmon (Totland et al., 1987). Endurance exercise training has been found to result in an increased aerobic capacity of the red muscle in some species such as the chub (*Leuciscus cephalus*) but not in others including brook trout (*Salvelinus fontinalis;* Johnston and Moon, 1980a) and Danube bleak (*Chalcalburnus Chalcoides mento;* Hinterleitner *et al.,* 1992).

VII. MECHANISMS UNDERLYING DIFFERENCES IN MUSCLE GROWTH PATTERNS

A. Number of Muscle Satellite Cells

Variation in the number of muscle satellite cells between populations is one possible explanation for differences in the relative importance of fiber recruitment to growth. Triploid Atlantic salmon have a lower density of satellite cells and recruit correspondingly fewer fibers to reach a given muscle cross-sectional area than diploid fish (Johnston *et al.,* 1999). Egg incubation temperature regimes that produced an increase in the density of muscle satellite cells also resulted in a greater contribution of fiber recruitment to muscle growth in Atlantic herring (Johnston, 1993; Johnston *et al.,* 1998) and Atlantic salmon (Johnston *et al.,* 2000c). Studies with other animals are also consistent with a relationship between fiber number and the number of satellite cells. In mammals, fiber numbers increase during the fetal period but remain constant after birth. Selection for high body mass in the mouse increased the numbers of satellite cells and muscle fibers present at birth, but not the ratio of satellite cells to myonuclei (Brown and Stickland, 1993, 1994).

B. The Regulation of Satellite Cell Behavior

The regulatory pathways controlling the behavior of satellite cells are known to be complex and there is evidence from *in vitro* studies with mammalian myoblast cell lines that different nuclear compartments are present in muscle fibers (Musarò *et al.,* 1999). Although the process of fiber recruitment is different in fish and mammals it would be surprising if the same regulatory molecules were not involved. A number of factors have been identified which influence the activation and proliferation of satellite cells in mammals and hence the supply of nuclei for fiber recruitment and hypertrophy, including IGF-1 (Florini *et al.,* 1991) and members of the fibroblast growth factor (FGF) family, FGF2 (Yablonka-Reuveni *et al.,* 1999), and FGF6 (Floss *et al.,* 1997). HGF/SF has been implicated in stimulating the first round of satellite cell proliferation in tissue culture studies with rat myocytes (Tatsumi *et al.,* 1998). HGF/SF is the putative receptor for c-met which is localized on the satellite cell membrane. Furthermore, HGF/SF factor has been shown to be present in mammalian muscles and to be released in response to crush

6. GENETIC AND ENVIRONMENTAL DETERMINANTS

injury, which suggests it may be the proximal signal for activating satellite cells *in vivo* in adult mammalian muscle (Tatsumi *et al.*, 1998). *In vitro* studies have implicated transforming growth factor-β (TGF-β) in inhibiting differentiation possibly due to a block in myogenic differentiation genes such as *myogenin* (Massague *et al.*, 1991). Yun *et al.* (1997) found that TGF-β1 inhibited both proliferation and differentiation of turkey satellite cells. In rainbow trout, growth hormone administration resulted in an increase in the percentage of small diameter muscle fibers relative to a control group, consistent with a role for this hormone in regulating recruitment either directly or indirectly through IGF-1 (Fauconneau *et al.*, 1997). The elucidation of the functions of the growth factors in fish muscle growth is only just beginning and will require a combination of tissue culture and *in vivo* studies and the development of suitable transgenic models as in birds and mammals.

C. Regulation of Muscle Fiber Hypertrophy

Insulin-like growth factors have been broadly implicated in regulating hypertrophic growth in mammals (Florini *et al.*, 1991). Induced expression of the non-circulating IGF-1 isoform in a post-mitotic rat myocyte line was shown to result in hypertrophy (Musarò and Rosenthal, 1999). There is evidence that IGF-1 mediated muscle fiber hypertrophy involves the activation of the calcineurin signal transduction pathway (Dunn *et al.*, 1999; Musarò *et al.*, 1999; Semsarian *et al.*, 1999). Calcineurin is a calcium-activated serum phosphatase comprising catalytic (CnA) and regulatory subunits (CnB) and the calcium sensor molecule, calmodulin. Transfected *IGF-1* gene expression in a post-mitotic cell line resulted in increased expression of CnA but not CnB transcripts (Musarò *et al.*, 1999). Cell cultures stably transfected with a calcium-independent form of CnA driven by a myogenin promoter had a pronounced hypertrophic phenotype. IGF-1 is thought to act in skeletal muscle by mobilizing intracellular calcium stores, which activate calcineurin, and induce the nuclear translocation of the transcription factor NF-ATc1 (nuclear factor of activated T cells) (Musarò *et al.*, 1999; Semsarian *et al.*, 1999). Three isoforms of NF-AT are present in mammalian muscle, and one of them, NF-ATc can preferentially translocate to a subset of nuclei within a single multinucleated myotube (Abbott *et al.*, 1998). IGF-1 and/or activated calcineurin have also been implicated in the expression of GATA-2, which also accumulates in a subset of myocyte nuclei (Musarò *et al.*, 1999). Although the details have not been established, CnA and GATA-2 expression may be involved in a signaling cascade associated with the upregulation of genes involved in muscle fiber hypertrophy (Musarò *et al.*, 1999; Semsarian *et al.*, 1999). The role of the calcineurin signaling pathway in stimulating the increase in fiber size observed with forced exercise in fish has yet to be investigated.

VIII. IMPLICATIONS FOR FLESH QUALITY IN FARMED FISH

A. Muscle Cellularity and Texture

Aquaculture provides an important alternative to the exploitation of wild stocks for the provision of fish muscle for human consumption. Fish muscle is low in saturated fats and is a valuable source of polyunsaturated fatty acids of the n-3 series (Henderson and Tocher, 1987). The fatty acid composition of the flesh and nutritional value of the product is largely determined by the fatty acid composition of the diet (Bell *et al.*, 1991). The amount and composition of lipid in the muscle can also potentially influence taste and texture (Sheehan *et al.*, 1996). The development of high fat feeds in the salmon farming industry has resulted in significantly enhanced growth rates. However, these growth advantages may have been at the expense of flesh quality, particularly for the smoked product.

Cultured fish tend to have softer flesh than their wild counterparts (Haard, 1992). Firm flesh is a valued sensory characteristic for consumers and also an important attribute for the mechanical processing of fillets by the food industry (Dunajski, 1979; Haard, 1992). The textural characteristics of the flesh are dependent on postmortem changes associated with the method of slaughter and storage (Fauconneau *et al.*, 1995). However, texture is also strongly influenced by the metabolic and structural characteristics of the muscle prior to slaughter. Potentially, the finding that muscle growth patterns can vary between different genetic strains and with diet and environmental factors has important consequences for the texture of the final product (Johnston, 1999).

Fish are harvested during the growth phase of the life cycle before sexual maturation and the muscle therefore comprises a wide range of diameters. Networks of collagen fibers surround individual muscle fibers and bundles of fibers (Love, 1988; Hallet and Bremner, 1988). The amount of connective tissue per unit volume would be expected to decrease as fiber diameter increases reflecting the lower surface to volume ratio. Fish muscle contains much lower concentrations of collagen than red meat and the collagen is much less cross-linked (Hallet and Bremner, 1988; Bracho and Haard, 1990). Collagen fibers are thought to contribute relatively little to the textural properties of the flesh after cooking (Hatae *et al.*, 1986). Several interspecific studies have found a negative correlation between the average muscle fiber diameter and "firmness" in cooked fish as assessed using both instrumental methods (Hatae *et al.*, 1990) and trained taste panels (Hurling *et al.*, 1996). In this case, the actomyosin component of the muscle fibers probably contributes most to the texture, and provides the main resistance to mastication. For smoked fish, the texture is a complex function of the structural characteristics of the muscle fibers and connective tissue and associated chemical changes during processing. Johnston *et al.* (2000d) found a positive correlation between the average muscle fiber density of the fillet (number of fibers per mm^2

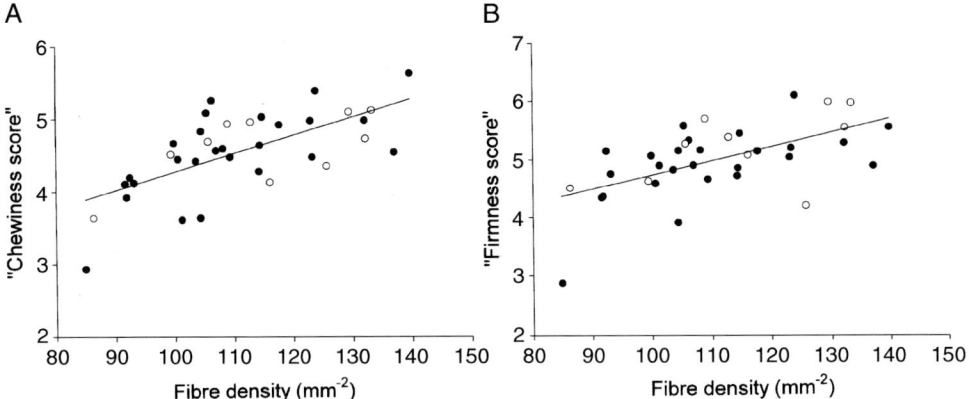

Fig. 14. The relationship between muscle fiber density (fibers mm^{-2} muscle cross-sectional area) in Atlantic salmon and textural characteristics measured by taste panels. (A) "chewiness," and (B) "firmness." Fish were from a predominantly early maturing population (closed circles) and a late maturing strain (open circles). From Johnston *et al.* (2000d).

muscle cross-sectional area) and the texture of the smoked product as assessed using trained taste panels. The strongest correlations were observed for "chewiness" and "firmness" scores explaining 40 and 30% of the variation in texture, respectively (Fig. 14A,B). It has been reported that sexual maturation may increase proteolytic enzyme activities and tenderize the flesh during storage and processing (Ando *et al.*, 1986). However, although the sensory evaluation of firmness in Atlantic salmon varied with muscle fiber density it was no higher in maturing than it was in immature fish (fig. 14A,B).

B. Color Visualization of the Fillet

In salmonid fish, a strong red color of the flesh is important for consumer acceptance. The color in wild salmonids comes from the absorption of oxygenated carotenoids from the diet and their deposition in the muscle tissue. In farmed Atlantic salmon, astaxanthin is added to the feed during seawater growth in order to achieve a final concentration in the flesh of 4–10 mg kg^{-1} wet mass (Torrissen *et al.*, 1989). Synthetic carotenoids pigments added to the feed constitute around 10% of total production costs (Torrissen *et al.*, 1995; Prendergast *et al.*, 1994). The deposition of astaxanthin in the muscle has a genetic component but also varies with age, growth rate, maturation, and season (Torrissen and Næval, 1988; Choubert *et al.*, 1997). Although muscle structure has little or no effect on pigment concentration it can influence the color visualization of the fillet (Johnston et al., 2000d). Flesh color in salmonid fish is influenced by lipid deposition and factors influencing the light scattering properties of the muscle (Torrissen *et al.*,

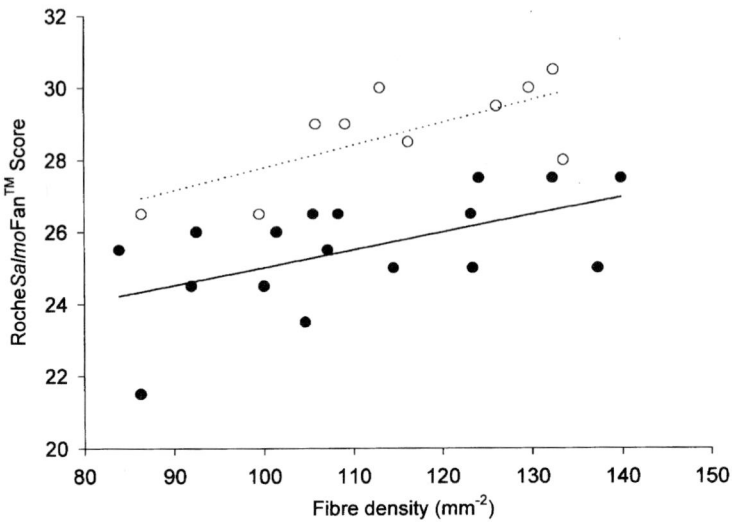

Fig. 15. The relationship between muscle fiber density (fibers mm^{-2} muscle cross-sectional area) and Roche *Salmo*Fan™ Score in Atlantic salmon (*Salmo salar* L.) from an early maturing (closed circles) and a late maturing strain (open circles). The lines were fitted by least-squares linear regression (RocheSalmoFan™ Score = a + b × fiber density: strain X: a = 20.10 ± 2.13, t = 9.39, p < 0.0001; b = 0.049 ± 0.019, t = 2.57, p < 0.05, n = 16; strain Y: a = 21.52 ± 2.669, t = 8.09, p < 0.0001; b = 0.063 ± 0.023, t = 2.73, p < 0.05, n = 9). From Johnston *et al.* (2000d).

1995). Color visualization for a given concentration of astaxanthin was found to be better in immature than maturing Atlantic salmon (Fig. 15). However, in both groups the color of the raw flesh as measured using the Roche SalmoFan™ was positively correlated with muscle fiber density, explaining 26% of the variation in grilse and 44% of the variation in immature fish (Fig. 15). There is evidence that the light scattering properties of beef and pork reside in the myofibrils or spaces between myofibrils (Offer *et al.,* 1989). One plausible explanation for the results in Fig. 15 is that muscle with a high density of muscle fibers scatters less of the incident light resulting in a deeper color. The Minolta color meter provides a measure of the reflectance (L), red color and yellow color of the fillet. Choubert *et al.* (1997) found that L values were higher in diploid than triploid Atlantic salmon for a given concentration of pigment. It is tempting to speculate that this was at least in part a function of the greater density of muscle fibers in diploid than triploid individuals (Johnston *et al.,* 1999).

C. Commercial Implications

Soft flesh and a pale uneven color are major contributions to downgrading losses in the salmon industry. Although soft flesh can easily result from incorrect slaughter and processing procedures, it is also strongly influenced by the number and size distributions of muscle fibers and associated connective tissue. The selection of strains and/or the adoption of production practices that are designed to maximize growth by fiber recruitment should result in flesh with a firmer texture and hence a higher value in the market place. However, before effective broodstock selection strategies can be devised to produce fish with a consistently firmer flesh it will be necessary to gain a deeper understanding of the cellular mechanisms underlying muscle fiber recruitment. However, one promising strategy might be to select for families with a high density of muscle satellite cells.

REFERENCES

Abbott, K. L., Friday, B. B., Thaloor, D., Murphy, T. J., and Pavlath, G. K. (1998). Activation and cellular localization of the cyclosporine A-sensitive transcription factor NF-AT in skeletal muscle cells. *Mol. Biol. Cell,* **9(10),** 2905–1916.

Alami-Durante, H., Fauconneau, B., Rouel, M., Escaffre, A. M., and Bergot, P. (1997). Growth and multiplication of white skeletal muscle fibres in carp larvae in relation to somatic growth rate. *J. Fish Biol.* **50,** 1285–1302.

Altringham, J. D., and Ellerby, D. J. (1999). Fish swimming: Patterns of muscle function. *J. Exp. Biol.* **202,** 3397–3403.

Altringham, J. D., and Johnston, I. A. (1988). The mechanical properties of polyneuronally myotomal muscle fibres isolated from a teleost fish (*Myoxocephalus scorpius*). *Pflügers Arch.* **412,** 524–529.

Ando, S., Hatano, M., and Zama, K. (1986). Protein degradation and protease activity of chum salmon (*Oncorhynchus keta*) during spawning migration. *Fish Physiol. Biochem.* **1,** 17–26.

Bell, J. G., McVicar, A. H., Park, M. T., and Sargent, J. R. (1991). High dietary linoleic acid affects the fatty acid composition of individual phospholipids from tissues of Atlantic salmon (*Salmo salar*) and cardiac lesion. *J. Nutr.* **121,** 1163–1172.

Blagden, C. S., Currie, P. D., Ingham, P. W., and Hughes, S. M. (1997). Notochord induction of zebrafish slow muscle mediated by Sonic hedgehog. *Genes Dev.* **11,** 2163–2175.

Blaxter, J. H. S. (1988). Pattern and variety in development. *In* "Fish Physiology, Vol. XIA" (Hoar, W. S., and Randall, D. J., eds.), pp. 1–58. Academic Press, New York.

Boddeke, R. E., Slijper, J., and van der Stelt, A. (1959). Histological characteristics of the body musculature of fishes in connection with their mode of life. *Koninkl. Ned. Akad. Weteenschap. Proc. Ser. C* **62,** 576–588.

Bone, Q. (1966). On the function of the two types of myotomal muscle fibre in Elasmobranch fish. *J. Mar. Biol. Assoc. U. K.* **46,** 321–349.

Bone, Q. (1978). Locomotor muscle. *In* "Fish Physiology Vol. VII," (Hoar, W. S., and Randall, D. J., eds.), pp. 361–424. Academic Press, New York.

Bowman, A. W., and Azzalini, A. (1997). "Applied Smoothing Techniques for Data Analysis. The Kernel Approach with S-Plus Illustrations," pp. 193. Oxford University Press, Oxford.

Bracho, G., and Haard, N. F. (1990). Determination of collagen crosslinks in rockfish skeletal muscle. *J. Food Biochem.* **14,** 435–451.

Brooks, S., and Johnston, I. A. (1993). Influence of development and rearing temperature on the distribution, ultrastructure and myosin sub-unit composition of myotomal muscle fibre types in the plaice *Pleuronectes platessa. Mar. Biol.* **117,** 501–513.

Brown, S. C., and Stickland, N. C. (1993). Satellite cell content in muscles of large and small mice. *J. Anat.* **183,** 91–96.

Brown, S. C., and Stickland, N. C. (1994). Muscle at birth in mice selected for large and small body size. *J. Anat.* **184,** 371–380.

Calvo, J. (1989). Sexual differences in the increase of white muscle fibres in Argentine hake, *Merluccius hubbsi,* from the San Matias Gulf (Argentina). *J. Fish Biol.* **35,** 207–214.

Carpenè, E., and Veggetti, A. (1981). Increase in muscle fibres in the *lateralis* muscle (white portion) of Mugilidae (Pisces, Teleostei). *Experientia* **37,** 191–194.

Carter, C. G., Mccarthy, I. D., Houlihan, D. F., Johnstone, R., Walsingham, M. V., and Mitchell, A. I. (1994). Food consumption, feeding behaviour, and growth of triploid and diploid Atlantic salmon, *Salmo salar* L., parr. *Can. J. Zool.* **72,** 609–617.

Cenciarelli, C., De Santa, F., Puri, P. L., Mattei, E., Ricci, L., Bucci, F., Felsani, A., and Caruso, M. (1999). Critical role played by Cyclin D3 in the MyoD-mediated arrest of cell cycle during myoblast differentiation. *Mol. Cell Biol.* **19,** 5203–5217.

Choubert, G., Blanc, J.-M., and Vallée, F. (1997). Colour measurement, using the CIELCH colour space, of muscle of rainbow trout, *Oncorhynchus mykiss* (Walbaum), fed astaxanthin: effects of family, ploidy, sex, and location of reading. *Aquaculture Res.* **28,** 15–22.

Cornelison, D. D. W., and Wold, B. J. (1997). Single-cell analysis of regulatory gene expression in quiescent and activated mouse skeletal muscle satellite cells. *Dev. Biol.* **191,** 270–283.

Davison, A. C., and Hinkley, D. V. (1997). "Bootstrap Methods and Their Applications." Cambridge University Press, Cambridge.

Delalande, J. M., and Rescan, P. Y. (1999). Differential expression of two nonallelic MyoD genes in developing and adult myotomal musculature of the trout (*Oncorhynchus mykiss*). *Dev. Genes Evol.* **209,** 432–437.

Devoto, S. H., Melancon, E., Eisen, J. S., and Westerfield, M. (1996). Identification of separate slow and fast muscle precursor cells *in vivo,* prior to somite formation. *Development* **122,** 3371–3380.

Dominov, J. A., Dunn, J. J., and Miller, J. B. (1998). Bcl-2 expression identifies an early stage of myogenesis and promotes clonal expansion of muscle cells. *J. Cell Biol.* **142,** 537–544.

Driedzic, W. R., and Hochachka, P. W. (1978). Metabolism of fish during exercise. In "Fish Physiology, Vol. VII" (Hoar, W. S., and Randall, D. J., eds.), pp. 503–543. Academic Press, New York.

Dunajski, E. (1979). Texture of fish muscle. *J. Texture Stud.* **10,** 301–318.

Dunn, S. E., Burns, J. L., and Michel, R. N. (1999). Calcineurin is required for skeletal muscle hypertrophy. *J. Biol. Chem.* **274,** 21908–21912.

Ennion, S., Gauvry, L., Butterworth, P., and Goldspink, G. (1995). Small-diameter white myotomal muscle fibres associated with growth hyperplasia in the carp (*Cyprinus carpio*) express a distinct myosin heavy chain gene. *J. Exp. Biol.* **198,** 1603–1611.

Fauconneau, B., Alami, Durante, H., Laroche, M., Marcel, J., and Vallot, D. (1995). Growth and meat quality relations in carp. *Aquaculture* **129,** 265–297.

Fauconneau, B., Andre, S., Chmaitilly, J., Le Bail, P.-Y., Krieg, F., and Kaushik, S. J. (1997). Control of skeletal muscle fibres and adipose cells in the flesh of rainbow trout. *J. Fish Biol.* **50,** 296–314.

Florini, J. R., Ewton, D. Z., and Magri, K. A. (1991). Hormones, growth factors, and myogenic differentiation. *Annu. Rev. Physiol.* **53,** 201–216.

Floss, T., Arnold, H.-H., and Braun, T. (1997). A role for FGF-6 in skeletal muscle regeneration. *Genes Dev.* **11,** 2040–2051.

Galloway, T. F., Kjørsvik, E., and Kryvi, H. (1998). Effect of temperature on viability and axial muscle development in embryos and yolk sac larvae of the Northeast arctic cod (*Gadus morhua*). *Mar. Biol.* **132,** 559–567.

Galloway, T. F., Kjørsvik, E., and Kryvi, H. (1999). Muscle growth and development in Atlantic cod larvae (*Gadus morhua* L.), related to different somatic growth rates. *J. Exp. Biol.* **202,** 2111–2120.

Gibson, S., and Johnston, I. A. (1995). Temperature and development in larvae of the turbot *Scophthalmus maximus*. *Mar. Biol.* **124,** 17–25.
Gorodilov, Y. N. (1983). Stadii embryonalnago razvitiya atlanticheskogo lososya (*Salmo salar* L.) II. Opisaniye I khronologiya. "Stages of embryonic development in Atlantic salmon, *Salmo salar* L. II. Description and chronology": Thorpe, J. E. *Scottish Fisheries Res. Transl.,* **19,** 1–34; GosNIORKh **200,** 107–126.
Greer-Walker, M. G. (1970). Growth and development of the skeletal muscle fibres of the cod (*Gadus morhua* L.). *J. Cons Perm. Int. Explor. Mer.* **33,** 228–244.
Greer-Walker, M. G., Bird, A. C., and Pull, G. A. (1972). The total number of white skeletal muscle fibres in cross section as a character for stock separation in North Sea herring (*Clupea harengus* L.). *J. Cons. Perm. Int. Explor. Mer.* **34,** 238–243.
Greer-Walker, M. G., and Pull, G. A. (1973). Skeletal muscle function and sustained swimming speeds in the coalfish (*Gadus virens* L.). *Comp. Biochem. Physiol. A,* **44,** 495–502.
Greenlee, A. R., Kersten, C. A., and Cloud, J. G. (1995). Effects of triploidy on rainbow trout myogenesis *in vitro*. *J. Fish Biol.* **46,** 381–388.
Hamor, T., and Garside, E. T. (1977). Size relations and yolk utilization in embryonated ova and alevins of Atlantic salmon *Salmo salar* L. in various combinations of temperature and dissolved oxygen. *Can. J. Zool.* **55,** 1892–1898.
Hanel, R., Karjalainen, J., and Wiesser, W. (1996). Growth of swimming muscles and its metabolic cost in larvae of whitefish at different temperatures. *J. Fish Biol.* **48,** 937–951.
Haard, N. F. (1992). Control of chemical composition and food quality attributes of cultured fish. *Food Res. Int.* **25,** 289–307.
Hallett, I. C., and Bremner, H. A. (1988). Fine structure of myocommata muscle fibre junction in Hoki (*Macruronus novaelandiae*). *J. Sci Food Agric.* **44,** 245–261.
Hatae, K., Tobimatsu, A., Takeyama, M., and Matsumoto, J. J. (1986). Contribution of the connective tissues on the texture differences of various fish species. *Bull. Jpn. Soc. Sci. Fish.* **52,** 2001–2008.
Hatae, K., Yoshimatsu, F., and Matsumoto, J. J. (1990). Role of muscle fibres in contributing firmness of cooked fish. *J. Food Sci.* **55,** 693–696.
Henderson, R. J., and Tocher, D. R. (1987). The lipid composition and biochemistry of freshwater fish. *Prog. Lipid Res.* **26,** 21–347.
Higgins, P. J., and Thorpe, J. E. (1990). Hyperplasia and hypertrophy in the growth of skeletal muscle in juvenile Atlantic salmon, *Salmo salar* L. *J. Fish Biol.* **37,** 505–519.
Hinterleitner, M., Huber, M., Lackner, R., and Weiser, W. (1992). Systemic and enzymatic responses to endurance training in two cyprinid species with different life styles (Teleostei: Cyprinidae). *Can J. Fish. Aquat. Sci.* **49,** 110–115.
Hurling, R., Rodell, J. B., and Hunt, H. D. (1996). Fibre diameter and fish texture. *J. Texture Stud.* **27,** 679–685.
Johnston, I. A. (1981). Structure and function of fish muscles. *In* "Vertebrate Locomotion" (Day, M. H., ed.). *Symp. Zool. Soc. Lond.* **48,** 71–113.
Johnston, I. A. (1993). Temperature influences muscle differentiation and the relative timing of organogenesis in herring (*Clupea harengus*) larvae. *Mar. Biol.* **116,** 363–379.
Johnston, I. A. (1999). Muscle development and growth: potential implications for flesh quality in fish. *Aquaculture* **177,** 99–115.
Johnston, I. A., Alderson, D., Sandeham, C., Mitchell, D., Selkirk, C., Dingwall, A., Nickell, D. C., Baker, R., Robertson, W., Whyte, D., and Springate, J. (2000a). Muscle fibre recruitment in early and late maturing strains of Atlantic salmon (*Salmo salar* L.). *Aquaculture* in press.
Johnston, I. A., McLay, H. A., Abercromby, M., and Robbins, D. (2000b). Phenotypic plasticity of early myogenesis and satellite cell numbers in Atlantic salmon spawning in upland and lowland tributaries of a river system. *J. Exp. Biol.* **203,** in press.
Johnston, I. A., McLay, H. A., Abercromby, M., and Robbins, D. (2000c). Egg incubation temperature

produces different effects on muscle fibre recruitment patterns in spring and autumn running Atlantic salmon populations. *J. Exp. Biol.* **203,** in press.

Johnston, I. A., Alderson, D., Sandham, C., Dingwall, A., Mitchell, D., Selkirk, C., Nickell, D., Baker, R., Robertson, B., Whyte, D., and Springate, J. (2000d). Muscle cellularity in relation to flesh quality in fresh and smoked Atlantic salmon (*Salmo salar* L.). *Aquaculture* in press.

Johnston, I. A., and Altringham, J. D. (1991). Movement in water: constraints and adaptations. In "Biochemistry and Molecular Biology of Fishes, Vol. 1" (Hochachka, P. W., and Mommsen, T. P., eds.), pp. 249–268. Elsevier, Amsterdam.

Johnston, I. A., Cole, N. J., Vieira, V. L. A., and Davidson, I. (1997). Temperature and developmental plasticity of muscle phenotype in herring larvae. *J. Exp. Biol.* **200,** 849–868.

Johnston, I. A., Cole, N. J., Abercromby, M., and Vieira, V. L. A. (1998). Embryonic temperature modulates muscle growth characteristics in larval and juvenile herring. *J. Exp. Biol.* **201,** 623–646.

Johnston, I. A., Davison, W., and Goldspink, G., (1977). Energy metabolism of carp swimming muscles. *J. Comp Physiol.* **114,** 203–216.

Johnston, I. A., and McLay, H. A. (1997). Temperature and family effects on muscle cellularity at hatch and first feeding in Atlantic salmon. *Can. J. Zool.* **75,** 64–74.

Johnston, I. A., and Moon, T. W. (1980a). Exercise training in skeletal muscle of the brook trout (*Salvelinus fontinalis*). *J. Exp. Biol.* **87,** 177–194.

Johnston, I. A., and Moon, T. W. (1980b). Endurance exercise training in the fast and slow muscles of a teleost fish (*Pollachius virens*). *J. Comp. Physiol. (B)* **135,** 147–156.

Johnston, I. A., Patterson, S., Ward, P. S., and Goldspink, G. (1974). The histochemical demonstration of myofibrillar adenosine triphosphatase activity in fish muscle. *Can. J. Zool.* **52,** 871–877.

Johnston, I. A., Patterson, S., and Goldspink, G. (1975). Studies on the swimming musculature of the rainbow trout. I. Fibre types. *J. Fish Biol.* **7,** 451–458.

Johnston, I. A., Strugnell, G., McCracken, M. L., and Jonstone, R. (1999). Muscle growth and development in normal-sex-ratio and all-female diploid and triploid Atlantic salmon. *J. Exp. Biol.* **202,** 1991–2016.

Johnston, I. A., Vieira, V. L. A., and Abercromby, M. (1995). Temperature and myogenesis in embryos of Atlantic herring *Clupea harengus*. *J. Exp. Biol.* **198,** 1389–1403.

Jolly, V. H. (1967). Observations on the smelt *Retropinna lacustris*. (Stokell). *N. Z. J. Sci.* **10,** 330–355.

Jørstad, K. E., King, D. P. F., and Nævdal, G. (1991). Population structure of Atlantic herring. *Clupea harengus* L. *J. Fish Biol.* **39** (Supplement A), 43–52.

Kiessling, A., Storebakken, T., Åsgård, T., and Kiessling, K.-H., (1991). Changes in the structure and function of the epaxial muscle of rainbow trout (*Oncorhynchus mykiss*) in relation to ration and age I. Growth dynamics. *Aquaculture* **93,** 335–355.

Killeen, J. R. (1999) The effects of temperature on development and growth in muscle of the trout (*Salmo trutta* L.) Ph.D. Thesis, University of St. Andrews, Fife, Scotland, pp. 240.

Killeen, J. R., McLay, H. A., and Johnston, I. A. (1999a). Temperature and neuromuscular development in embryos of the trout (*Salmo trutta* L.). *Comp. Biochem. Physiol. (A)* **122,** 53–64.

Killeen, J. R., McLay, H. A., and Johnston, I. A. (1999b). Development in *Salmo trutta* at different temperatures, with a quantitative scoring method for intraspecific comparisons. *J. Fish Biol.* **55,** 382–404.

Koumans, J. T. M., Akster, H. A., Brooms, G. H. R., Lemmens, C. J. J., and Osse, J. W. M., (1991). Numbers of myosatellite cells in white axial muscle of growing fish. *Cyprinus carpio* L. (Teleostei). *Am. J. Anat.* **192,** 418–424.

Koumans, J. T. M., Akster, H. A., Brooms, G. H. R., Lemmens, C. J. J., and Osse, J. W. M., (1994). Numbers of myonuclei and of myosatellite cells in red and white axial muscle during growth of carp *Cyprinus carpio* L. (Teleostei). *J. Fish Biol.* **44,** 391–408.

Koumans, J. T. M., and Akster, H. A. (1995). Myogenic cells in development and growth of fish. *Comp. Biochem. Physiol.* **110A**, 3–20.
Love, R. M. (1988). "The Food Fishes: Their Intrinsic Variation and Practical Implications" pp. 276. Farrand Press, London.
Mascarello, F., Rowlerson, A., Radaelli, G., Scaolo, P.-A., and Veggetti, A. (1995). Differentiation and growth of muscle in the fish *Sparus aurata* (L.). I. Myosin expression and organisation of fibre types in lateral muscle from hatching to adult. *J. Muscle Res. Cell Motil.* **16**, 213–222.
Massague, J., Heino, J., and Laiho, M. (1991). Mechanisms in TGT-beta action. *Ciba Found. Symp.* **157**, 51–59.
Musarò, A., and Rosenthal, N. (1999). Maturation of the myogenic program is induced by post-mitotic expression of IGF-1. *Mol. Cell Biol.* **19**, 3115–3124.
Musarò, A., McCullagh, K. J. A., Naya, F. J., Olson, E. N., and Rosenthal, N. (1999). IGF-1 induces skeletal myocyte hypertrophy through calcineurin in association with GATA-2 and NF-Atcl. *Nature* **400**, 581–585.
Martinez, I., and Christiansen, J. S. (1994). Myofibrillar proteins in developing white muscle of the Arctic charr, *Salvelinus alpinus* (L.). *Comp. Biochem. Physiol.* **107B**, 11–20.
Matschak, T. W., Stickland, N. C., Crook, A. R., and Hopcroft, T. (1995). Is physiological hypoxia the driving force behind temperature effects on muscle development in embryonic Atlantic salmon (*Salmo salar* L.)? *Differentiation* **59**, 71–77.
Matschak, T. W., Stickland, N. C., Mason, P. S., and Crook, A. R. (1997). Oxygen availability and temperature affect embryonic muscle development in Atlantic salmon (*Salmo salar* L.). *Differentiation* **61**, 229–235.
Matschak, T. W., Hopcroft, T., Mason, P. S., Crook, A. R., and Stickland, N. C., (1998). Temperature and oxygen tension influence the development of muscle cellularity in embryonic rainbow trout. *J. Fish Biol.* **53**, 51–590.
Matsuoka, M., and Iwai, T. (1984). Development of the myotomal musculature in the Red Sea bream. *Bull. Jpn. Soc. Sci. Fish.* **50(1)**, 29–35.
Mauro, A. (1961). Satellite cells of skeletal muscle fibers. *J. Biophys. Biochem. Cytol.* **9**, 493–495.
McLay, H. A., Johnston, I. A., Webb, J. H., and Robbins, D. (2000). The effects of temperature and egg size on the early development of spring and autumn running Atlantic salmon (*Salmo salar* L.) populations. *Can J. Aquat. Fish. Sci.* in press.
Metcalfe, N. B., Huntingford, F. A., and Thorpe, J. E. (1988). Feeding intensity, growth rates and the establishment of life history patterns in juvenile Atlantic salmon. *J. Anim. Ecol.* **57**, 463–474.
Meyer-Rochow, V. B., and Ingram, J. R. (1993). Red-white muscle distribution and fibre growth dynamics: a comparison between lacustrine and riverine populations of the Southern smelt *Retropinna retropinna* Richardson. *Proc. R. Soc. Lond. B.* **25**, 85–92.
Nag, A. (1972). Ultrastructure and adenosine triphosphatase activity of red and white muscle fibres of the caudal region of a fish, *Salmo gairdneri*. *J. Cell Biol.* **55**, 42–57.
Nag, A. C., and Nursall, J. R. (1972). Histogenesis of white and red muscle fibres of trunk muscles of a fish, *Salmo gairdneri*. *Cytobios* **6**, 227–246.
Nathanailides, C., Lopez-Albors, O., and Stickland, N. C. (1995). Influence of prehatch temperature on the development of muscle cellularity in posthatch Atlantic salmon (*Salmo salar*). *Can. J. Fish. Aquat. Sci.* **52**, 675–680.
Nicieza, A. G., Reyes-Gavián F. G., and Braña, F. (1994). Differentiation in juvenile growth and bimodality patterns between northern and southern populations of Atlantic salmon (*Salmo salar* L.). *Can J. Zool.* **72**, 1603–1610.
Offer, G., Knight, P., Jeacocke, R., Almond, R., Cousins, T., Elsey, J., Parsons, N., Sharp, A., Starr, R., and Purslow, P. (1989). The structural basis of the water-holding, appearance and toughness of meat and meat products. *Food Microstruc.* **8**, 151–170.
Podhorska-Okolow, M., Sandri, M., Brun, B., Rossini, K., and Carraro, U. (1998). Apoptosis of myo-

fibres and satellite cells: exercise-induced damage in skeletal muscle of the mouse. *Neuropathol. Appl. Neurobiol.* **24,** 518–531.
Prendergast, A. F., Higgs, D. A., Beames, D. M., Dosanjh, B., and Deacon, G. (1994). Searching for substitutes: canola. *North. Aquaculture* **10,** 15–19.
Proctor, C., Mosse, P. R. L., and Hudson, R. C. L. (1980). A histochemical and ultrastructural study of the development of the propulsive musculature of the brown trout, *Salmo trutta* L., in relation to its swimming behaviour. *J. Fish Biol.* **30,** 161–172.
Quinn, L. S., Norwood, T. H., and Nameroff, M. (1988). Myogenic stem cell commitment probability remains constant as a function of organismal and mitotic age. *J. Cell Physiol.* **134,** 324–336.
Ramírez-Zarzosa, G., Gil, F., Vázquez, A., Arenciba, A., Latorre, R., López-Albors, O., Ortega, A., and Moreno, F. (1998). The post-larval development of lateral musculature in gilthead sea bream *Sparus aurata* (L.) and Sea Bass *Dicentrarchus labrax* (L.). *Anat. Histol. Embryol.* **27,** 21–29.
Rescan, P. Y., and Gauvry, L. (1996). Genome of the rainbow trout (*Oncorhynchus mykiss*) encodes two distinct muscle regulatory factors with homology to MyoD. *Comp. Biochem. Physiol. B* **113,** 711–715.
Riddell, B. E., and Leggett, W. C. (1981). Evidence of an adaptive basis for geographic variation of body morphology, and time of downstream migration of juvenile Atlantic salmon (*Salmo salar*). *Can. J. Fish. Aquat. Sci.* **38,** 308–320.
Rowlerson, A., Mascarello, F., Radaelli, G., and Veggetti, A. (1995). Differentiation and growth of muscle in the fish *Sparus aurata* (L): II. Hyperplastic and hypertrophic growth of lateral muscle from hatching to adult. *J. Muscle Res. Cell Motil.* **16,** 223–236.
Rudnicki, M. A., and Jaenisch, R. (1995). The MyoD family of transcription factors and skeletal muscle myogenesis. *Bioessays* **17,** 203–209.
Ryman, N. U., Lagercrantz, L., Anderson, R., Chakraborky, R., and Rosenberg, R. (1984). Lack of correspondence between genetic and morphological variability patterns in Atlantic herring (*Clupea harengus*). *Heredity* **53,** 687–704.
Sänger, A. M. (1992). Effects of training on axial muscle of two cyprinid species: *Chondrostoma nasus* (L.) and *Leuciscus cephalus* (L.). *J. Fish Biol.* **40,** 637–646.
Scapolo, P. A., and Rowlerson, A. (1987). Pink lateral muscle in the carp (*Cyprinus carpio* L.): histochemical properties and myosin composition. *Experentia* **43,** 384–386.
Schultz, E. (1996). Satellite cell proliferative compartments in growing skeletal muscles. *Dev. Biol.* **175,** 84–94.
Semsarian, C., Wu, M.-J., Ju, Y.-K., Marciniec, T., Yeoh, T., Allen, D. G., Harvey, R. P., and Graham, R. M. (1999). Skeletal muscle hypertrophy is mediated by a Ca^{2+}-dependent calcineurin signalling pathway. *Nature* **400,** 576–581.
Sheehan, E. M., Connor, T. P. O., Sheehy, P. J. A., Buckley, D. J., and FitzGerald, R. (1996). Effect of dietary fat intake on the quality of raw and smoked salmon. *Irish J. Agric. Food Res.* **35,** 37–42.
Silverman, D. (1986). "Density Estimation for Statistics and Data Analysis" pp. 175. Chapman & Hall, New York.
Small, S. A., and Benfey, T. J. (1987). Cell size in triploid salmon. *J. Exp. Zool.* **241,** 339–342.
Smith, P., and Jamieson, A. (1986). Allozyme data and stock discreteness in herrings: A conceptual revolution. *Fish. Res.* **4,** 223–234.
Stickland, N. C. (1983). Growth and development of muscle fibres in the rainbow trout (*Salmo gairdneri*). *J. Anat.* **137,** 323–333.
Stickland, N. C., White, R. N., Mescall, P. E., Crook, A. R., and Thorpe, J. E. (1988). The effect of temperature on myogenesis in embryonic development of the Atlantic salmon (*Salmo salar* L.) *Anat. Embryol.* **178,** 253–257.
Stoiber, W., and Sänger, A. M. (1996). An electron microscope investigation into the possible source of new muscle fibres in teleost fish. *Anat. Embryol.* **194,** 569–579.
Stoiber, W., Haslett, J. R., Goldschmid, A., and Sänger, A. M. (1998). Patterns of superficial fibre

formation in the european pearlfish (*Rutilus frisii meidingeri*) provide a general template for slow muscle development in teleost fish. *Anat. Embryol.* (Berl) **197,** 485–496.
Suresh, A. V., and Sheehan, R. J. (1998). Muscle fibre growth dynamics in diploid and triploid rainbow trout. *J. Fish Biol.* **52,** 570–587.
Tåning, A. V. (1952). Experimental study of meristic characters in fishes. *Biol. Rev.* **27,** 169–193.
Tatsumi, R., Anderson, J. E., Nevoret, C. J., Harvey, O., and Allen, R. E. (1998). HGF/SF is present in normal adult skeletal muscle and is capable of activating satellite cells. *Dev. Biol.* **194,** 114–128.
Taylor, E. B. (1981). A review of local adaptation in Salmonidae, with special reference to Pacific and Atlantic salmon. *Aquaculture* **98,** 185–207.
Thorgaard, G. H. (1986). Ploidy manipulation and performance. *Aquaculture* **57,** 57–64.
Thorpe, J. E. (1987). *In* "Age and Growth in Fishes" (Summerfelt, R. C., and Hall, G. E., eds.), pp. 463–474. Iowa State University Press, Ames, IA.
Thorpe, J. E. (1989). Developmental variation in salmonid populations. *J. Fish Biol.* **35** (Supplement A), 295–303.
Thorpe, J. E., and Mitchell, K. A. (1981). Stocks of Atlantic salmon (*Salmo salar*) in Britain and Ireland: discreteness, and current management. *Can. J. Fish. Aquat. Sci.* **38,** 1576–1590.
Thorpe, J. E., Morgan, R. I. G., Talbot, C., and Miles, M. S. (1983). Inheritance of developmental rates in Atlantic salmon, *Salmo salar* L. *Aquaculture* **33,** 119–128.
Torrissen, O. J., Christiansen, R., Struksnæs, G., and Estermann, R. (1995). Astaxanthin deposition in the flesh of Atlantic salmon, *Salmo salar* L., in relation to dietary astaxanthin concentration and feeding period. *Aquaculture Nutr.* **1,** 77–84.
Torrissen, O. J., Hardy, R. W., and Shearer, K. D. (1989). Pigmentation of salmonids—carotenoid deposition and metabolism. *CRC Crit. Rev. Aquat. Sci.* **1,** 209–225.
Torrissen, O. J., and Nævdal, G. (1984). Pigmentation of salmonids—genetical variation in carotenoid deposition in rainbow trout. *Aquaculture* **38,** 59–66.
Torrissen, O. J., and Nævdal, G. (1988). Pigmentation of salmonids—variation in flesh carotenoids of Atlantic salmon. *Aquaculture* **68,** 305–310.
Totland, G. K., Kryvi, H., Jødestol, K. A., Christiansen, E. N., Tangerås, A., and Slinde, E. (1987). Growth and composition of the swimming muscle of adult Atlantic salmon (*Salmo salar* L) during long-term sustained swimming. *Aquaculture* **66,** 299–313.
Usher, M. L., Stickland, N. C., and Thorpe, J. E. (1994). The role of temperature in the development of muscle cellularity in Atlantic salmon (*Salmo salar* L.) embryos. *J. Fish Biol.* **44,** 953–964.
Valente, L. M. P., Gomes, E. F. S., and Fauconneau, B. (1998). Biochemical growth characterisation of fast and slow-growing rainbow trout strains: effect of cell proliferation and size. *Fish Physiol. Biochem.* **18,** 213–224.
Valente, L. M. P., Rocha, E., Gomes, E. F. S., Silva, M. W., Oliveira, M. H., Monteiro, R. A. F., and Fauconneau, B. (1999). Growth dynamics of white and red muscles in fast and slow growing strains of rainbow trout. *J. Fish Biol.* **55,** 675–691.
van Raamsdonk, W., van der stelt, A., Diegenbach, P. C., van den berg, W., De beryn, H., Van dijk, J., and Mijzen, P. (1974). Differentiation of the musculature of the teleost *Brachydanio rerio. Anat. Embryol.* **153,** 137–155.
van Raamsdonk, W., Mos, W., Smit-Onel, M. J., van der Laarse, W. J., and Fehres, R. (1983). The development of the spinal motor column in relation to the myotomal muscle fibres in the zebrafish (*Brachydanio rerio*). I. Posthatching development. *Anat. Embryol.* **167,** 125–139.
Verspoor, E. (1997). Genetic diversity among Atlantic salmon (*Salmo salar* L.) populations. *ICES J. Mar. Sci.* **54,** 965–973.
Veggetti, A., Mascarello, F., Scapolo, P. A., and Rowlerson, A. (1990). Hyperplastic and hypertrophic growth of lateral muscle in *Dicentrarchus labrax* (L.). An ultrastructural and morphometric study. *Anat. Embryol.* **182,** 1–10.

Vieira, V. L. A., and Johnston, I. A. (1992). Influence of temperature on muscle-fibre development in larvae of the herring *Clupea harengus. Mar. Biol.* **112**, 333–341.

Vieira, V. L. A., and Johnston, I. A. (1996). Muscle development in the tambaqui, an important Amazonian food fish. *J. Fish Biol.* **49**, 842–52.

Vieira, V. L. A., and Johnston, I. A. (1999). Temperature and neuromuscular development in the tambaqui, Colossoma macropomum. *J. Fish Biol.* **55** (Supplement A), 66–83.

Watabe, S. (2001). *In* "Muscle Development and Growth," Ch. 2. (Johnston, I., ed.), San Diego, California.

Weatherley, A. H., Gill, H. S., and Rogers, S. C. (1979). Growth dynamics of muscle fibres, dry weight, and condition in relation to somatic growth rate in yearling rainbow trout (*Salmo gairdneri*). *Can. J. Zool.*, **57**, 2385–2392.

Weatherley, A. H., and Gill, H. S. (1985). Dynamics of increase in muscle fibers in fishes in relation to size and growth. *Experentia* **41**, 353–354.

Weatherley, A. H., Gill, H. S., and Lobo, A. F. (1988). Recruitment and maximal diameter of axial muscle fibres in teleosts and their relationship to somatic growth and ultimate size. *J. Fish Biol.* **33**, 851–859.

Webb, J. H., and McLay, H. A. (1996). Variation in the time of spawning of Atlantic salmon (*Salmo salar*) and its relationship to temperature in the Aberdeenshire Dee, Scotland. *Can. J. Fish Aquat. Sci.* **53**, 2739–2744.

Weintraub, H., Genetta, T., and Kadesch, T. (1994). Tissue-specific gene activation by MyoD: determination of specificity by cis-acting repression elements. *Genes Dev.* **8**, 2203–2211.

Yablonka-Reuveni, Z., and Rivera, A. J. (1994). Temporal expression of regulatory and structural muscle proteins during myogenesis of satellite cells on isolated adult rat fibers. *Dev. Biol.* **164**, 588–603.

Yablonka-Reuveni, Z., Seger, R., and Rivera, A. J. (1999). Fibroblast growth factor promotes recruitment of skeletal muscle satellite cells in young and old rats. *J. Histochem. Cytochem.* **47**, 23–42.

Yang, Q., Basel-Duby, R., and Williams, R. S. (1997). Transient expression of a winged helix protein, MNF-β, during myogenesis. *Mol. Cell. Biol.* **17**, 5236–5243.

Youngson, A. F., Jordan, W. C., and Hay, D. W. 1994. Homing of Atlantic salmon (*Salmo salar* L.) to a tributary spawning stream in a major river catchment. *Aquaculture*, **121**, 259–267.

Yun, Y. D. C., McFarland, D. C., Pesall, J. E., Gilkerson, K. K., and Pesall, J. E. (1997). Variation in response to growth factor stimuli in satellite cell populations. *Comp. Biochem. Physiol.* **117A**, 463–470.

7

MUSCLE FIBER DIVERSITY AND PLASTICITY

A. M. SÄNGER
W. STOIBER

I. Introduction
II. Muscle Fiber Diversity
   A. Red Muscle Fibers
   B. White Muscle Fibers
   C. Intermediate or Pink Muscle Fibers
   D. So-Called Tonic Muscle Fibers
   E. Red Muscle Rim Fibers
   F. Muscle Fibers of the Transitional Zone
   G. Scattered Dorsal and Ventral Muscle Fibers
III. Plasticity of Muscle Phenotype
   A. Temperature and Season
   B. Oxygen Tension
   C. Exercise Training
IV. Summary
   Acknowledgments
   References

I. INTRODUCTION

The swimming muscle of teleost fish, comprising 60% of the total body mass, consists of a number of almost identical units, the myotomes. They are separated from each other by connective tissue sheets, the myosepta through which the force of contraction of myotomal muscle fibers is transmitted, via tendons, to the axial skeleton and caudal fin, resulting in undulation of the body and forward propulsion (Videler, 1993). Each myotome contains a superficial, wedge-shaped region lying directly beneath the lateral line, where the muscle fibers run parallel to the body axis, and a deeper part where the muscle fibers are arranged in a helical fashion, forming angles of up to 40° (Fig. 1). This typical orientation of muscle fibers is associated with the need for constant amounts of sarcomere shortening at

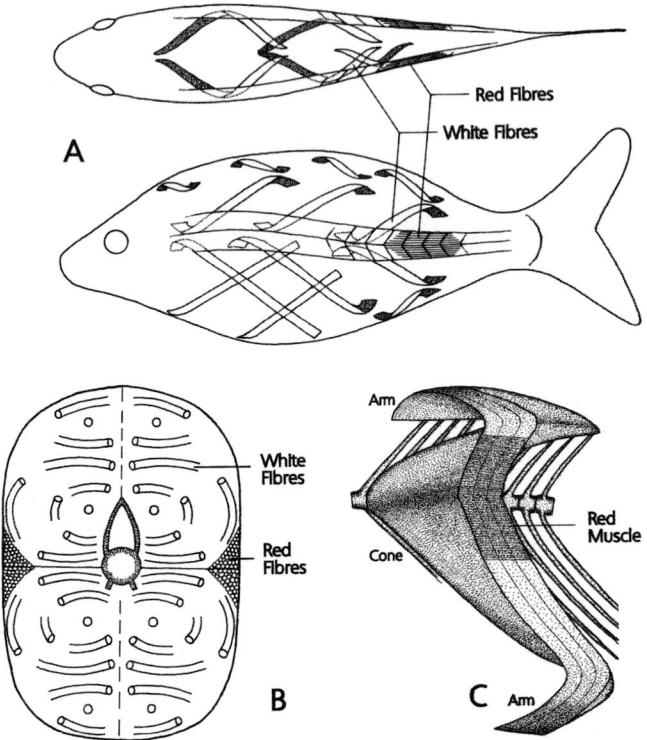

Fig. 1. Schematic diagrams of muscle fiber arrangement within the teleost myotome: (A) dorsal and lateral view, (B) cross-section. Red fibers are confined to a wedge-shaped superficial strip along the lateral line and run parallel to the longitudinal axis of the trunk. White fiber paths within successive myotomes follow helical trajectories. Modified after Alexander (1969) and Rome *et al.* (1988). (C) Lateral view of three mid-body myotomes of king salmon *Oncorhynchus tschawytscha*. W-shaped teleost muscle segments consist of a central cone and two arm regions. Modified after Greene and Greene (1913) and Winterbottom (1974).

different body flexures (Alexander, 1969; van Raamsdonk *et al.*, 1980; Rome and Sosnicki, 1990).

The locomotor muscles of all groups of fish are very highly specialized to meet the wide range of force production that is required from the muscular system both during sustained economical cruising and also in high velocity bursts. The solution adopted universally to these conflicting requirements has been to divide the locomotor system into several parts containing different muscle fiber types. These different muscle fiber types are designed to operate at different oscillatory frequencies (Rome *et al.*, 1988). The organization which is found in both elasmobranch and teleost fish is as follows: axial muscle consists mainly of fast-white

7. MUSCLE FIBER DIVERSITY AND PLASTICITY

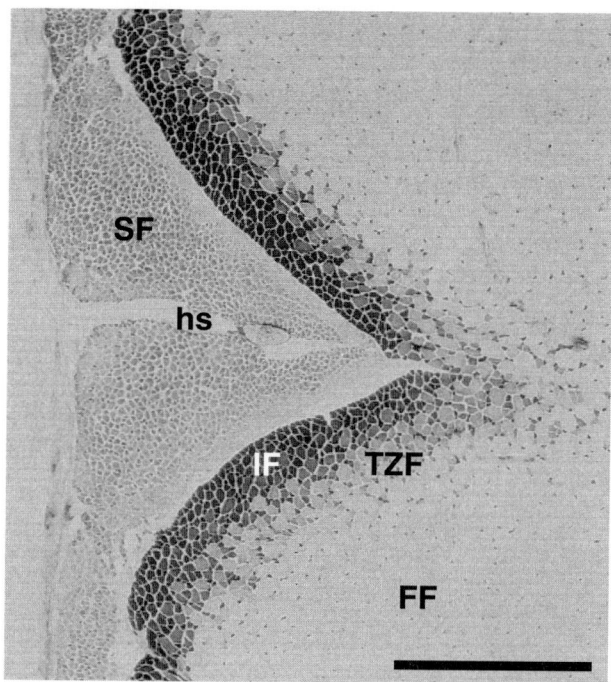

Fig. 2. Chub adult, age 1+. Transverse section at the horizontal septum (*hs*) stained for mATPase activity (acid preincubation pH 4.4). *SF*, slow (red) fibers; *IF*, intermediate fibers; *TZF*, transitional zone fibers; *FF*, fast (white) fibers. Lateral is to the left. Scale bar 1 mm.

fibers, covered by a thin layer of slow-red muscle fibers, and a layer of pink or intermediate muscle fibers in between them (Fig. 2). Muscle color is indicative of the degree of vascularization of each muscle type. Red muscle appears dark because of its high myoglobin contents, mitochondrial density, and degree of capillarization in contrast to the low values of these parameters in white muscle. Pink muscle possesses intermediate characteristics. The corresponding nomenclature of slow oxidative (slow aerobic), fast oxidative glycolytic (fast aerobic), and fast glycolytic fibers combines information on both contractile and metabolic properties. The use of other qualitative and quantitative analytical methods (e.g., histochemistry, immunocytochemistry, electronmicroscopy, morphometry, *in situ* hybridization) can further help to characterize and identify the mentioned muscle fiber types and to distinguish additional ones (Table I).

Fiber types as distinguished by histochemical staining for mATPase activity contain a characteristic isoform or mixture of isoforms of myosin which can be shown by immunostaining with isoform-specific antibodies (Billeter *et al.*,

Table I.
Histochemically Characterized Myotomal Muscle Fiber Types in Various Fish Species

Fish species	Fiber types	Author
Carassius auratus Tinca tinca	Red = slow Pink = fast aerobic White = fast glycolytic	Johnston and Lucking, 1978 Johnston and Bernard, 1982a
Cyprinus carpio	Red a Red b Small pink Large pink White	Akster, 1983
Esox lucius	Red Tonic Intermediate White	Zawadowska and Kilarski, 1984
Barbatula barbatula Various cyprinid species	Red Tonic Intermediate Transitional White	Kilarski and Kozlowska, 1985 Sänger et al., 1988, 1989
Brachydanio rerio	Red Red muscle rim Intermediate White Scattered dorsal Scattered ventral	Van Raamsdonk et al., 1980, 1982

1980, 1981; Pierobon Bormioli et al., 1980; Rowlerson et al., 1985; Scapolo and Rowlerson, 1987). In common with all other vertebrates, fish myosins have a subunit composition of two heavy chains and four light chains. The light chain patterns of myosins from fast contracting white fibers and from pink fibers (contracting at intermediate speeds) are identical, but different from those of slow contracting red fibers (Focant and Huriaux, 1976; Focant et al., 1976; Johnston et al., 1977). The differences in light chain electric charge and/or molecular weight provide a useful tool for the characterization of fish species-specific fiber types (Huriaux and Focant, 1985; Rowlerson et al., 1985; Martinez et al., 1989). In regard to myosin heavy chains (MHCs), the red fibers, the so-called tonic fibers, the red muscle rim fibers, the intermediate fibers, and the white fibers all contain distinct types. The fibers of the transitional zone and the scattered dorsal and ventral fibers contain mixtures of white and intermediate MHCs (van Raamsdonk et al., 1980; Karasinski et al., 1994). Further, during ontogeny there is a sequential expression of developmental-stage-specific myosin isoforms, and of other muscle proteins, resulting in related transformations of fiber type (Scapolo et al., 1988; Johnston, 1994).

A further muscle fiber type specific protein which also changes with development, is parvalbumin, a low molecular weight sarcoplasmic Ca^{2+}-binding protein. This protein has several isoforms, which vary in amount and distribution in a fiber type specific manner (Hamoir et al., 1972; Gerday, 1982). It has been shown in several species including rainbow trout (*Oncorhynchus mykiss* W.), brown trout (*Salmo trutta* L.), and barbel (*Barbus barbus* L.), that during fish development transitions occur in isotype expression with predominantly larval isoforms replaced by adult forms. In sea bass (*Dicentrarchus labrax* L.), however, the larval form remains the principal isotype in adult fish indicating that the polymorphism of parvalbumins in fish constitutes a subtle mechanism modulating the speed and power of muscle contraction. It has been suggested that each isotype plays a specific role in relation to the muscle activity required in fish at a given developmental stage or a given trunk level in the adult (Huriaux et al., 1996; Focant et al., 1999).

Fiber types also differ in their ultrastructural properties like Z-line thickness, volume density and type of mitochondria, volume density and distribution of sarcoplasmic reticulum (SR), and T-tubule system (Bone et al., 1986; Sänger, 1992a; Luther et al., 1995).

Cells of muscle tissues are extremely plastic and adaptable (Sänger, 1993). Unlike other tissues that respond to mechanical stimuli, muscle fibers create the mechanical stresses as well as respond to them, so that it is sometimes difficult to discern which is the cause or the effect when studying muscle adaptation (Goldspink, 1985). Thus the highly plastic nature of skeletal muscle tissue based on its heterogeneity is particularly demonstrated by its adaptability. Even in the adult animal, skeletal muscle fibers are not static structures but are capable of altering their contractile and metabolic properties by altering protein composition to adapt to altered functional demands, environmental changes, hormonal signals, and changes in neural input. Extreme examples of phenotypic plasticity in muscle include the transformation of muscle tissue to thermogenic organs (Carey, 1982; Block, 1986; Tullis and Block, 1997), electric organs (Wachtel, 1964; Patterson and Zakon, 1997; Unguez and Zakon, 1998), and bioluminescent tissue (Johnston and Herring, 1985).

The remainder of this chapter is used to further pursue many of the above topics in a manner that we hope is both constructive and a timely stimulation for new research. We use the characteristics of the various muscle fiber types and their development (Section II) and the plasticity of muscle phenotype in relation to environmental factors—temperature, oxygen concentration, etc. (Section III)—as the framework for our discussion. Attention is directed toward both the anatomical aspects of muscle fiber types and their functional significance.

II. MUSCLE FIBER DIVERSITY

A. Red Muscle Fibers

1. POSITION AND OCCURRENCE

Slow or red muscle fibers are commonly confined to a narrow superficial strip along the lateral line with a wedge-like insertion in the region of the horizontal septum (Fig. 2). Among all teleost species investigated, so far only the stickleback (*Gasterosteus aculeatus* L.) seems to lack true slow (red) fibers (Kilarski and Kozlowska, 1983; Te Kronnie *et al.*, 1983). Slow muscle fibers generally display a parallel alignment to the body axis suited to slow speed body movements (Rome, 1994).

Red muscle fibers, being small in diameter (25–45 μm), usually constitute less than 10% and never more than 30% of the myotomal musculature (Boddeke *et al.*, 1959; Greer-Walker and Pull, 1975). Similar to the 4.73% found in nase (*Chondrostoma nasus* L.) by our group, red muscle fibers in goldfish (*Carassius auratus* L.) constitute around 4–5% of the total cross-sectional area except in the last two myotomes adjacent to the tail where the proportion is about 15–17% (Johnston and Lucking, 1978; 11.5–15%, Gill *et al.*, 1989). Zhang *et al.* (1996), examining the distribution of red, pink, and white muscle along the length of scup (*Stenotomus chrysops*), report the largest cross-sectional area of red fibers to occur at 60% of total fish body length measured in the rostral to caudal direction. The proportion of the fishes' cross-section occupied by red fibers increases from 1.37–8.42% moving caudally along the length of the fish.

2. DEVELOPMENT AND GROWTH

Slow fiber formation in the teleost embryo is directed by molecular signals from the notochord (Du *et al.*, 1997). The myogenic cells within each somite are destined to have different fates. Slow (red) fiber precursors originate from a group of cells next to the notochord (adaxial cells). A subset of these cells (muscle pioneer cells) differentiate at the site of the prospective horizontal septum. Other slow fiber precursors migrate laterally to form a slow fiber monolayer at the surface of the myotome (Devoto *et al.*, 1996; Stoiber *et al.*, 1998), while the dorsal and ventral bulks of somitic cells develop into fast (white) fibers. Apart from the pre-differentiated muscle pioneers, slow fiber maturation in many teleosts (e.g., zebrafish, rainbow trout, Atlantic salmon, herring) is similar to that of the most developed fibers within fast muscle (Waterman, 1969; Nag and Nursall, 1972; Johnston, 1993a; Usher *et al.*, 1994), but may also be much faster, as observed in the curimata-pacu (*Prochilodus marggravii*; Brooks *et al.*, 1995) and the pearlfish (*Rutilus frisii meidingeri* Heckel; Stoiber *et al.*, 1998). By the time of hatching, the state of slow fiber differentiation varies considerably between

7. MUSCLE FIBER DIVERSITY AND PLASTICITY

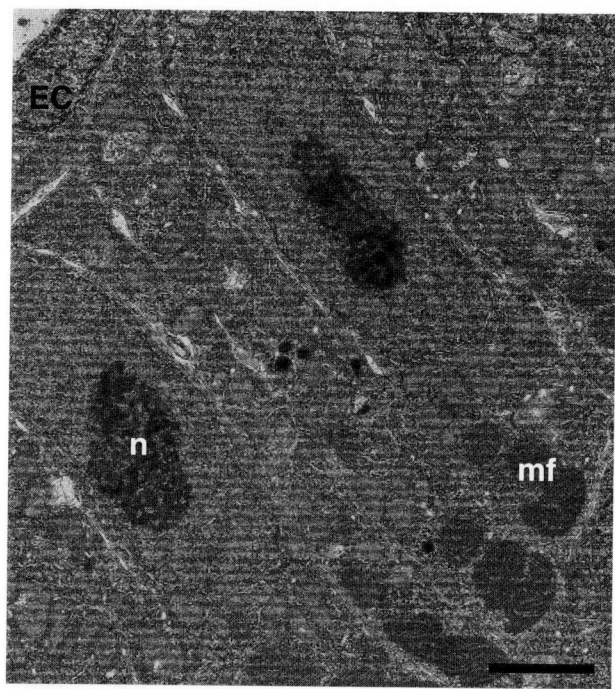

Fig. 3. 30-somite pearlfish embryo. Transverse section of slow fibers at the surface of a midbody myotome. Initial myofibrils (*mf*) run medially while nuclei (*n*) and most cytoplasm follow laterally. *EC* so-called external cells. Lateral is top left. Scale bar 2 μm.

species, from rather advanced, as in roach (*Rutilus rutilus* L.), bleak (*Alburnus alburnus* L.), chub (*Leuciscus cephalus* L.), pearlfish, lake whitefish (*Coregonus lavaretus* L.), turbot (*Scophthalmus maximus* L.), sea bass, and herring (*Clupea harengus* L.; El-Fiky *et al.*, 1987; Scapolo *et al.*, 1988; Calvo and Johnston, 1992; Vieira and Johnston, 1992, 1996; Johnston and Horne, 1994; Hanel *et al.*, 1996; Stoiber *et al.*, 1998) to more retarded, as in the red sea bream (*Pagrus major;* Matsuoka and Iwai, 1984). An exceptionally long delay in slow fiber differentiation has been reported for the northern anchovy (*Engraulis mordax;* O'Connell, 1981). In cyprinid embryos, slow fiber myofibrillogenesis begins medially with a fibril cluster while nuclei and cytoplasm with abundant mitochondria occupy the lateral domains of the cells (Fig. 3). The cluster is expanded into circular or garland-like myofibril chains until the animals start free swimming (Stoiber *et al.*, 1998) (Fig. 4). Concurrently, the tubules of the SR attain their adult shape with a multiple (often double row) arrangement in cross-sectional view (Fig. 4, inset) and multiple overlaps in frontal (*en face*) view (Stoiber, 1991, 1996; Luther *et al.*,

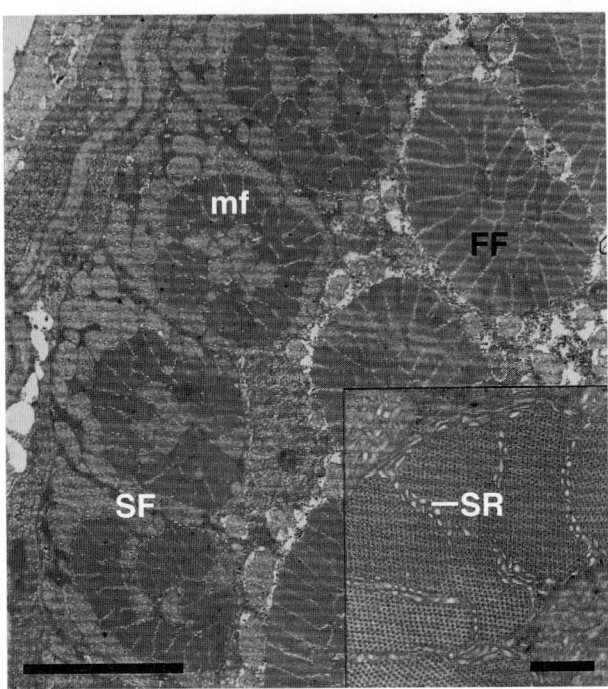

Fig. 4. Roach larva 2 days after the transition to free swimming, transverse section. Slow fiber (*SF*) myofibrils (*mf*) are joined to circular patterns. *FF*, fast fibers. Lateral is to the left. Scale bar 5 μm. Inset: Multiple arrangement of sarcoplasmic reticulum (*SR*) tubules between slow fiber myofibrils of a pearlfish larva. Bar 0.5 μm.

1995). Similar patterns of early myofibril formation have been found in other teleost taxa (e.g., Veggetti *et al.,* 1990, sea bass; Johnston, 1993a, herring; Brooks *et al.,* 1995, curimata-pacu; Stoiber, 1996, rainbow trout). Partitioning of intracellular space within young slow fibers may also appear less orderly, as shown for rainbow trout from a different source (Nag and Nursall, 1972) and for the gilthead sea bream (*Sparus aurata*) (Ramirez-Zarzosa *et al.,* 1995). From the late embryo stage, cyprinid slow fibers differ from their faster contracting counterparts by having higher relative volumes of mitochondria and broader Z-lines in their sarcomeres (Stoiber, 1991). The differentiation of the vascular system in the slow muscle (as in all other axial muscle) of cyprinid fry is free of capillaries until after the transition to free swimming, but develops a dense capillary network toward the end of the larval period (Stoiber, 1991, 1996) (for adult state see Fig. 5). In the zebrafish (*Brachydanio rerio* Hamilton Buchanan), the initial motor nerve contact with slow fibers is established medially with the muscle pioneer fiber subpopulation at the horizontal septum (Felsenfeld and Curry, 1988; Liu and Westerfield,

7. MUSCLE FIBER DIVERSITY AND PLASTICITY

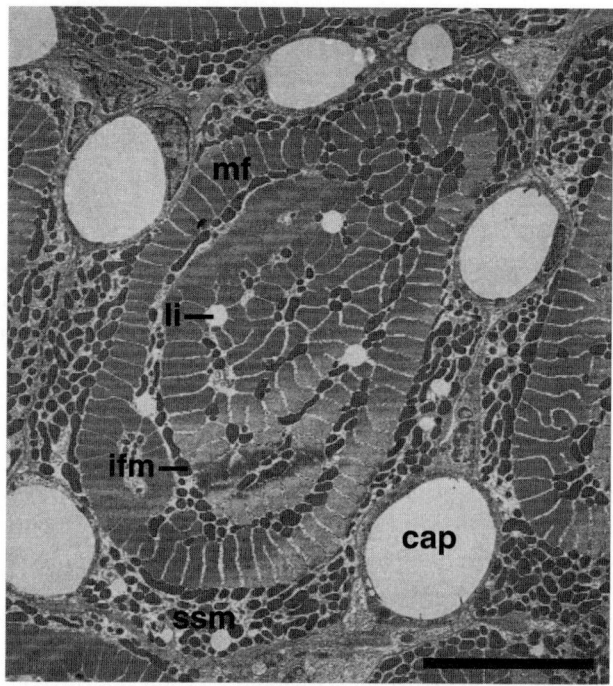

Fig. 5. Nase adult. Transverse section of a slow (red) fiber surrounded by capillaries (*cap*). *mf* myofibrils, *ssm* and *ifm* subsarcolemmal and intermyofibrillar mitochondria, respectively, *li*, lipid droplets. Scale bar 10 µm.

1990). In the pearlfish, the innervation of the lateral slow fibers has a focal pattern in the hatchlings, with neuromuscular terminations next to the fibers' myoseptal insertions. This scheme is maintained far into larval life and then becomes gradually expanded into the adult-type multiterminal pattern (Stoiber, 1996; Stoiber et al., 1998).

3. Histochemical Analysis and Ultrastructural Features

Histochemical analysis of larval and adult red muscle shows intense staining for succinic dehydrogenase (SDH) activity and an alkaline-labile mATPase activity (Johnston and Lucking, 1978; Akster, 1983; Sänger et al., 1988, 1989; Stoiber, 1996). The lability of mATPase activity to acid preincubation pH values is species dependent. Acid-stability is, for example, seen in some cyprinid fish (Fig. 2) and European eel (*Anguilla anguilla* L.) (Johnston and Bernard, 1982a; Rowlerson et al., 1985; Sänger et al., 1988), but acid-lability is seen in catfish (*Ictalurus melas*) and common carp (*Cyprinus carpio* L.) (Rowlerson *et al.*, 1985). The distinct phenotype of red fibers is demonstrated by the presence of two "slow" myosin light chains (Rowlerson *et al.*, 1985). Titin, also known as connectin, a striated muscle-specific protein spanning the distance between the Z- and M-lines of a sarcomere, has a larger molecular mass in red than in white muscle fibers. On the other hand, comparing fibers from anterior and posterior axial muscle, for both white and red fibers the molecular mass of titin in posterior muscle fibers is larger than in anterior ones demonstrating that the same fiber type can express different titin isoforms depending on its location along the body axis. As red fibers are exposed to larger changes in sarcomere strain during continuous swimming than white fibers, it is proposed, that sarcomere strain is one of the functional parameters that modulates the expression of different titin isoforms in axial muscle fibers (Spierts *et al.*, 1997).

The major ultrastructural features of this muscle type are a good capillary supply, high amount of subsarcolemmal and intermyofibrillar mitochondria of the lamellar type, and lipid droplets (Fig. 5). The relative amount of these component features differs with fish species (Table II, III, V) and also in response to environmental factors (see Section III). Concentrations of myoglobin and cytochromes are high. Red muscle fibers are rich in glycogen which is mainly found in close association with the sarcolemma but also between myofibrils. The intermyofibrillar aggregates are generally observed to comprise single, double, or even multiple chains of mitochondria interspersed with lipid droplets and glycogen granules. The myofibrils are approximately polygonal in shape with the exception of the peripheral ones which exhibit radial alignment (Fig. 5). As illustrated in the inset of Fig. 4, sometimes such peripheral myofibrils are partially separated by an ingrowing sarcotubular system. The SR is well developed but the relative length of the T-SR junction, in contrast to white muscle fibers, is small (Akster, 1985). The triad is situated at the level of the Z-line.

4. ENERGY METABOLISM

Energy metabolism in embryos and early larvae of fish is almost entirely aerobic. Mitochondrial density in the swimming muscles of a species of fish, from larva to adult, covers about the same range as mitochondrial density in the skeletal muscles of mammals. However, depending on species, the aerobic capacity (power density) of mitochondria may be one order of magnitude lower in fish than in mammals (Wieser, 1995; Moyes *et al.*, 1992; Johnston *et al.*, 1998). Mitochondria of red slow muscle are able to utilize both carbohydrates and lipids as fuels and possibly also certain amino acids (Jones and Sidell, 1982; van Waarde and De Wilde-van Berg Henegouwen, 1982; Moyes *et al.*, 1989). Carbohydrates play a minor role in sustained/prolonged swimming (Weber and Haman, 1996) with long term high level utilization of this fuel probably incompatible with its very

Table II.
Total Mitochondrial Volume Densities in Percentages in Red and White Muscle Fibers of Different Fish Species

Fish species	Red fibers	White fibers	Author
Leuciscus cephalus laboratory reared	21.3 ± 0.5	1.1 ± 0.1	Sänger, 1992b
Anguilla anguilla	21.7 ± 1.7	8.9 ± 0.8	Egginton and Johnston, 1982b
Aspius aspius	22.5 ± 1.9	1.7 ± 0.3	Sänger, 1992a
Tinca tinca	22.9 ± 1.1	4.5 ± 0.5	Johnston and Bernard, 1982
Salmo trutta	23.1 ± 1.6	2.8 ± 0.6	Davison, 1983
Rutilus rutilus	24.5 ± 1.4	1.4 ± 0.3	Sänger, 1992a
Pleuronectes platessa	24.6 ± 0.8	2.0 ± 0.3	Johnston and Moon, 1981
Cyprinus carpio	24.8 ± 1.6		Akster, 1985
Notothenia gibberifrons	24.9 ± 0.7	0.6 ± 0.3	Londraville and Sidell, 1990
Barbatula barbatula	25.0 ± 2.7	9.0 ± 0.4	Kilarski and Kozlowska, 1987
Carassius carassius	25.5 ± 1.0	4.6 ± 0.1	Johnston and Moon, 1981
Leuciscus cephalus wild population	26.7 ± 2.6	1.5 ± 0.2	Sänger, 1992a
Oncorhynchus mykiss winter-acclimatized	26.9 ± 0.9		St.-Pierre et al., 1998
Oncorhynchus mykiss summer-acclimatized	27.0 ± 1.0		St.-Pierre et al., 1998
Alburnus alburnus	27.9 ± 2.5		Sänger, 1992a
Chondrostoma nasus laboratory reared	28.3 ± 1.1	1.5 ± 0.1	Sänger, 1992b
Morone saxatilis acclimated to 25°C	28.6 ± 1.8	2.7 ± 0.3	Egginton and Sidell, 1989
Chondrostoma nasus wild population	29.9 ± 0.9	2.5 ± 0.2	Sänger, 1992a
Salvelinus fontinalis	31.3 ± 0.9	9.3 ± 0.7	Johnston and Moon, 1981
Chalcalburnus chalcoides mento	32.7 ± 1.3	2.8 ± 0.4	Sänger, 1992a
Trematomus newnesi	34.8 ± 1.2	1.4 ± 0.5	Londraville and Sidell, 1990
Morone saxatilis acclimated to 5°C	44.8 ± 2.4	4.0 ± 0.4	Egginton and Sidell, 1989
Engraulis encrasicolus	45.5 ± 1.1		Johnston, 1982c

Note: Values are given as mean ± SEM.

low availability in fish diets (Halver, 1972). By contrast, protein is commonly believed to be a major aerobic energy source for fish swimming, particulary during exercise (Brett and Groves, 1979; van Waarde, 1983; van den Thillart, 1986; Jobling, 1994; Weber and Haman, 1996). However, the work by Lauff and Wood (1996) and Alsop and Wood (1997) is not in support of an increasing importance of this fuel as sustained swimming speeds increase. Their results suggest that the reliance on protein as a fuel for red muscle is greatly dependent

Table III.
Lipid Volume Densities of Red Muscle Fibers in Percentages of Different Fish Species

Fish species	Red fibers	Author
Blicca bjoerkna	0.4 ± 0.1	Sänger, 1992a
Abramis ballerus	0.5 ± 0.1	Sänger, 1992a
Morone saxatilis acclimated to 25°C	0.6 ± 0.3	Egginton and Sidell, 1989
Notothenia gibberifrons	1.2 ± 0.4	Londraville and Sidell, 1990
Scardinius erythrophthalmus	1.6 ± 0.3	Sänger, 1992a
Chondrostoma nasus	1.7 ± 0.2	Sänger, 1992a
Aspius aspius	2.0 ± 0.5	Sänger, 1992a
Trematomus newnesi	2.6 ± 0.2	Londraville and Sidell, 1990
Alburnus alburnus	3.2 ± 0.9	Sänger, 1992a
Leuciscus cephalus	3.3 ± 0.4	Sänger, 1992a
Salmo trutta	3.5 ± 1.3	Davison, 1983
Rutilus rutilus	3.7 ± 0.7	Sänger, 1992a
Leuciscus leuciscus	5.5 ± 0.6	Sänger, 1992a
Oncorhynchus mykiss summer-acclimatized	7.5 ± 0.6	St.-Pierre et al., 1998
Morone saxatilis acclimated to 5°C	7.9 ± 1.4	Egginton and Sidell, 1989
Oncorhynchus mykiss winter-acclimatized	10.0 ± 1.0	St-Pierre et al., 1998
Chalcalburnus chalcoides mento	10.1 ± 0.7	Sänger, 1992a
Salvelinus fontinalis	10.9 ± 0.4	Johnston and Moon, 1980b

Note: Values are given as mean ± SEM.

on feeding quantity (protein intake) and that protein does not become more important as a fuel during exercise, even when abundantly available in the diet. Indeed, to the contrary, as demonstrated in unfed rainbow trout, lipid becomes increasingly important as the major energy source as swimming speed increases. Carbohydrates are of secondary importance and protein makes the smallest

Table IV.
Volume Densities (in Percentages) of Various Cell Components in Different Muscle Fiber Types

	Red fibers	Tonic fibers	Intermediate fibers	White fibers
Mitochondria	21–41	4–10	5.7–14.4	0.6–10
Lipid	0.3–11	0.07–0.9	0.05–0.2	0
Sarcotubular system	2.8–5.1	7.3–7.6	5.9–13.6	8.2–16.2
Myofibrils	41–65	49–71	65.8–73.5	63–89
Subsarcolemmal cytoplasm	6.1–11.6	30.0–30.7	5.8–11.2	3.3–3.7
Nuclei	0.08–1.5	3.2–3.5	0.5–0.8	0.02–0.6

Note: Based on data of the following authors: Johnston and Moon, 1980b; Johnston, 1981; Davison, 1983; Kilarski and Kozlowska, 1987; Londraville and Sidell, 1990; Sänger, 1992a,b, 1997.

Table V.
Numerical Capillary to Fiber Ratio (N/N(C,F)) and Numerical Capillary Density (N/A(C,F)) of Red and White Muscle Tissue

Fish species	N/N(C/F) N/A(C,F), mm^{-2}		Author
	Red fibers	White fibers	
Trematomus newnesi	0.62 ± 0.66	1.20 ± 0.20	Londraville and Sidell, 1990
	296 ± 30	74 ± 11	
Chalcalburnus chalcoides mento	0.79 ± 0.08		Sänger, 1999
wild population	998 ± 87		
Salmo trutta	0.85 ± 0.06	0.39 ± 0.05	Davison, 1983
Leuciscus cephalus	0.89 ± 0.04	0.43 ± 0.04	Sänger, 1999
laboratory reared	1959 ± 54		
Notothenia gibberifrons	0.91 ± 0.07	0.71 ± 0.10	
	438 ± 9	76 ± 14	
Chalcalburnus chalcoides mento	0.97 ± 0.06		Sänger, 1999
laboratory reared	806 ± 64		
Morone saxatilis	0.99 ± 0.04	0.40 ± 0.04	Egginton and Sidell, 1989
acclimated to 25°C	1937 ± 329	176 ± 35	
Chondrostoma nasus	1.01 ± 0.08	0.33 ± 0.02	Sänger, 1999
laboratory reared	1966 ± 79		
Leuciscus cephalus	1.27 ± 0.07		Sänger, 1999
wild population	1697 ± 36		
Chondrostoma nasus	1.32 ± 0.11		Sänger, 1999
wild population	2337 ± 30		
Morone saxatilis	1.41 ± 0.07	0.65 ± 0.03	Egginton and Sidell, 1989
acclimated to 5°C	1898 ± 63	269 ± 28	
Carassius carassius	1.50 ± 0.09	0.35 ± 0.05	Johnston and Bernard, 1984
Oncorhynchus mykiss	1.73 ± 0.09		Egginton and Cordiner, 1997
summer-acclimatized	2178 ± 105		
Oncorhynchus mykiss	2.10 ± 0.08	0.95 ± 0.08	Davie *et al.*, 1986
Oncorhynchus mykiss	2.50 ± 0.15		Egginton and Cordiner, 1997
winter-acclimatized	1371 ± 36		
Salvelinus fontinalis	3.42 ± 0.20		Johnston and Moon, 1980b
Engraulis encrasicolus	12.90 ± 0.50		Johnston, 1982b

Note: Values are given as mean ± SEM.

contribution overall. Steady-state exercise by the red muscle is fueled mainly by mitochondrial oxidation of pyruvate, fatty acids and, to a lesser extent, ketone bodies. This spares glycogen reserves for tissues that rely primarily on glucose such as the brain (Moyes *et al.*, 1989). In contrast, in elasmobranchs carbohydrates and ketone bodies are the predominant fuel for slow muscle metabolism (Moyes *et al.*, 1990).

5. FUNCTIONAL CONSIDERATIONS

Species differ in the ways in which they employ their red fibers in sustained slow and moderate speed swimming. Coughlin and Rome (1996) concluded that most power for steady swimming at moderate speeds comes from posterior rather than anterior musculature. However, in mackerel (*Scomber scombrus*) cruise swimming, red muscle net positive power output is fairly uniform over the entire length of the trunk (Shadwick *et al.*, 1998; see also Katz and Shadwick, 1998). Different again, in the American eel (*Anguilla rostrata*), anteriorly located red muscle is not active at the lowest swimming speeds but is recruited in a posterior-to-anterior direction as swimming speed increases (Gillis, 1998). This may also be true for taxa with less elongated bodies and thus provide a more general mechanism for fish to vary their swimming speeds (Jayne and Lauder, 1995; Gillis, 1998).

These functional variations are at least partly paralleled by differences in the relative proportion of red muscle in the anterior and posterior regions of the trunk. The proportion of red muscle in the posterior (caudal) region reflects the extent to which sustained swimming is part of a particular lifestyle. Active pelagic species such as the Scombridae, Clupeidae, and Carangidae, which share a carangiform swimming style, possess the greatest amount of red muscle (18–30%). In contrast "sit and wait" predators, such as Esocidae and Congridae, have no more than 5% red muscle in the caudal region (Luther *et al.*, 1995).

Red muscle fibers show a high density of nerve terminations (Akster, 1983). They are multiply innervated (multiterminally as well as polyneuronally) in both teleosts and elasmobranchs (Barets, 1961; Bone, 1978), slow contracting and fatigue-resistant (van Raamsdonk *et al.*, 1978). Red muscle has a relatively low maximum velocity of shortening and slow activation-relaxation kinetics (during oscillatory contractions, the rate of activation and particularly the rate of relaxation are important determinants of power production) and are used, accordingly, during steady slow-to-medium swimming speeds (low shortening velocities and low tailbeat frequencies; Bone, 1966; Johnston *et al.*, 1977; Curtin and Woledge, 1993; Rome *et al.*, 1984, 1985, 1988, 1992a; Altringham and Johnston, 1990; Rome and Swank, 1992; Woledge, 1992; Jayne and Lauder, 1994; Gillis, 1998). Posterior red muscle undergoes relatively high strains during swimming and produces high levels of mass-specific power when compared to the anterior red muscle. The relatively slow relaxation rate of red muscle contributes to the lower power production associated with low strain during oscillatory activity (Rome *et al.*, 1993).

With regard to these contractile and metabolic characteristics red muscle is recruited for long term, low intensity sustained swimming activities with maximal economy of operation (Bone, 1975). The cost of locomotion is low at those speeds and slow muscle fibers, as mentioned above, typically make up just a few percent of the total myotomal muscle mass. A comparison of red fiber function with that of white fibers is provided as part of the discussion of white fibers (see below).

B. White Muscle Fibers

1. Position and Occurrence

A fish is made up predominantly of white muscle. The white fibers represent the bulk of myotomal muscle (Fig. 2), never less than 70%, and show the largest fiber diameters ranging between 50 and 100 μm or even more. The proportion of the cross-sectional area of the myotomal muscle that is comprised of white muscle varies along the length of the fish being greatest in the anterior of the animal and declining caudally. In the scup the maximal cross-sectional area has been reported to be 82.5% with a decline to a minimal value of 66.3% (Zhang et al., 1996).

2. Development and Growth

In accordance with the histochemical fast fiber characteristics of most adult teleost fish (e.g., Rowlerson et al., 1985; Sänger et al., 1988, 1989), post-hatch fast fibers are characterized by an acid-labile and alkali-stable mATPase activity (Stoiber, 1996, pearlfish). The myosin isoform pattern of the small fibers generated through rapid fast muscle hyperplasia of teleost larvae differs from that of the larger fast fibers and from the small fibers within adult fast muscle. This indicates that fast fiber formation may rely on successive activation of separate stem cell populations (Scapolo et al., 1988; Koumans and Akster, 1995). The helical deep fiber paths running through successive myotomes of the adults (Alexander, 1969) start to form during the late embryonic stage (Stoiber, 1991, 1996). The fibers of the original (somite-cell derived) fast fiber stock initially contain a small number of myofibril clusters which split up in a roughly radial pattern and eventually become fused (Figs. 6 and 7). By contrast, the later established new fast fibers derived from myosatellite cells usually contain only a single group of myofibrils (Fig. 8). As myofibril growth (apart from terminal elongation by addition of new sarcomeres) is largely based on peripheral addition of new filaments (Patterson and Goldspink, 1976), myofibril cross-sectional profiles (except those in the centers of the fibers) become progressively ribbon-shaped and arborized (Figs. 7, 9), the latter due to incomplete splitting (Stoiber, 1991, 1996). Relative volumes of fast fiber mitochondria are quite high at the late embryonic stage, but decrease with further development to very low levels. Differences in mitochondrial content of larval fast fibers occur in parallel with the activity patterns of the larvae; for example, values for the rather inactive larvae of the plaice (*Pleuronectes platessa*) are 6–8% compared to 15–26% for the more active larvae of herring (Brooks and Johnston, 1993; Johnston, 1994).

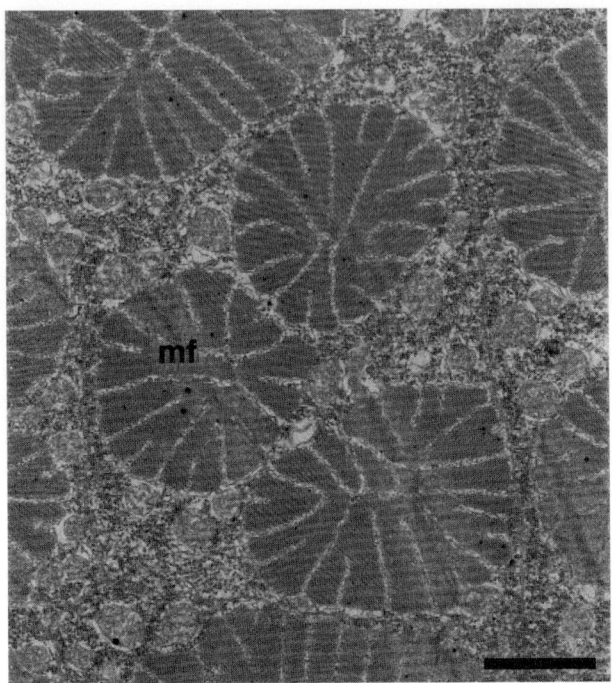

Fig. 6. Free embryo of roach 2 days post-hatching. Transverse section of a fast (white) fiber with three groups of myofibrils (*mf*). Scale bar 2 μm.

3. Histochemical Analysis and Ultrastructural Features

Adult white muscle fibers show no SDH activity. Species differences are found in the pH lability of the mATPase activity of white muscle myosins: alkaline-lability (Kilarski and Kozlowska, 1985; Sänger *et al.*, 1988, 1989) or alkaline-stability (Johnston and Lucking, 1978; Akster, 1983; Zawadowska and Kilarski, 1984; Rowlerson *et al.*, 1985). Three "fast" light chains and in trout also one "slow" light chain can be observed in white muscle (Rowlerson *et al.*, 1985).

White fibers (Fig. 9) are tightly packed with myofibrils occupying between 75 and 95% of fiber volume (Johnston, 1980; Sänger *et al.*, 1990; Sänger, 1992a). The myofibrils show a marked radial orientation and retain a relatively uniform width (Fig. 9). Organelles such as mitochondria (tubular ones; Sänger, 1992a) which interrupt the arrays of myofibrils are few and both lipid droplets and myoglobin are present only at very low levels and the vascularization is poor. Glycogen content is also low with granules mainly located between the myofibrils. As

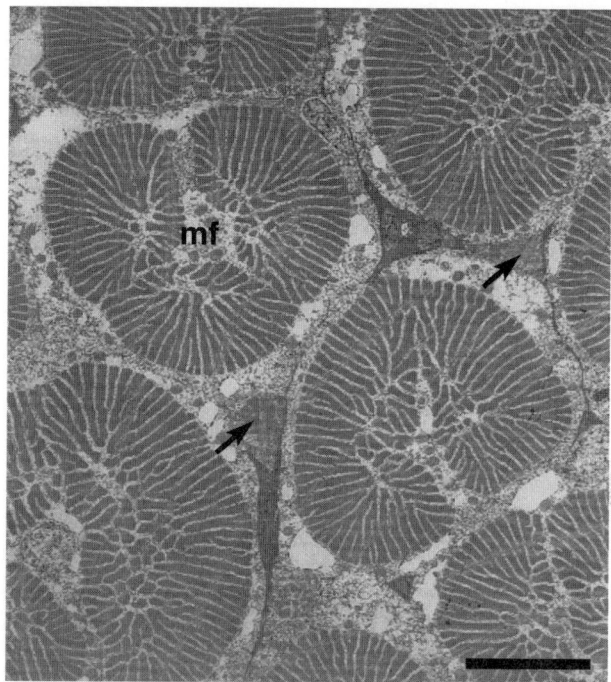

Fig. 7. Roach larva after 15 days of free swimming. Transverse section of fast muscle next to the horizontal septum. Myofibril (*mf*) clusters within the fibers of the original stock have become fused, while small new fibers (*arrows*) insert between the larger fibers. Scale bar 10 μm.

with red muscle fibers, white muscle fibers exhibit a well-developed sarcotubular system, but in contrast to the former they have a relatively longer T-SR junction which influences the rate of calcium release and thereby the rate of tension development (Akster, 1985). Rapid relaxation is aided both by an extensive network of SR occupying 6–14% of fiber volume (Nag, 1972; McArdle and Johnston, 1981; Sänger, unpublished) and by the presence of high concentrations of parvalbumins. The triad is situated at the level of the Z-line.

4. ENERGY METABOLISM

Generally, there are two pathways for supplying energy for contraction in fast muscle, phosphagen hydrolysis and anaerobic glycolysis. Differences in the energy metabolism probably reflect varied utilization of burst swimming activity (Dunn *et al.*, 1989, and see functional considerations below). The energy for white muscle, operating as a closed system, comes almost exclusively from the anaerobic breakdown of intramuscular glycogen with small contributions from cytosolic

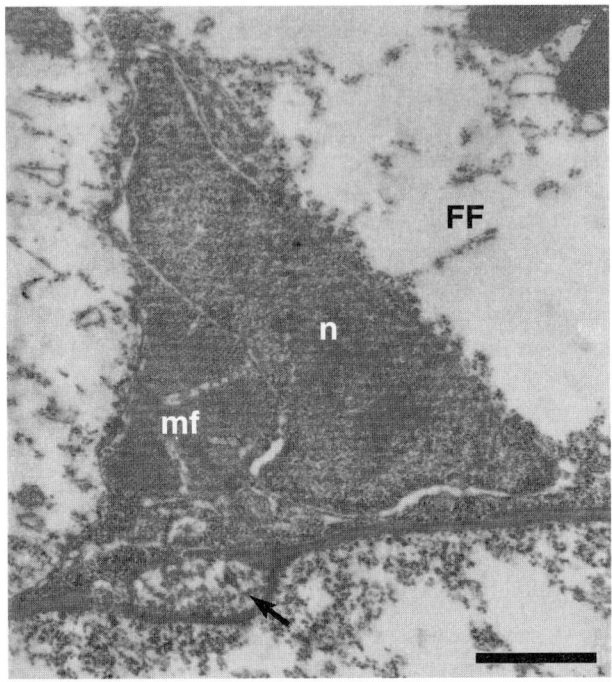

Fig. 8. Danube bleak at the larva/juvenile transition. Transverse section of a satellite-cell derived small new fast fiber with a single myofibril (*mf*) cluster. The attached myoneural synapse is indicated by an *arrow*. *n*, nucleus; *FF*, large fast fiber. Scale bar 1 μm.

phosphocreatine (PCr) and ATP (Bone, 1975; Weber and Haman, 1996). This holds particularly for white fiber contraction in so-called "fast starts" (e.g., Eaton *et al.*, 1977; Webb, 1978; Morley and Batty, 1996; Wakeling and Johnston, 1998) which rely entirely on local fuels. Glycogen is accumulated inside the muscle during periods of recovery (Arthur *et al.*, 1992; Schulte *et al.*, 1992) and is, therefore, placed at its final site of catabolism before exercise is started. High performance fish such as tuna store about eight times more glycogen in their white muscle than sedentary carp, and over four times more than rainbow trout (Arthur *et al.*, 1993). Fish white muscle has the capacity for very rapidly increasing its glycolytic flux (Driedzic and Hochachka, 1976). In notothenioid fish *Notothenia neglecta* and *Chaenocephalus aceratus* Lönnberg fast twitch fibers, having relatively low activities of glycolytic enzymes, rely on PCr hydrolysis as an anaerobic energy supply pathway limiting burst endurance (Dunn and Johnston, 1986; Johnston, 1987). An inability to generate energy via anaerobic glycolysis for burst swimming is also demonstrated in *Pagothenia borchgrevinki*, an active Antarctic teleost (Davison *et al.*, 1988). In contrast, another notothenioid

7. MUSCLE FIBER DIVERSITY AND PLASTICITY 205

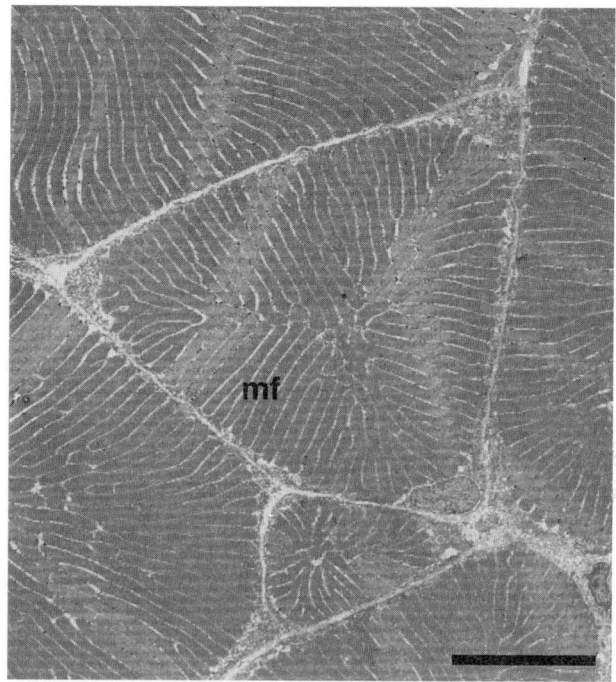

Fig. 9. Nase adult, age 1+. Transverse section of fast (white) fibers. The ribbon-shaped arborized profiles of fast fiber myofibrils (*mf*) have a radial orientation. Scale bar 10 μm.

N. gibberifrons Lönnberg has high activities of glycolytic enzymes in its fast muscles which is consistent with it having a greater burst swimming endurance than the above species (Dunn *et al.*, 1989). Short periods at burst speed result in the mobilization of a significant proportion of muscle glycogen stores and a concomitant buildup of lactic acid. The lactate, produced in white muscle during burst exercise, is retained (particularly in those fish with high proportions of white muscle) and apparently metabolized *in situ* (Turner *et al.*, 1983a,b; Milligan and McDonald, 1988). Tissue lactate levels approach resting levels after 8–12 hr of recovery (Turner *et al.*, 1983a,b). The recovery time is dependent on several factors including LDH activity and the rates of pyruvate transport and mitochondrial oxidation (Moyes *et al.*, 1989). Here white muscle mitochondria may play a critical role in lactate metabolism after burst exercise. Mitochondrial content and enzyme activities found in the skipjack tuna (*Katsuwonus pelamis*) support the idea that in high performance swimmers such as the scombroids, fast (white) muscle may be used aerobically to convert lactate into glycogen *in situ* (Arthur *et al.*, 1992; Moyes *et al.*, 1992). Studies on the oxidative properties of carp white muscle (Moyes *et al.*, 1989) showed that white muscle mitochondria appear to be specialized in using pyruvate (deriving from glucose and lactate, generated during high intensity exercise). The low capacity for ketone body oxidation of teleost muscle mitochondria (red as well as white muscle) probably reflects low levels of ketolytic enzymes (Moyes et al., 1989). Teleosts are thought not to rely on ketone bodies as fuel, even after extended periods of food deprivation (Zammit and Newsholme, 1979).

Finally it may be noted that energy metabolism in embryos and early fish larvae differs from the adult pattern by being almost entirely aerobic (Wieser, 1995).

5. Functional Considerations (with particular reference to comparison with red fibers)

In white muscle of most fish, it is maximum power output which is important rather than economy when compared to red muscle. Generally, this muscle type is used at high swimming speeds (high tailbeat frequencies), e.g., in fast-start burst swimming for prey capture and escape response. White muscle fibers are fast contracting (high maximum velocities of shortening, fast activation-relaxation kinetics) and fatigue rapidly. During fast-starts, they experience lower sarcomere strains than red fibers (Bone, 1966; Johnston et al., 1977; van Raamsdonk et al., 1978; Curtin and Woledge, 1988; Rome et al., 1984, 1985, 1988, 1992a,b; Altringham and Johnston, 1990; Spierts and van Leeuwen, 1999).

The functional role of white fibers is reflected in their ultrastructure (see above), and in the innervation. With the exception of elasmobranchs, holosteans, chondrosteans, and some less advanced teleosts (eels, herrings) which have focally innervated white fibers (Bone, 1964, 1978), most fish have multiply and polyneuronally innervated white fibers with propagation of action potentials (Bone, 1978; Johnston, 1983; Ono, 1983). This difference in the innervation pattern (focal vs. multiple) is reflected in some contractile and electrophysiological properties of white muscle. White muscle fibers show the lowest nerve termination density of all known fiber types (Akster, 1983).

In contrast to elasmobranchs and some primitive teleosts where only red muscle fibers are active during sustained swimming and white fibers are not used during sustained/intermediate speed swimming (Bone, 1966; Bone et al., 1978), such a simple division of labor between red and white muscle is thought not to be available for the majority of fish. Generally, in juvenile and adult but not in larval fish red and white muscle forms a two-gear system which powers very different movements. The red muscle powers slow movements, while the white muscle powers very fast movements, both while working at the appropriate V/V_{max} (where V is the velocity of shortening; Rome, 1994). In many teleosts, electromyography (EMG) evidence suggests that both red and white muscle, specially the more superficial fast fibers, have a role to play in sustained/intermediate speed swimming (Hudson, 1973, rainbow trout; Johnston et al., 1977 and Bone et al., 1978, common carp; Johnston and Moon, 1980a, coalfish *Pollachius virens;* Dickson, 1996, tunas), although this may not be the case in all species (e.g., Rome *et al.,* 1985). There is evidence from the mackerel that red and white muscles at the same axial location might contract out of phase with each other when both are active (Shadwick *et al.,* 1998). In anguilliform swimmers, the activation of white fibers at high swimming speeds occurs in a posterior-to-anterior direction similar to red fibers (Gillis, 1998). EMG data from the largemouth bass (*Micropterus salmoides*) further suggest that in fast-starts, white fiber activity within a single myotome can be spatially regionalized (Thys, 1997).

The recruitment of white muscle fibers for sustained swimming would suggest a certain aerobic capacity of some fibers in the white myotomal muscle and this may be reflected by the finding of one "slow" myosin light chain (MLC) besides the three "fast" ones in some fish species (Rowlerson *et al.,* 1985). However, the normally low mitochondrial volume fractions (1.4–1.7%) and the very low capillary densities of white muscle fibers (Table V) (Kryvi *et al.,* 1980; Totland *et al.,* 1981; Egginton and Johnston, 1982a,b) together with the findings of training studies (Sänger and Lackner, 1991; Sänger, 1992b) raise the question as to whether white fibers are really recruited during sustained swimming. The low mitochondrial content is probably only sufficient to meet the energetic requirements of basal metabolism (for example, protein synthesis) but is inadequate to meet the demands of sustained swimming activities. The mitochondrial metabolism in white muscle may have a major role in recovery and rest (lactate metabolism as mentioned above). During sustained swimming the bulk of white muscle is probably moving entirely passively, with the fibers of the transition zone being activated to overcome this passive resistance (Johnston, 1983). On the other hand it is possible that, during bursts of acceleration powered by the bulk of white muscle, the red and intermediate fibers will be moving passively (Johnston *et al.,* 1977). Under the conditions of higher swimming speeds red fibers are probably mechanically ineffective since they will be shortening at velocities far in excess of their optimum power output. At these speeds of shortening the myosin crossbridges become detached from the actin filaments before the full cycle of force production is complete and, hence, there is very little ATP breakdown (Curtin and Davies, 1975; Goldspink, 1975). At enhanced swimming speeds there is evidence for a sequential recruitment of faster contracting fiber types (Hudson, 1973; Bone, 1978; Sisson and Sidell, 1987; Rome and Sosnicki, 1990). Freadman (1979) has shown that in striped bass (*Morone saxatilis* Walbaum) sustained activity is entirely supported by slow fibers until quite high swimming speeds are reached (more than 4 body lengths per second) when fast fibers are recruited. Johnston and Bernard (1982a) supposed that this variation in fiber recruitment between different fish species may reflect differences in either the number of neuromuscular endplates per fiber and/or the degree of motor axon branching or multiterminal innervation. Tuna white muscle has a very high anaerobic as well as aerobic metabolic capacity allowing the endothermic tunas to accelerate and then rapidly restore intracellular acid-base balance and glycogen stores and to do this repeatedly. White muscle anaerobic capacity is significantly greater in tunas than in the related ectothermic scombrid fishes, indicating that energy generation for bursts may be greater in tunas (Dickson, 1996). Because of their high capacities to buffer metabolic acids within the white muscle and blood, and to recover from exhaustive exercise rapidly, they may be able to burst repeatedly. The high aerobic capacity (again higher than that of scombrids) may indicate that at least some tuna white muscle fibers are recruited at sustainable speeds or it may be related to rapid lactate turnover or high growth rates and rapid rates of food processing (Brill, 1996; Dickson, 1996).

In conclusion, the problem of fiber type recruitment may be explained by the large diversity of fish, their swimming modes, and lifestyles. It is hardly surprising then, that there are differences in their locomotor musculature and fiber type recruitment.

C. Intermediate or Pink Muscle Fibers

1. POSITION AND OCCURRENCE

In accordance to their name, intermediate or pink fibers are not only intermediate in position between red and white muscle fibers but also in many other aspects, as detailed below. The general ultrastructural characteristics of an intermediate fiber are to be seen in Fig. 12.

In juveniles and adults of most teleost species, a zone of intermediate or pink fibers is inserted between slow and fast fiber domains (Fig. 2), having the same longitudinal fiber orientation as the red ones. The intermediate fibers do not intersperse with the slow fibers but are, to some extent, intermingled with the most lateral fast fibers (Waterman, 1969; van Raamsdonk et al., 1980, 1982; Akster, 1983; Kilarski and Kozlowska, 1983, 1985; Scapolo and Rowlerson, 1987; Gill et al., 1989; Ramirez-Zarzoza et al., 1991). Because of differences in the histochemical profile, different authors have classified the "intermediate" fibers of this transitional zone of some species as a separate category (Kilarski and Kozlowska, 1985; Sänger et al., 1988, 1989). The European eel is totally lacking fibers of this intermediate type (Egginton and Johnston, 1982b).

The relative amount of intermediate muscle differs both between fish species and developmental stage. In common carp, these fibers normally comprise about 10% of the myotomal bulk making them slightly more numerous than the red fibers (Davison et al., 1976; 2–4 cells deep, Gill et al., 1989). In the cyprinids that we have examined (Sänger et al., 1988) the red muscle population is always larger than the intermediate one. In contrast to red and white muscle, it has been shown in the scup that the proportion of cross-sectional area occupied by pink fibers is constant along the length of the fish (Zhang et al., 1996). The mean fiber diameters lie between those of red and white muscle fibers.

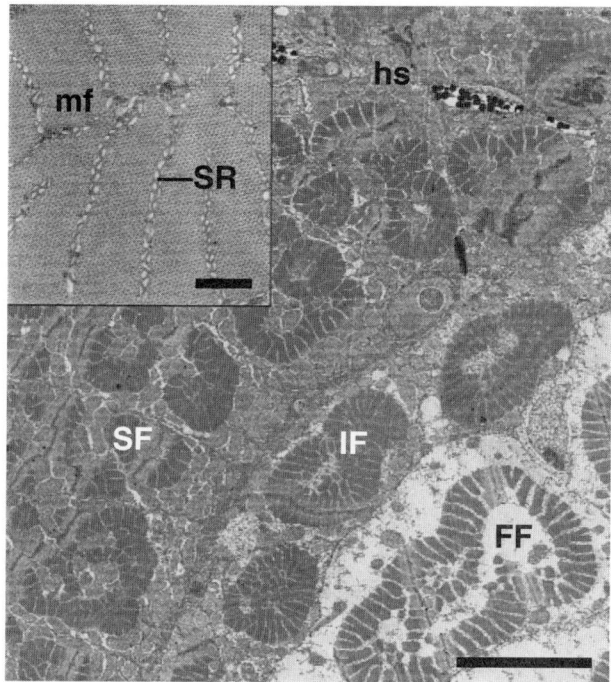

Fig. 10. Roach larva after 20 days of free swimming. Transverse section of the horizontal septum (*hs*) area (ventral part) showing an initial layer of intermediate fibers (*IF*) along the medial boundary of the slow fiber (*SF*). *FF,* fast fibers. Lateral is to the left. Scale bar 10 μm. Inset: Myofibril (*mf*) and *SR* tubule arrangement within a young intermediate fiber from a pearlfish larva. Bar 0.5 μm.

2. Development and Growth

Compared to both slow and fast fibers, the intermediate fibers commonly appear relatively late in development. In some species like zebrafish and red sea bream their appearance coincides with the end of yolk resorption and the switch to exogenous nutrients (van Raamsdonk *et al.,* 1982; Matsuoka and Iwai, 1984). Sometimes indeed, in certain cyprinid species (roach, pearlfish, bleak), the first intermediate fibers are only visible with the aid of electron microscopy and histochemical techniques long after the animals have switched to free-swimming life (El-Fiky *et al.,* 1987; Stoiber, 1996) (Fig. 10). In the sea bass, intermediate fibers are detectable with histochemical techniques only at the end of the larval period (Scapolo *et al.,* 1988). It is still a matter of debate as to where the intermediate fibers originate. On the basis of his fine structural work in zebrafish, Waterman (1969) concluded that the origin of the intermediate fiber population may be associated with the formation of the fascia separating the superficial slow from the deep portion of axial muscle. Similarly, fine structural findings in

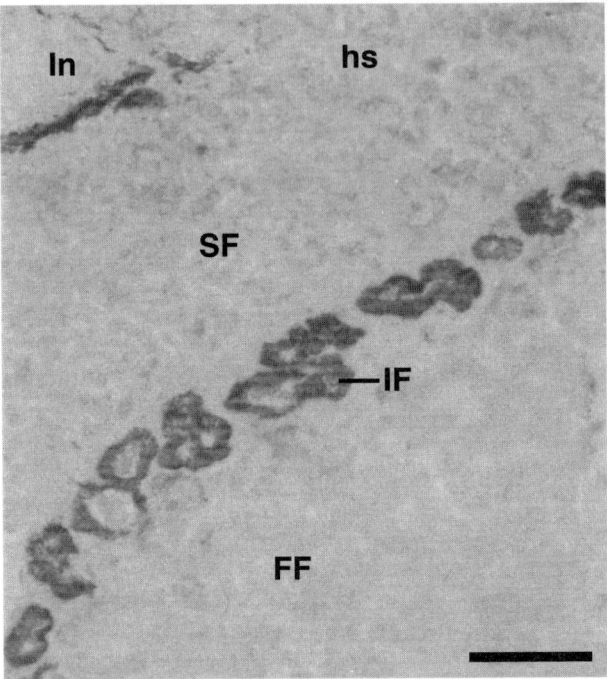

Fig. 11. Pearlfish larva after 35 days of free swimming. Transverse section stained for mATPase activity. Alkaline preincubation at pH 10.35 does not affect the single layer of intermediate fibers (*IF*) while slow fibers (*SF*) and fast fibers (*FF*) are inactivated. *hs* horizontal septum; *ln*, lateral line nerve. Lateral is to the left. Scale bar 50 μm.

roach, pearlfish, and Danube bleak (*Chalcalburnus chalcoides mento* Agassiz) (1832) suggest that small new fibers with intermediate fiber characteristics (radial cleavage of myofibrils combined to double row arrangement of SR tubules; see Fig. 10, inset) are formed from myogenic cells arising at the lateral confines of the deep fast fiber bulk (Stoiber, 1991, 1996). However, again in the pearlfish, well-differentiated, larger intermediate fibers are the first to be labeled by mATPase histochemistry (Fig. 11) and by myosin antisera (Stoiber, 1996). This supports the assumption of other workers that teleost intermediate fibers differentiate from (probably immature) fast fibers (van Raamsdonk *et al.*, 1982; Scapolo *et al.*, 1988; Veggetti *et al.*, 1993). Such contradictory findings have at least two interpretations. First, the intermediate fibers may acquire their fine structural identity prior to their enzymatic and immunological one, or second, intermediate fiber generation. Starting with some mATPase-active reacting fibers along the lateral rim of the deep muscle portion close to the horizontal septum in pearlfish larvae beyond 12 mm SL, intermediate fiber identification via resistance against high alkaline preincubation improves greatly with fish growing larger, accentuating a single layer of fibers demarcating the lateral boundary of the deep muscle portion. By retaining high and quite homogeneous mATPase activity, these fibers create a sharp contrast to the completely inactive superficial and deep fibers immediately above and beneath. Toward the juvenile state the monolayer arrangement gives rise to a two- or three-layered system that merges into the bulk of unreactive deep fibers without sharp confines. As for pearlfish, roach, and Danube bleak, intermediate fibers represent a mixture of superficial and deep fiber characteristics with regard to their cell morphology, fine structure, and arrangement. Thus they resemble the former by an axially parallel orientation and a double-row arrangement of SR tubules, while polygonal cross-sections and radial cleavage of myofibrils is similar to the latter.

3. Histochemical Analysis and Ultrastructural Features

With their intermediate SDH activity and acid as well as alkaline-stabile mATPase activity (Figs. 2 and 11) cyprinid intermediate muscle fibers resemble the pink or fast red fibers described by other workers (Johnston et al., 1975b, 1977; Kryvi et al., 1981; Akster, 1983). Pink or fast red fibers are also observed to have intermediate levels of oxidative and glycolytic enzymatic activities, with higher oxidative activity levels found in red muscle and higher glycolytic activity levels found in white muscle (Patterson et al., 1975; Johnston et al., 1977; Kryvi et al., 1981; Mascarello et al., 1986). mATPase activity level again is intermediate between red and white muscle fibers, with a reported red:pink:white ratio of 1:2:4 (Johnston et al., 1977). Due to their distinct MHC isoform (Scapolo and Rowlerson, 1987), the pink or intermediate muscle fibers of most fish exhibit a distinct histochemical profile, identifiable on the basis of a characteristic pH sensitivity of the mATPase. Immunohistochemically these fibers all show a positive reaction with anti-fast sera and no reaction with anti-slow sera (Rowlerson et al., 1985). Intermediate or pink fibers are therefore classified as a fast aerobic type (Johnston et al., 1974; Johnston, 1982a).

Two further kinds of pink muscle have been described on the basis of their histochemical and immunohistochemical characteristics. In mullet (*Mugil capito*) the pink muscle consists of a mosaic of small and medium diameter fibers; the smaller fibers having the histochemical and immunohistochemical properties typical of red muscle, whereas the larger ones resemble those of white muscle (Rowlerson et al., 1985). Similarly Akster (1983) also found small and large pink fibers in common carp, the smaller fibers having a higher SDH activity and glycogen content than the larger ones. As small fibers often have higher activities of metabolic enzymes or more glycogen than the larger ones (Patterson et al., 1975; Mosse and Hudson, 1977; Korneliussen et al., 1978; Egginton and Johnston, 1982a), a functional diversity between the two pink fiber types is suggested (Boddeke et al., 1959; Bone et al., 1978). Alternatively small pink fibers may be explained as fibers of an early growth stage similar to those of

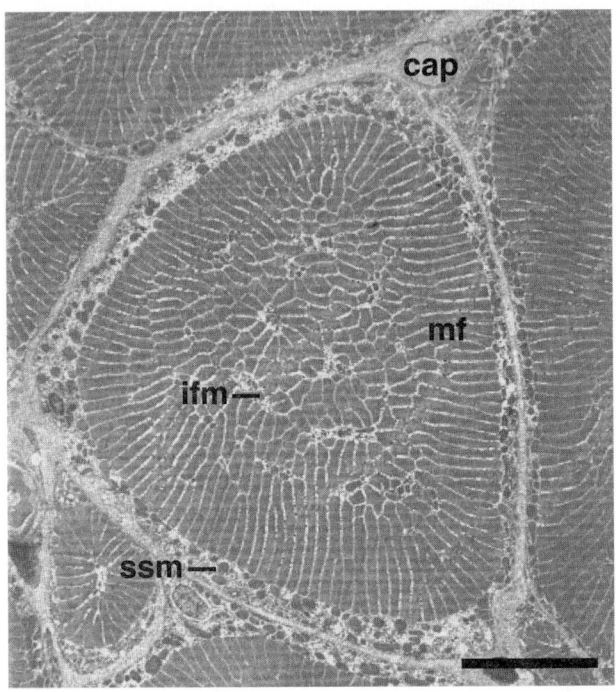

Fig. 12. Nase adult, age 1+. Transverse section of an intermediate fiber. *mf,* myofibrils; *ssm* and *ifm* subsarcolemmal and intermyofibrillar mitochondria, respectively; *cap,* capillary. Scale bar 10 μm.

the transitional zone (see below). This is supported by the gradual transition of both the fiber size and pH stability of the mATPase activity from small to large fibers (Carpene and Veggetti, 1981; Akster, 1983), an increase in fiber number in growing fish (Greer-Walker, 1970; Willemse and van den Berg, 1978; Weatherley and Gill, 1981), and greater fractional volume of nuclei in these small fibers (Egginton and Johnston, 1982b). Since morphological studies on these pink fiber populations did not show differences (Akster, 1985), the latter explanation of a developmental role may be favored. In the guppy *Lebistes reticulatus* the pink muscle near the lateral line resembles a transition zone between red and white muscle; adjacent to the red fibers the pink ones have an alkali- as well as acid-stable mATPase activity and react only with anti-slow myosin sera, whereas the deeper fibers are only alkali-stable and react only with the anti-fast sera (Rowlerson *et al.,* 1985).

Ultrastructural aspects also display a mixture of white and red fiber features; for example, arrangement of myofibrils in cross-section, storage of lipid and glycogen granules, amounts of subsarcolemmal and intermyofibrillar mitochondria (tubular and lamellar), and relative length of T-SR junction and sarcoplasm (Johnston *et al.,* 1975b, 1977; Akster, 1983, 1985; Sänger, 1992a) (Fig. 12).

4. ENERGY METABOLISM

Pyruvate and lactate concentrations and activities of phosphorylase, pyruvate kinase, and lactate dehydrogenase are, however, significantly higher in pink fibers than in either red or white muscle (Johnston *et al.*, 1977). These authors suggest that this high capacity for anaerobic glycolysis of the pink muscle is associated with its recruitment for sustained effort at swimming speeds above which the fish can no longer meet all its energy requirements by gas exchange at the gills.

5. FUNCTIONAL CONSIDERATIONS

On the basis of physiological and histochemical properties, the intermediate fibers are characterized as fast contracting with intermediate resistance to fatigue and intermediate speed of shortening between red and white muscles (Johnston *et al.*, 1977; van Raamsdonk *et al.*, 1978). The activation and relaxation rates as well as the maximum velocity of shortening of pink fibers are faster than those of red fibers (Coughlin *et al.*, 1996). These properties facilitate higher mass-specific maximum oscillatory power production relative to that of red muscle at frequencies similar to the tailbeat frequency at maximum sustained swimming speeds. The force generated per unit pink fiber cross-sectional area is about the same as for red fibers (Granzier *et al.*, 1983; Rome and Swank, 1992). Additionally, pink muscle is found in anatomical positions in which red muscle is quantitatively lowest and produces very little power during swimming; the anterior region of the fish, which undergoes the lowest strain during swimming. Pink muscle lies closer to the backbone and will therefore undergo lower strain during swimming than the overlying red muscle. This lower strain increases the need for faster relaxation rates to permit positive power production which is met by the pink muscle fiber population. Pink muscle produces more oscillatory power than red muscle under low-strain conditions ($\pm 2-3\%$) and this may allow pink muscle to supplement the relatively low power generated by red muscle in the anterior regions of swimming fish (Rome *et al.*, 1993; Coughlin *et al.*, 1996). As shown with training studies on cyprinids (Sänger, 1992b), intermediate muscle fibers are able to adapt to an endurance exercise program by increasing their aerobic capacity supporting the suggestion that intermediate or pink fibers are recruited for sustained swimming at low and higher speeds.

D. So-Called Tonic Muscle Fibers

1. POSITION AND OCCURRENCE

The so-called tonic fibers are a population of small-diameter fibers (therefore also sometimes referred to as such) with an apparently species-specific location within the peripheral part of the myotome next to the red muscle fibers. To date, they have been described in the stone loach (*Barbatula barbatula* L.) (Kilarski and Kozlowska, 1985), the pike (*Esox lucius* L.) (Zawadowska and Kilarski, 1984), a series of cyprinids (Tatarczuch and Kilarski, 1982; Sänger *et al.,* 1988, 1989; Kilarski, 1990; Sänger, 1997), and the pond loach (*Misgurnus fossilis* L.) (Karasinski *et al.,* 1994). In the cyprinids studied by our group, this fiber population is situated between red and intermediate muscle fibers, where they form groups near the horizontal septum which become an almost continuous layer more laterally. A similar belt-like distribution between red and intermediate fibers is reported for common carp and goldfish (Kilarski and Zawadowska, 1985). In contrast to that, in stone loach (Kilarski and Kozlowska, 1985) and pond loach (Karasinski *et al.,* 1994) groups of tonic fibers are found among the red fibers or they are scattered throughout the red fiber zone as in the myotome of pike (Zawadowska and Kilarski, 1984). A third type of distribution is frequently found in small, fast-swimming fish like zebrafish (van Raamsdonk *et al.,* 1980) and stickleback (Kilarski and Kozlowska, 1983; te Kronnie *et al.,* 1983), where they are located just under the skin, dispersed in a single layer among the red muscle fibers. A similar superficial layer of tonic fibers located immediately beneath the skin is found in the elasmobranch species *Scyliorhinus canicula* L., the dogfish, and the closely related *S. stellaris* (Bone *et al.,* 1986) as well as in a series of notothenioid fish (Dunn *et al.,* 1989). These diverse locations and distributions of the tonic muscle fibers make it rather unlikely that they are merely growing or developing fibers. This raises the question as to whether these features are related to swimming activity and body size (see below) which could support the concept of a distinct fiber type (Sänger, 1997).

Tonic muscle fibers are very small. The diameter of the fibers ranges from 8 up to 24 μm (Kilarski and Kozlowska, 1987; Kilarski, 1990; Sänger, 1997). This fiber type constitutes only a very small proportion of the myotomal muscle (probably less than 3% in cyprinids; Sänger, unpublished observations).

2. DEVELOPMENT AND GROWTH

The ontogeny of tonic fibers has been described in some cyprinids, namely the roach, pearlfish, and Danube bleak (Stoiber, 1991, 1996). These fibers are absent in pre-hatch embryos and early hatchlings and appear shortly before the animals switch from bottom-resting to free-swimming life. Occurrence of tonic fibers during the larval period is reported for turbot fry at metamorphosis (Calvo and Johnston, 1992). Expression of "tonic" characteristics always begins with a few fibers of the dorsal part of the slow muscle insertion at the horizontal septum (Fig. 13). Tonic-like fine structure appears to develop gradually, prior to the differentiation of histochemical and immunohistochemical characteristics (Figs. 14 and 15). The developing tonic fibers, similarly to the adult tonic fibers, show an extremely low or even no SDH activity stain, low but alkali-resistant mATPase reaction (Fig. 13), and characteristic sarcomeric structure and T-tubule arrange-

7. MUSCLE FIBER DIVERSITY AND PLASTICITY

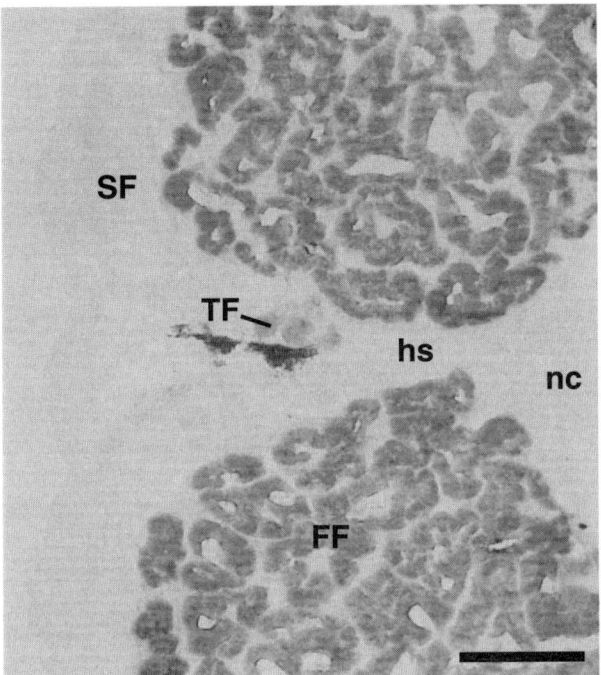

Fig. 13. Pearlfish larva 4 days after transition to free swimming, transverse section. Staining for mATPase activity (alkaline preincubation pH 10.2) visualizes initial tonic fibers (*TF*) within the dorsal part of the slow fiber (*SF*) layer. *FF*, fast fibers; *hs*, horizontal septum; *nc*, notochord. Lateral is to the left. Scale bar 25 μm.

ment (Fig. 16). In contrast to the adult tonic fibers (as characterized by Kilarski and Kozlowska, 1985, 1987; Sänger, 1997), the developing fibers may have a higher abundance of mitochondria. The first tonic fibers probably develop *de novo* from myosatellite-like cells, which are present in the horizontal septum area from the late embryonic stage onward, but could conceivably also arise by re-differentiation of a few superficial fibers (Stoiber, 1996).

3. Histochemical Analysis and Ultrastructural Features

Adult tonic fibers always exhibit negative SDH activity but their mATPase activity again is species specific. Similar to the true tonic fibers, which are known to be moderately stable to both acid and alkali, tonic fibers of cyprinid fish and stone loach show a low mATPase activity for both preincubation pH values, acid as well as alkaline (Kilarski and Kozlowska, 1987; Sänger, 1997). The clusters of

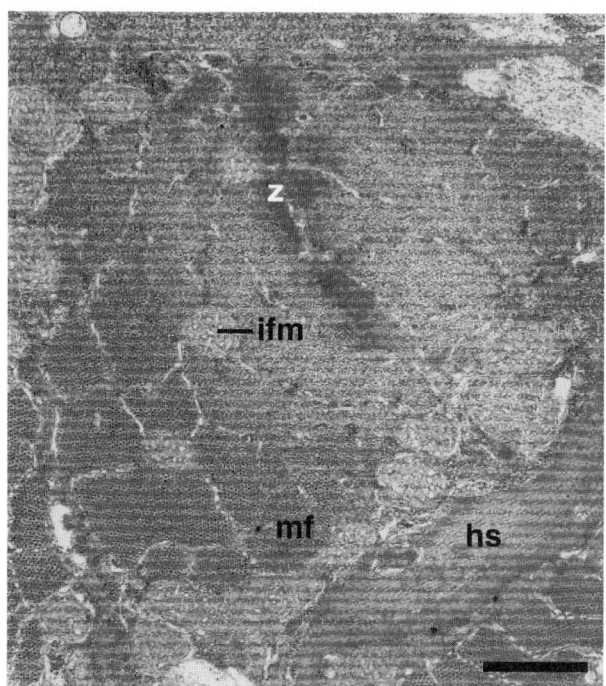

Fig. 14. Pearlfish larva on the day of transition to free swimming, transverse section. Electron micrograph of a nascent tonic fiber. Myofibrils (*mf*) are of irregular shape but less cleaved than in older stages, intermyofibrillar mitochondria (*ifm*) are frequent though less abundant than in slow fibers. *hs*, horizontal septum; Z, Z-disc. Dorsal is top left. Scale bar 1 μm.

small-diameter fibers in pond loach show a different histochemical reaction that is mATPase labile after acid and stable after alkaline pretreatment (Karasinski *et al.*, 1994). In pike the scattered tonic fibers are negative after alkaline preincubation (Zawadowska and Kilarski, 1984). Regarding the histochemical findings, these so-called tonic muscle fibers resemble the nuclear-bag-two fibers of the muscle spindles in mammals and mammalian extraocular muscle slow tonic fibers (Pierobon Bormiolo *et al.*, 1980). These findings indicate once more the problems of classifying muscle fiber types solely on the basis of histochemistry. The high degree of heterogeneity may support the assumption that tonic fiber localization and properties may vary between fish species according to size, swimming speed, and/or other lifestyle-related parameters (see also Kilarski and Kozlowska, 1985). As anuran tonic fibers reflect similar differences by providing a moderately acid- and alkali-stable mATPase in *Xenopus laevis* but only an acid-stable in *Rana temporaria, R. pipiens,* and *R. esculenta* (Rowlerson and Spurway,

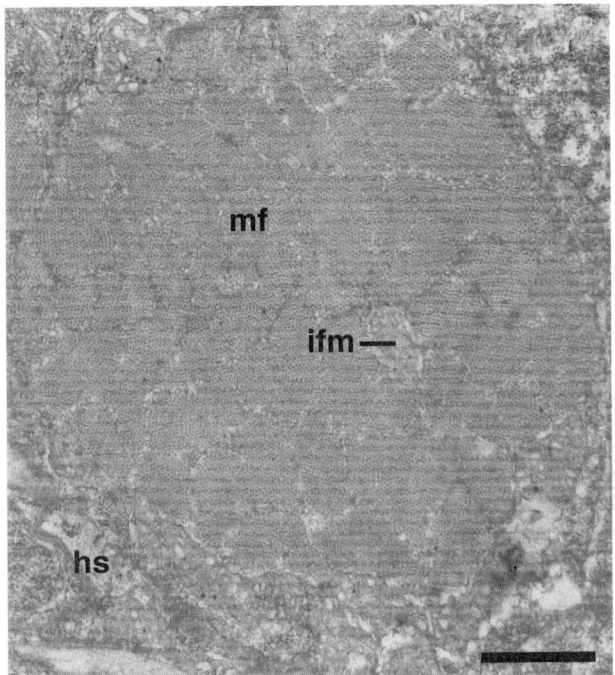

Fig. 15. Danube bleak larva 5 days after transition to free swimming. Transverse section of a tonic fiber. Myofibril (*mf*) cleavage is more complex than in the less mature tonic fiber shown in Fig. 6. Intermyofibrillar mitochondria (*ifm*) as in adult tonic fibers are scarce. *hs*, horizontal septum. Dorsal is top right. Scale bar 1 μm.

1988), it may be inferred that such tonic fiber heterogeneity is a general phenomenon among lower vertebrates. Further, it raises the question as to whether these different fish species have different MHC isoforms in their small-diameter fibers with respect to the above-mentioned differences in mATPase activity, as it has been shown in amphibian and mammalian muscle that the contractile properties and mATPase activity of muscle fibers correlate well with the composition of MHC isoforms (Billeter *et al.,* 1981; Staron and Pette, 1986; Lännergren, 1987). It has been shown in pond loach that tonic muscle fibers are not just distinct in mATPase activity, but seem to contain a unique type of MHC (Karasinski *et al.,* 1994). With immunostaining, tonic fibers show positive reaction with the antibody against tonic myosin as do the true tonic fibers which indicates this fiber population to be very probably tonic in nature (Sänger, 1997).

The ultrastructural characteristics of this fiber type are similar to those of the slow contracting tonic fibers of other vertebrates (Kilarski and Kozlowska, 1985,

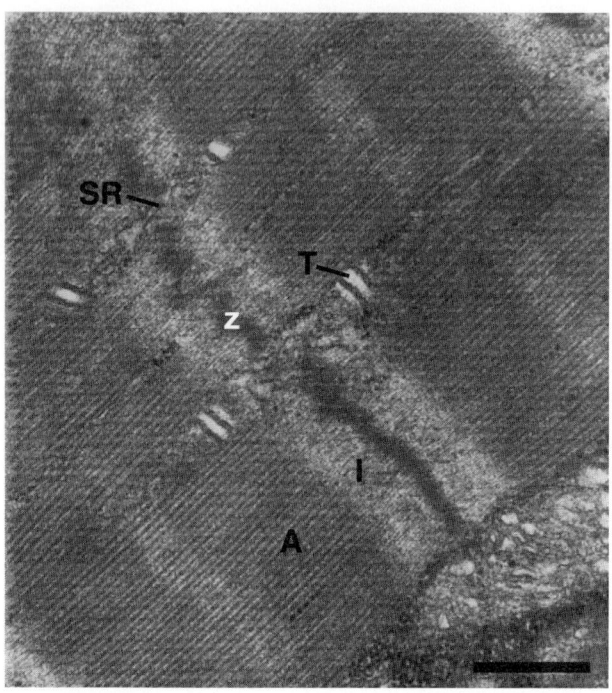

Fig. 16. Pearlfish larva after 25 days of free swimming. Horizontal section of tonic fiber myofibrils. Z-lines are thicker than those of the other fiber types and frequently lack straight orientation. *T*-system tubules travel at the *A*-band/*I*-band transitions. *SR* components have very small lumina. Scale bar 0.5 μm.

1987; Zawadowska and Kilarski, 1984) and also resemble the superficial fibers of the dogfish (Bone *et al.*, 1986; Kilarski, 1990). Z-lines are broad and wavy, M-lines are ill-defined, and triads, partially incomplete, are at the A/I junctions (Fig. 16). The nuclei are situated more proximally and intermyofibrillar mitochondria are rare (Fig. 15). Tonic muscle fibers show the smallest mitochondria in size compared to the other muscle fiber types. The inner mitochondrial architecture of these fibers is similar to that of mitochondria from white and intermediate muscle fibers (Sänger, 1997). Although most adult tonic fibers in cyprinids have a relatively high sarcoplasmic volume with loosely distributed subsarcolemmal mitochondria, some are distinctly different at an ultrastructural level with a dense packing of myofibrils and little sarcoplasm. These two populations have also been described in the muscle of developing fish. With the exception of myofibrillar and sarcoplasmic content, the volume densities of other cell components like mitochondria, lipid droplets or sarcotubular system in tonic fibers are most similar to those of intermediate muscle fibers (Sänger, 1997) (Table IV).

4. Functional Considerations

The nature and function of the so-called tonic muscle fiber type is still a matter of discussion. With respect to their similarities to intermediate muscle fibers in quantitative and qualitative fine structural characteristics (see above) one could suppose that they respresent a developmental stage of the latter. Immunohistochemical studies on goldfish have demonstrated very small fibers between the red and intermediate fibers with different staining characteristics—a more intense reaction with some anti-slow sera than the typical red fibers (Rowlerson et al., 1985). Does this indicate that tonic fibers are more similar to red than to intermediate fibers? The fine structural features do not show any of the characteristics of red muscle fibers, however, and not only do they contain a unique slow MHC distinct from that of the intermediate and red muscle fibers (Karasinski et al., 1994), but they are also tonic in nature (Sänger, 1997). This would all support the supposition that tonic fibers are a separate fiber type.

The uncertain classification of teleost tonic muscle fibers and the lack of data on excitation/contraction properties means that little is known about the role of these fibers during myotomal power generation. Different hypotheses regarding the function of this fiber type have been put forward. Similar histochemical features to those described above were observed in amphibian muscle fibers, mammalian extraocular muscle fibers, and intrafusal bag fibers (Pierobon Bormioli et al., 1980). Thus similar to the red muscle rim fibers a sensory or proprioceptive function is favored (van Raamsdonk et al., 1980; Zawadowska and Kilarski, 1984). The small histochemically weakly staining fibers observed on the periphery of the myotomes of notothenioid fish have been suggested to have a postural or tonic role, holding the trunk in a fixed position during labriform locomotion and when stationary (Dunn et al., 1989). The superficial fibers observed in the myotomes of an elasmobranch, the dogfish, are suggested to have a tonic (postural) rather than a locomotory role (Bone et al., 1986). Similarly, Kilarski and Kozlowska (1987) supposed a high capacity for sustained force generation involved in supporting tonicity of relaxed twitch muscle systems based on the high protein content of tonic fibers represented by the filamentary system. In contrast, our own studies generally show quite low myofibrillar content of tonic muscle fibers. Furthermore, if tonic muscle fibers function in maintenance of body posture, why should there be differences in mATPase activity in various fish species? These findings do not support the idea of tonic muscle fibers supporting posture nor does the observation that tonic fibers respond to exercise training programs by increasing their aerobic capacity (Sänger, 1997). Although Karasinski et al. (1994) suggest unique contractile properties for this fiber type depending on the distinct MHC expressed, we do not think that a functional role in maintaining tonicity of a fish's relaxed muscle system can be favored. Ontogenic studies demonstrating the timing of tonic fiber first appearance with the onset of free-swimming life in the pearlfish, roach, and Danube bleak suggest that, in these species, the function of such fibers may be to allow full control of the axial muscle locomotory system, a prerequisite for efficient prey capture, shoaling, and predator avoidance. It may therefore be essential to have an initial cluster of tonic fibers established dorsal to the horizontal septum, while the ventral part stays devoid of such fibers until the fish have grown to juvenile size.

In summary, there are characteristics of this fiber type which support its tonic nature, but others which make such a postural role unlikely.

E. Red Muscle Rim Fibers

Red muscle rim fibers are found in the zebrafish (van Raamsdonk et al., 1980). They are located between red and intermediate fibers and have a small diameter. Using immunohistochemistry these red muscle rim fibers have been found in zebrafish larvae about one week after yolk store depletion and it is suggested that they are generated by longitudinal splitting of slow fibers (van Raamsdonk et al., 1982). Because of the relatively low number of these fibers, van Raamsdonk et al. (1980) suggested that they may not play a role in force generation but could possibly act as a sensory system. If a myotome contains fibers with a sensory function then one would expect them to be located in the same region as the most intensively utilized fibers, i.e., near or among the red muscle fibers. Their histochemical properties most closely resemble the so-called tonic fibers.

F. Muscle Fibers of the Transitional Zone

This fiber population lies between intermediate and white muscle fibers and is without distinct boundaries (Fig. 2). For example, in more juvenile pearlfish fibers exhibiting a high-to-moderate mATPase activity are observed in the region of transition of intermediate to the deep muscle zone. This is suggested as the onset of the transitional muscle fiber zone including an even higher degree of fiber heterogeneity in the adults of other cyprinid species (Sänger et al., 1988, 1989; Zawadowska and Karasinski, 1988; Kilarski, 1990, Stoiber, 1996), as well as in the pike (Zawadowska and Kilarski, 1984) and in the stone loach (Kilarski and Kozlowska, 1985).

These fibers show a comparable variety of reaction intensities for mATPase activity with both acid as well as alkaline preincubation pH values (Fig. 17) and the SDH activity is relatively low (Kilarski and Kozlowska, 1985; Sänger et al., 1988, 1989). Van Raamsdonk et al. (1980) call the part where the intermediate muscle portion merges with the white part of the musculature, the transition zone. No distinct boundary between fast aerobic and fast glycolytic fibers is observed in tench (*Tinca tinca* L.), but Johnston and Bernard (1982a) show a mosaic appearance of heavily (fast aerobic) and lightly stained (fast glycolytic) fibers along

7. MUSCLE FIBER DIVERSITY AND PLASTICITY

Fig. 17. Roach adult. Transverse section of the transitional zone stained for mATPase activity (acid preincubation pH 4.4). Lateral is to the left. Scale bar 100 μm.

the inner boundary of the fast aerobic zone. Although the same range of staining intensities is not found, there are similarities to the fibers of the transitional zone found in other cyprinids (Sänger *et al.,* 1988, 1989), in the stone loach (Kilarski and Kozlowska, 1985), and in the pike (Zawadowska and Kilarski, 1984).

The histochemical staining reactions and the range of fiber diameters found in the transitional zone suggest that they may represent a growing stage or fiber "pool" (see below) rather than satellite cells as suggested by Zawadowska and Kilarski (1984). In fact the ultrastructural characteristics of these supposed muscle fibers are not in agreement with those determined for satellite cells (Bischoff, 1990; Koumans *et al.,* 1990). Support for the hypothesis that these fibers function as a "pool" is obtained from the findings of training studies in cyprinids (Sänger, 1992b) together with their MHC composition (mixture of white and intermediate MHC). The increase in number of intermediate muscle fibers in response to endurance training might be brought about by recruitment and transformation of the small dark fibers of the transitional zone. They may increase their fiber size and aerobic potential to transform to intermediate fibers.

G. Scattered Dorsal and Ventral Muscle Fibers

These fibers, described for zebrafish, are scattered throughout the dorsal and ventral part of the white muscle area. These also have mixtures of white and intermediate heavy chain myosins (van Raamsdonk *et al.,* 1980). Similar to the fibers of the transitional zone they show intermediate myofibrillar characteristics and vary in diameter. As these scattered fibers appear last of the muscle fiber types during post-embryonic development and are related to the girth of the myotomal muscle, they are thought to play an important role in overcoming the passive resistance of the inactive bulk of the helicoid muscle mass when only the lateral fibers are contracted (van Raamsdonk *et al.,* 1980, 1982).

III. PLASTICITY OF MUSCLE PHENOTYPE

A. Temperature and Season

Seasonal effects on the morphology and physiology of muscle tissue are mainly connected with changes in environmental temperature. Intracellular lipid storage in red muscle fibers differs with season, being higher in winter which is clearly connected with lower ambient temperature (Sänger *et al.,* 1990). Similar seasonal changes in oxidative activity (Johnston and Lucking, 1978) and mATPase activity (Chayen *et al.,* 1993) are detectable histochemically in the muscle of some fish. Many fish are exposed to large seasonal fluctuations in temperature and, as a fundamental element of the environment, temperature clearly has influences on many aspects including muscle development (Chapters 4 and 6), morphology, and function. As far as their metabolism and activity are concerned, poikilotherms are assumed to be completely dependent on their thermal environment. It has become evident, however, that certain poikilotherms have developed strategies for reducing this dependence and for sustaining locomotory ability at decreased temperatures. Freshwater fish, for example, experience large seasonal fluctuations in water temperature ranging from around $0-4°C$ in winter to $25-30°C$ in summer and these animals frequently exhibit a remarkable maintenance of biological activity and metabolism. Extreme environmental stresses such as those experienced by polar fish can lead to a rapid and transient expression of a set of genes, whose products code for heat shock proteins. In addition there is a wide range of adaptational changes in the muscular system for maintaining the necessary activity patterns which have been demonstrated by cellular and biochemical studies on the effects of thermal acclimation/acclimatization. Note, "acclimation" refers to responses to an environmental variable in the laboratory while "acclimatization" refers to similar responses occurring under natural conditions.

Rome and Sosnicki (1990) estimate a 12- to 14-cm carp swimming at 2.4 body lengths per second would require a 50% greater cross-sectional area of red muscle fibers to generate the same power at 10°C as at 20°C. As discussed in Section II.B., in most fish there is a sequential recruitment of slow red, fast red (pink or intermediate) to fast white muscle fibers with increasing speed. Just before the speed of white muscle recruitment all the red fibers are recruited and maximally activated. The order of recruitment is the same for low and high temperatures (Rome, 1990). However, Rome et al. (1984) have shown that a reduction in water temperature is accompanied by a decrease in the threshold speed for recruitment of fast motor units, and that recruitment of fast anaerobic muscle fibers compensates for the reduced power output of aerobic muscle fiber types. To generate the same power and force to locomotor at a given speed at low temperature as at the high ones, the fish compress their recruitment order into a narrow speed range. More muscle fibers and faster fiber types are recruited at low temperatures (Rome, 1990; Rome et al., 1992b). Non-acclimated fish have to recruit their white muscle even at low-swimming speeds resulting in a lower maximum sustainable swimming speed. In species that can acclimate to low temperatures by increasing the mATPase activity (hence increasing the instrinsic rate of contraction) of their red fibers, the maximum cruising speed is greater than in fish acutely cooled to the same temperature (Heap and Goldspink, 1986) and the recruitment thresholds of white muscle fibers are also considerably higher (Rome et al., 1985). Performance at high temperatures in cold-acclimated animals is reduced (Fig. 18; Fry and Hart,

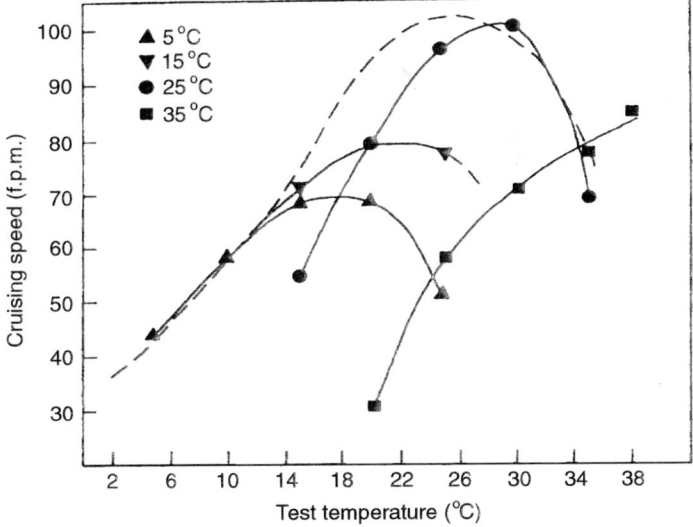

Fig. 18. Effect of acclimation temperature on cruising speed. From Temple and Johnston (1997) *The Thermal Dependence of Fast-Start Performance in Fish;* page 393, with permission from Elsevier Science.

1948; Johnston, 1993b; Temple and Johnston, 1997). This increase in locomotor performance in the cold with acclimation is primarily due to an increase in contractile properties such as maximum isometric force, maximum contraction velocities V_{max} (large increase for fast and more modest in slow muscle fibers), and maximum power output which enables the fish to swim at higher speeds (Johnston et al., 1985, 1990; Rome, 1990).

Thermal acclimation also influences the twitch contraction kinetics of muscle fibers. An acute drop in temperature from 20 – 8°C leads to approximately two- to threefold increases in the half-times for activation and relaxation of muscle. The cold-induced decrease in the average diameter of myofibrils together with increased crossbridge cycle times and a higher surface density of SR, respectively, may account for these changes in activation and relaxation kinetics (Penney and Goldspink, 1980; Johnston, 1993b). However, alterations in the SR-surface area are not a common feature according to data reported by Fleming et al. (1990), suggesting that changes in the density and/or properties of calcium pumps (Ca^{2+}-ATPase activity of the SR is about 60% higher in cold-acclimated fish) may contribute to this capacity adaptation in relaxation rate with temperature acclimation. Parvalbumins obviously play no role in this context, as concentrations and relative proportions of parvalbumin isoforms are unaffected by acclimation temperature (Fleming et al., 1990; Rodnick and Sidell, 1995).

The observed changes in recruitment patterns and swimming performance with thermal acclimation reflect adaptations in the phenotype of myotomal muscle. Improved sustained swimming performance with cold acclimation/acclimatization is partly due to increases in the proportion of red and intermediate muscle fibers (number and diameter) (Fig. 19; Johnston and Lucking, 1978; Sidell, 1980; Jones and Sidell, 1982; Sisson and Sidell, 1987; Egginton and Sidell, 1989; Taylor et al., 1996; Zhang et al., 1996; Egginton and Cordiner, 1997). A similar effect on white muscle is not reported. Evidence suggests that muscle fibers in fish larvae are even more sensitive to temperature change. Only a 5°C difference in acclimation temperature significantly alters the proportion and distribution of tonic and red muscle fibers of larvae. Specimens reared at 17°C exhibit a 43% higher total number of tonic fibers than their 22°C-reared counterparts (Calvo and Johnston, 1992). In adult carp the acclimation to different temperatures has profound effects on muscle contractile properties (Johnston et al., 1990; Langfeld et al., 1991) and recent work points to a change in myosin expression which could account for this (Gerlach et al., 1990; Johnston, 1994; Johnson and Bennett, 1995). In some species, the expression of contractile protein isoforms (myosins) varies according to the acclimation temperature and correlates with adaptive changes in locomotory performance (Johnston et al., 1990). mATPase activity and muscle-shortening speed are predominantly determined by the particular MHC composition, but are modulated by other proteins, particularly the MLC (Greaser et al., 1988; Bottinelli et al., 1991; Lowey et al., 1993). Changes in mATPase activity might be based

7. MUSCLE FIBER DIVERSITY AND PLASTICITY

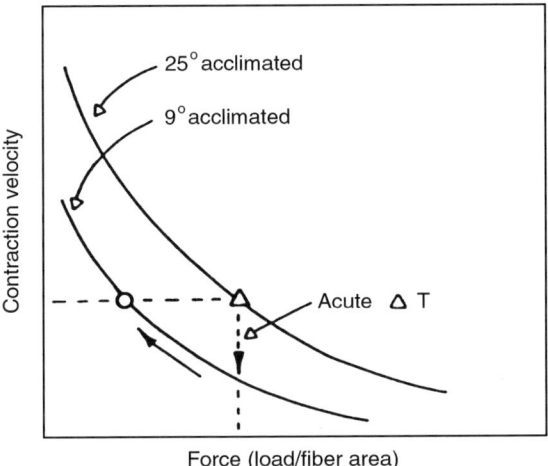

Fig. 19. How proliferation of red muscle mass at cold temperature may contribute to maintenance of sustained swimming. From Sisson and Sidell (1987). With permission from The University of Chicago Press.

on significant changes at the molecular level involving the synthesis of different myosin isoforms. Johnston *et al.* (1975a) and Penney and Goldspink (1981a,b) show that cyprinid and percid muscle produce a different set of myofibrils for low temperature swimming. The contractile proteins of these species change so that the specific mATPase is much higher at low temperatures and muscle contraction can proceed at a faster rate. Similar increases in mATPase at low temperature are reported for other cyprinid species (Heap *et al.*, 1985; Crockford and Johnston, 1990; Hwang *et al.*, 1990) suggesting that this is an adaptive strategy to preserve burst speed (Gerlach *et al.*, 1990).

In many fish, the acclimatory change to low temperatures seems to involve a switch in myosin gene expression resulting in the appearance of different myosin isoforms which are held responsible for the variation of muscle mATPase and contractile properties (e.g., increase in V_{max}). This has been demonstrated for white muscle MHC in carp (Gerlach *et al.*, 1990, Hwang *et al.*, 1990; Goldspink *et al.*, 1992) and goldfish (Johnson and Bennett, 1995), for white muscle MLC again in the carp (Crockford and Johnston, 1990, Johnston *et al.*, 1990, Langfeld *et al.*, 1991), and in the short-horn sculpin (*Myoxocephalus scorpius*) (Ball and Johnston, 1996). Similarly, Langfeld *et al.* (1991) studying carp red muscle show that cold acclimation results in the expression of MLC isoforms normally associated with faster contracting fiber types. They suggest that this isoform change is involved in the improvement in the contractile performance of the slow muscle fibers at low temperatures. Further evidence for myosin isoform involvement comes from the finding that the relative volume of myofibrils in slow fibers (i.e., the proportion of contractile proteins that produce force) show only small and variable changes following temperature acclimation (Johnston and Maitland, 1980). The temperature-specific myosin isoform change is thought to depend on the polyploidy of the related genes (Sidell and Johnston, 1985; Goldspink et al., 1992). However, acclimation strategies are likely to vary between species and do not affect MHC and MLC in the same ways. Accordingly, no temperature-induced changes in the expression of fast muscle MHC have been observed in killifish (*Fundulus heteroclitus*), short-horn sculpin, or rainbow trout (Johnson and Bennett, 1995; Ball and Johnston, 1996; Johnson et al., 1996). This supports the view that acclimatory adjustments in fish muscle also depend on mechanisms other than differential expression of myofibrillar protein isoforms. For example, temperature acclimation can also affect non-myofibrillar proteins, such as myoglobin. Myoglobin content increases with colder temperatures (Johnston and Lucking, 1978; Sidell, 1980; Jones and Sidell, 1982; Egginton and Sidell, 1989). Recent data indicate that myoglobins from fish bind and release oxygen more rapidly at cold temperature than do those from mammals (Sidell, 1998).

As revealed by ultrastructural studies, cold acclimation is associated with increases in mitochondrial volume density of slow and fast muscle cells (see Table II). Numerous studies have supported the idea that increases in volume density and/or cristae surface density of mitochondria are primary mechanisms for enhancing the aerobic capacity of muscle of fish inhabiting cold water environments (e.g., Johnston and Maitland, 1980; Tyler and Sidell, 1984; Egginton and Sidell, 1989; Kilarski et al., 1996, Johnston et al., 1998, St. Pierre et al., 1998). This reduces the mean diffusional distance for oxygen between capillaries and mitochondria, which in turn conserves or increases the rates of cellular ATP regeneration/synthesis. Furthermore, this also reduces the diffusion distance between mitochondria and myofibrils and increases the exchange surface area between mitochondria and cytoplasm, compensating for decreases in diffusivity of aqueous solutes (Johnston and Maitland, 1980; Tyler and Sidell, 1984; Egginton and Sidell, 1989; Sidell, 1998). However, investigating diffusion coefficients of ATP and PCr, Hubley et al. (1997) found that "the proximal stiumulus for temperature-induced changes in mitochondrial volume density in muscle is not a disruption in intracellular diffusion of high energy phosphates."

In fish larvae, however, mitochondrial volume density of superficial fibers is greater with higher rearing temperatures (Johnston, 1994).

A very high mitochondrial content is a common feature in Arctic and Antarctic fish (Table II), which can be regarded as ultrastructurally cold adapted (Johnston, 1987; Johnston et al., 1988; Londraville and Sidell, 1990). These fish have been also regarded as metabolically cold adapted, but recent work suggests that metabolic cold adaptation MCA is an artifact due to problems with methods and/or interpretations (Steffensen, 1997).

The thermal response of the microvasculature is correlated with muscle fiber growth dynamics and varies between species. For example, in the crucian carp (*Carassius carassius* L.), cold acclimation leads to an increase of the capillary to fiber ratio (C/F-ratio), while muscle fiber cross-sectional area is unaffected (Johnston, 1982b). This means that there is a parallel increase in the physiologically more relevant measure of capillarization, the capillary density (= number of capillaries per unit of muscle cross-sectional area). By contrast, in other species the cold-induced increase in C/F-ratio is accompanied by an unchanged or even reduced capillary density due to hypertrophy of the muscle fibers (Table V) (Egginton and Sidell, 1989; Egginton and Cordiner, 1997).

There is evidence that these differences may be connected with differences in intracellular lipid content. In adult fish an increase in intracellular lipid droplets in red muscle fibers is a well-known effect with cold acclimation (Egginton and Sidell, 1989; Table III) and regarded as one of the key mechanisms to ensure superficial muscle oxygen supply under cold acclimation and phylogenetic cold adaptation in some species. This proliferation in intracellular lipid is suggested to accelerate oxygen flux (greater solubility and roughly equivalent diffusion coefficient of oxygen in lipid vs. aqueous compartments) and to provide a significant intracellular oxygen store in cold-acclimated fish. The enhanced diffusion of oxygen in lipid may obviate the need for increases in capillary density (see also Section III.B.), which is supported by the finding that white muscle of striped bass lack lipid accumulation but show an increase in both capillary per fiber ratio and capillary density at low temperatures. This vascular expansion and mitochondrial proliferation in white muscle fibers is suggested to be a compensation for the lack of lipid accumulation in this tissue during acclimation to cold (Egginton and Sidell, 1989; Londraville and Sidell, 1990).

In fish larvae cold-acclimated specimens lack lipid droplets in superficial muscle fibers in contrast to their 20°C-reared counterparts. Possibly, lipid accumulation within superficial muscle plays a less important role during cold acclimation in fish larvae (Stoiber, 1991, 1996).

Interestingly, the red fibers of some Antarctic fish living at permanently cold conditions have relatively low lipid content (Table III). Nevertheless, lipid is thought to play a substantial role in enhancing the oxygen supply to these fibers which are devoid of or extremely poor in myoglobin and/or badly capillarized (Table V, Londraville and Sidell, 1990).

Fast-start behavior (see also pages 202–206) changes with temperature, over acute, seasonal, developmental, and evolutionary time scales. The extent to which temperature acclimation compensates for such changes appears to be dependent on species (see below), type of fast-start, and stage of development (Temple and Johnston, 1997, 1998). The link between changes in the instrinsic muscle properties with thermal acclimation and fast-start performance can be demonstrated with work loop experiments (Josephson, 1985). Work loops estimate the power

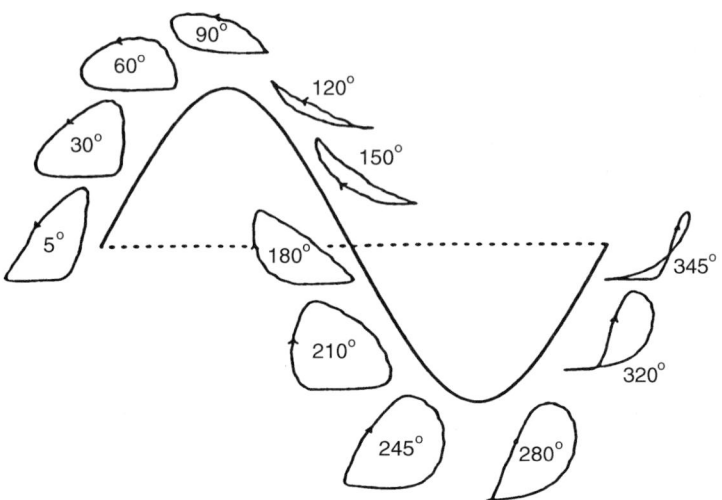

Fig. 20. The effect of varying the phase of muscle activation relative to muscle strain in work loop experiments. From Johnson and Johnston (1991). With permission from The Company of Biologists LTD.

available from a muscle during cycles of contraction and relaxation, as occur during locomotion (Fig. 20). In the short-horn sculpin, improved fast-start performance with warm acclimation is indicated by increased maximum power output of fast muscle undergoing sinusoidal length changes (Johnson and Johnston, 1991). Johnston et al. (1995), again using this method in the same species, found near perfect compensation of muscle performance with temperature acclimation. Fast-start performances and escape responses vary between cold water (Antarctic) and warmer water species, with a low capacity to compensate for the cold effects in the Antarctic fish (Archer and Johnston, 1989; Franklin and Johnston, 1997; Temple and Johnston, 1997).

A thermal response is also seen with the relative importance of different metabolic fuels in muscle. An increase in the utilization of fatty acids (Jones and Sidell, 1982) as also observed in human muscle (Timmons et al., 1985) is associated with cold acclimation. Furthermore, cold acclimation also results in an increase of glycogen content (Johnston and Maitland, 1980) and an increase in the activities of enzymes of the electron transport system, citric acid cycle, and aerobic glucose utilization for both slow and fast skeletal muscles, whereas glycolytic enzymes remain unchanged or decline slightly (Sidell, 1980; Johnston et al., 1985, St.-Pierre et al., 1998). This may be due to corresponding changes in the frequency and/or duration of burst swimming activities. The recruitment of fast motor units at sustained swimming speeds may be facilitated by the enhanced aerobic potential of these fibers following cold acclimation (Johnston et al., 1985). The compensatory increases in pathways of aerobic energy production in the muscles of cold-acclimated fish may represent an adaptation to increase the maximum rate of ATP supply at low temperatures (Hazel and Prosser, 1974; Johnston et al., 1985). In fast muscle fibers ATP supply depends largely upon energy sources already present within the fibers themselves: initially ATP levels are maintained constant by the breakdown of phosphoryl creatine and with prolonged contraction anaerobic glycogenolysis is activated, resulting in the accumulation of lactic acid.

The activation of the glycolytic pathway in white muscle has been shown to occur at similar rates in Antarctic and temperate water species. However, lactate clearance after exercise is faster in the cold water species (Hardewig et al., 1998).

Larvae of a species of freshwater fish, the Danube bleak, expend considerably more metabolic energy at 15°C than at 20°C at a defined speed (Kaufmann and Wieser, 1992). Wieser and Kaufmann (1998) suggest that the higher cost of swimming at the lower temperature might be because the two-gear system of the swimming muscles operating in juvenile and adult fish is not yet functional in the larvae. This has the consequence that, when the larvae are swimming at high speeds in cold water, the muscle fibers have to operate over an increasingly inefficient range of shortening velocities.

All these factors result in an increase in the mechanical power output of the more economical aerobic muscle fibers at low temperatures. The red fibers are aerobic and it is an advantage to the fish to derive as much power from aerobic pathways as possible before it has to recruit its anaerobic fast fibers. The increase in the power production of the red fibers is derived from the above-mentioned change in the mATPase properties and also from an increase in the number of red fibers. The consequence is that cold-acclimated fish can swim faster than warm-acclimated ones with their aerobic muscle at low temperatures and can derive more aerobic sustainable power from its myotomal musculature (also some white muscle fibers increase their aerobic potential; Johnston et al., 1985) (Goldspink, 1985; Rome et al., 1985). Such compensatory increases of muscle aerobic capacity occur when fish are acclimated to a temperature greater than 8°C below their optimum for locomotory performance (Guderley, 1990).

The data for temperature acclimation suggest a general trend toward a greater proportion of aerobic fibers and a greater power production by these fibers in cold-adapted fish. During cold acclimation, red muscle seems to adapt mainly by increasing its mass, mitochondrial content, capillarization, and by slightly changing its biochemical/metabolic properties. White muscle, on the other hand, does not increase in size but markedly changes its biochemical properties, in particular its specific myofibrillar activity. Locomotor performance in different groups possesses the attributes necessary for evolutionary adaptation and only partial (but incomplete) adaptation to environmental temperature can occur (Bennett, 1990).

B. Oxygen Tension

Environmental oxygen levels affect not just swimming performance, gills, and systemic parameters but also muscle fiber development and characteristics. Oxygen levels play a major role in the development of muscle cellularity in a temperature-dependent way. With regard to muscle morphology well-oxygenated waters are inhabited by fish with not just efficient gills, larger hearts, and blood volume capacities but also large amounts of red muscle. Environmental oxygen tension in water is affected by a variety of different factors. Low oxygen tensions, for example, can be a product of pollution, high altitude, the metabolic activity of plants, or due to the restricted volume of the water body. In addition to hypoxia induced by such environmental constraints, limitations at specific points in the respiratory chain pathway can also give rise to hypoxia at the cellular level.

Many fish live in habitats with highly variable oxygen content due to above-mentioned factors. To ensure a sufficient oxygen supply to the tissues, large respiratory surfaces, good capillarization, high concentrations of hemoglobin (Hb) and myoglobin, serving as oxygen stores, are required. Recent studies in our laboratory provide evidence that intracellular lipid storage too may play a role for sufficient oxygen supply of muscle tissue. Those cyprinid fish with a high lipid content turn out to have a poorly capillarized muscle which supports the idea that the oxygen needs of the tissue are met by lipid, facilitating oxygen transport from capillaries to mitochondria, accelerating oxygen diffusion, and providing an intracellular oxygen store (Sänger, 1999).

Adaptations to either stable or unstable hypoxic environments are variable. Childress and Seibel (1998) show that the primary adaptations of animals living in zones of minimum oxygen level are those that support aerobic metabolism by enhancing the organism's abilities to extract oxygen from water. These abilities are notably better than those of animals adapted to unstable hypoxic environments. Further mechanisms involve either a reduction of metabolic rates or the use of anaerobic metabolism to make up the difference between the aerobic capacity and total metabolic demand. A further response to hypoxic conditions might be expressed in changed mode of life (locomotion) or alteration of one of the above parameters involved in the respiratory cascade to increase oxygen supply. There is evidence for all of these mechanisms depending upon the severity of the oxygen depletion and also the species (different behavioral, cardiovascular, and respiratory adjustments made to low oxygen partial pressure PO_2).

Some species, as mentioned above, can further increase their oxygen extraction efficiency at low PO_2 following a period of acclimation by enhanced ventilation at the gills, enhanced removal of oxygen, short diffusion distances from the water to the blood, and an increase in utilization of blood circulated oxygen by the tissues (Hughes, 1973; Lomholt and Johansen, 1979; Johnston et al., 1983; Childress and Seibel, 1998). Other fish reduce their spontaneous locomotory activity thus decreasing the demand for oxygen (Hughes, 1973; Lomholt and Johansen, 1979) or, as in fish larvae, swim more economically in hypoxic water (i.e., their cost of transport is lower) than in normoxic water (Wieser, 1995). Hypoxia can result in an increase in the effective gill surface area, altered perfusion of branchial capillaries, increased blood pressure (Holeton and Randall, 1967), increase in the cardiac output (Hughes, 1973), and changes in the respiratory properties of the blood (Johnston and Bernard, 1984). This includes changes in the properties of Hb resulting in an increase in oxygen-carrying capacity and in oxygen affinity associated with a reduction in the ATP content of the red cells (Wood and Johansen, 1972; Wells *et al.*, 1989). One modulating factor of the O_2-affinity of Hb is the concentration of red cell nucleotide triphosphates, for example, ATP, which can decrease Hb-O_2 affinity.

Apart from raising the concentrations of Hb and myoglobin, an additional possibility for an adaptation to oxygen deficiency is to increase another oxygen carrier or user, i.e., the mitochondria. The volume density of mitochondria and the activity and number of the respiratory chains on the inner mitochondrial membrane are among those factors determining the maximum aerobic capacity of the locomotory muscles, together with the capacity of the capillary circulation to deliver oxygen and substrates. Though there are contradictory data (Johnston and Bernard, 1982a,b), the same authors show crucian carp mitochondrial number (subsarcolemmal as well as intermyofibrillar ones) and capillarization of slow muscle to be higher under hypoxic conditions (Johnston and Bernard, 1984). Again oxygen deficiency is thought to explain the observed differences in mitochondrial content (subsarcolemmal as well as intermyofibrillar mitochondria) and capillarization in red muscle fibers of roach inhabiting two different water bodies, the eutrophic backwaters of the River Danube (Stopfenreuth) and the oligotrophic Seefelder See (Sänger *et al.*, 1990). The proliferation of mainly slow or red muscle mitochondria together with the higher capillary supply in roach from the eutrophic water body may be seen to compensate for oxygen deficiency and represent an adaptation to improve the utilization of circulating oxygen stores at low PO_2. The supposition that such changes are due to differences in environmental oxygen concentration is supported by the findings of Palzenberger and Pohla (1989). Handling the same population as mentioned above, they found higher values for total filament length and number of lamellae of gills in roach from the backwaters of the River Danube compared with those from the Seefelder See. The resulting higher oxygen extraction by the muscle tissue will reduce venous PO_2 and thus raise the rate of oxygen transfer across the gills. The higher mitochondrial abundance in goldfish and common carp that encounter hypoxia and anoxia during winter dormancy is suggested to facilitate not only ethanol production during anoxia but also the diffusion of oxygen to mitochondria during hypoxia (Guderley, 1990).

As mentioned above, there is also evidence that fish respond metabolically to hypoxia or anoxia and that the capacity to store large amounts of glycogen is one of the factors determining resistance to anoxia. Goldfish and crucian carp possess very large glycogen stores in comparison to other fish species and these stores are increased with acclimation to hypoxia (van den Thillart et al., 1980; Holopainen and Hyvärinen, 1985). There are some cyprinid species, like goldfish (Shoubridge and Hochachka, 1980), crucian carp (Johnston and Bernard, 1983) and *Rhodeus amarus*, a small cyprinid (Wissing and Zebe, 1988), which are able to survive long periods of anoxia by utilizing an anaerobic pathway leading to the production of ethanol. Mitochondrial pyruvate dehydrogenase is able to decarboxylate pyruvate to form acetaldehyde under anaerobic conditions (Mourik et al., 1982). In the cytoplasm acetaldehyde is subsequently reduced to ethanol by alcohol dehydrogenase (van den Thillart, 1982). Red muscle of crucian carp accumulates ethanol at nearly twice the rate of white muscle (Johnston and Bernard, 1983) and the relative activities of alcohol dehydrogenase are correspondingly higher in goldfish red than white muscle tissue (van den Thillart, 1982). The utilization of the ethanol pathway allows glycolysis to proceed without accumulation of acid endproducts thus avoiding lactic acidosis. In contrast, common carp try to meet a high anoxic energy demand by conversion of glycogen to lactic acid (van den Thillart et al., 1989). A combination of metabolic suppression with such a modified metabolism is another strategy for hypoxic or anoxic survival. It further extends the period during which hypoxia or anoxia can be resisted as shown for goldfish (van den Thillart et al., 1989) and crucian carp (Holopainen and Hyvärinen, 1985), the latter being able to survive for up to 4–5 months in the absence of oxygen at low temperature.

C. Exercise Training

Two basic responses to exercise training have been observed in humans and some other mammals. Activities such as running, swimming, and bicycling result in an increased capacity of the muscle to do aerobic work and metabolic adjustments which improve endurance (Holloszy and Booth, 1976; Hoppeler et al., 1985). In contrast, training which includes a significant component of isometric exercise (for example, weightlifting) results in fiber hypertrophy and an increase in muscle strength (Goldspink et al., 1976; Hoppeler, 1986).

Few studies have been carried out on exercise in "lower" tetrapods mainly due to difficulties involved in forcing these animals to exercise at known, repeatable work rates and the highly anaerobic nature of any exercise (Bennett, 1978). But fish, unlike amphibia and reptiles, can be exercised with greater ease by forcing them to swim against a water current in a flume or respirometer, making them an ideal subject for so-called exercise training studies. However, comparison between training studies in fish are difficult because of the variety of exercise apparatus, the exercise conditions, the program used during training, and in the analytical tests performed to detect training responses. Further, training responses are species specific relating to the swimming activity and lifestyle of each fish species. It is therefore not surprising that data on the effect of fish exercising are sometimes contradictory and this makes it difficult to get a general overview on the effects of training in fish. Although most of the training studies have been and are still done on salmonids (mainly because of their easy availability from hatcheries, their naturally running water habitat being a good pre-condition to training studies and, last but not least, their commercial importance) there is a good deal of recent work on species of fish other than the "athletic" salmonids.

Besides growth rates, food conversion efficiencies, anaerobic recovery, and systemic parameters of aerobic capacity endurance, training also affects muscle morphology and phenotype. An increase in cell diameter and number of fibers in aerobic muscle in response to training seems to be a general feature in teleost fish. Generally, training at sustainable swimming speeds (speeds of 2 body lengths per second or less) leads to an increase in red muscle proportion with an increase in cell diameter and cell number, respectively (Hinterleitner et al., 1992; Sänger, 1992b; Meyer-Rochow and Ingram, 1993; Young and Cech, 1993). Such an increase in red muscle mass can be seen also with "natural training": roach from a running water habitat contain significantly more red muscle, particularly in the tail region, than fish of the same age from still water. Subsequently fish conditioned to running water can swim faster in exercise tests (Broughton et al., 1981). In salmonids, however, such changes are not reported (Houlihan and Laurent, 1987; Totland et al., 1987). Changes in pink or intermediate fiber diameter of salmonids depend on swimming speed: no changes could be shown at low swimming speeds, whereas training at higher speeds resulted in increased cell diameter (Davison and Goldspink, 1977). In cyprinids intermediate muscle fibers increased their diameters at all swimming speeds (Davison and Goldspink, 1978; Hinterleitner et al., 1992). Increases in the number of these fibers can be obtained at intermediate speeds (Sänger, 1992b). Contradictory results are reported in white muscle. In most species, the proportion of white muscle is unaffected under endurance training conditions (Davie et al., 1986; Sänger, 1992b), but a number of investigations have shown increases in white fiber diameter at all swimming speeds, these increases being directly correlated to the specific speed (Greer-Walker, 1971; Greer-Walker and Pull, 1973; Davison and Goldspink, 1977, 1978; Greer-Walker and Emerson, 1978; Johnston and Moon, 1980a; Hinterleitner et al., 1992; Davison, 1994).

The data regarding the effects of training upon capillarization in muscle are also contradictory. No changes were observed in salmonids studied by Johnston and Moon (1980b) and Davison (1983). In contrast increases in the capillary to muscle fiber ratio were observed in the red and white muscle of rainbow trout studied by Davie et al. (1986) and studies by our group showed similar increases in these muscle types in the Danube bleak (Sänger, unpublished) and increases in the red and intermediate muscle of some cyprinids (Sänger, 1992b). These differing results could be due partly to different species-specific capillarization (Table V) and, again, to different durations of training. Capillaries transport not only oxygen but also lipid and glucose and, moreover, remove waste products such as lactate and metabolic heat. As a consequence, the limiting factor in oxidative (aerobic) muscle fibers is oxygen supply whereas in glycolytic muscle fibers it is the removal of waste products. Increased capillarization of red and intermediate muscle fibers means improved oxygen supply and accounts for higher aerobic capacity of these muscle fiber types. Increased capillary supply to white muscle may be seen to improve glycogen synthesis pathways, enhance removal of metabolic products such as lactate, and heat and/or prevent local fatigue.

Biochemical work done on swimming myotomal muscle is somewhat contradictory again. Johnston and Moon (1980a,b), looking at enzyme activities, suggest that the oxidative capacities of red and white muscle of coalfish (Gadidae) and brook trout (*Salvelinus fontinalis,* Salmonidae) were not affected much by training, whereas Love *et al.* (1977) show that myoglobin concentrations of red muscle increased as a result of training which would allow better transport of oxygen within muscle tissue. Based on shifts in the ratios of LDH (lactate dehydrogenase) isoenzymes with training favoring lactate oxidation, Davie *et al.* (1986) also assume that exercise produced an increased aerobic capacity. A change toward lipid metabolism to fuel swimming and increase aerobic capacity of red muscle is suggested (see below). It is known that training increases protein turnover in most tissues in rainbow trout (Houlihan and Laurent, 1987). Trained rainbow trout have higher levels of enzymes associated with aerobic metabolism in red and white muscle indicating a shift to lipid metabolism (Farrell *et al.,* 1990, 1991; Lauff and Wood, 1996; Alsop and Wood, 1997). Regarding SDH activity contradictory results have been obtained so far. In all muscle fiber types of zebrafish (De Graaf *et al.,* 1990) SDH activity increases, whereas in trained Danube bleak, chub, and nase the enzymes of energy metabolism are generally unaffected. In the white fibers of Danube bleak three glycolytic enzymes display a parallel trend indicating an increase of the glycolytic potential with training of about 25% (Hinterleitner *et al.,* 1992). Lackner *et al.* (1988) show PCr levels dropping much more in control than in trained chub. Lactate and many glycolytic intermediates peak at exhaustion in controls, but continue to rise in trained fish. It has been shown in crucian carp that recovery to base lactate levels are much faster in trained fish (Johnston and Goldspink, 1973). It is suggested that in trained fish lactate production continues in muscle tissue following exercise, allowing other tissues to recover faster.

Although earlier work has failed to show major changes in muscle fiber ultrastructure more recent work has shown training responses also on the fine structural level. In the cyprinid species chub, we observed an increase in mitochondrial and lipid volume density of red muscle, indicating, together with increases in muscle type proportions and capillarization, an improved aerobic potential of trained fish (Sänger, 1992b). The second cyprinid examined in this study, the nase, shows little change in mitochondria and intracellular lipid, due to its higher content of these cell components per se (Tables II and III) which in turn is very probably related to the swimming activity. Nase, a herbivorous cyprinid, is known as a sustained swimmer, undertaking long spawning migrations. In contrast the chub is an omnivorous-carnivorous fish, showing burst-swimming activity. A recent study in our laboratory on a third cyprinid species, the above-mentioned Danube bleak, has shown no training response at the ultrastructural level. This fish maintained an irregular burst-like mode of swimming under the training program. Three different members of the same fish family show three different responses, indicating a significant influence of swimming performance and lifestyle on any training response pattern. However, a higher lipid content in red muscle has also been observed in salmonids together with increases in glycogen content (Davison and Goldspink, 1977). In addition, Davison and Goldspink showed that differences between anterior and posterior regions of the myotome of untrained fish disappear following training and suggested that fish forced to swim against a water current use more of the myotome compared with fish in still water. The different responses of salmonids and cyprinids to training can be explained by the different mitochondrial and lipid volume densities of these fish (Tables II and III) and the differences in the training regimes (duration and threshold level of exercise being reached) used in these studies. Similarly, so-called tonic muscle fibers increase not only their mitochondrial and lipid but also their myofibrillar content with training, indicating increased efficiency for sustained swimming locomotion (Sänger, 1997).

In summary, there is good evidence that endurance exercise training leads to increased aerobic potential in red, intermediate, and possibly, in some cases, even in white muscle. This improved aerobic capacity is due to better tissue capillary supply, increased mitochondrial and lipid content, and increases in enzyme activity. Enzymes involved in lipid metabolism are particularly affected, indicating that exercising fish use mainly lipid as a fuel. Increases in glycolytic potential in white muscle suggests increased lactate production. The corresponding improved capillary supply, as shown in the Danube bleak suggests enhanced ability to clear this lactate and to deliver substrates.

Most of the studies to date have employed sustainable swimming speeds with few using higher speeds such as those leading to rapid exhaustion. As Davison (1997) points out in his recent review, one should bear in mind that a training regime leading to rapid exhaustion must produce severe stress reactions and corresponding high levels of stress hormones which might outweigh any effects derived from the training regime itself. Exercise which leads to exhaustion involves short bouts of high intensity swimming and sprinting and is powered primarily by white myotomal muscle and supported by anaerobic metabolism. High speed training thus should have some effect on this muscle population. Gamperl *et al.*

(1988) have shown that anaerobic (sprint) training decreases growth and food consumption without altering body composition. Trained fish grow at 81% of the rate of control fish and have smaller white muscle fiber diameters, although it is not clear whether the latter is a training response or due to the much smaller size of the trained fish (Gamperl and Stevens, 1991). Pearson *et al.* (1990) report that sprint training enhances burst-swimming stamina. In white muscle fibers levels of glycogen and glycolytic enzymes are increased after sprint training. Trained fish, swimming faster and further, accumulate more lactate, but show no differences in glycogen or PCr depletion and less ATP depletion, suggesting that training increases uptake of exogenous glucose by white muscle in post-exercise gluconeogenesis. Such a system is beneficial in that glycogen synthesis from glucose requires an investment of only two ATP per glucosyl unit, while three are required when lactate is the substrate. Training enhances lactate clearance between 1 and 3 hr after exercise, allowing recovery from the higher loads to be completed in the same time as in untrained fish. Higher energy stores are maintained in muscle of trained fish after exercise and especially throughout recovery. Overall, sprint training minimizes endogenous fuel depletion during exhaustive swimming, even though swim speed (and distance) increases, and enhances the rate of metabolic recovery following the swim. Sprint training increases endurance in trout at short-time high swimming speeds. Glycogen repletion may be enhanced by the use of exogeneous glucose early in recovery (Pearson *et al.,* 1990).

Sprint training failed to alter volume densities of terminal cisternae, T-tubules, mitochondria, or lipid content (Gamperl and Stevens, 1991).

With regard to the data reviewed here, it is difficult to draw overall conclusions in terms of training effects on fish muscle. There is a great deal of variability in the results, often with conflicting data which may be a result of comparing fish species with different morphological and physiological properties (for example, different energy requirements) as well as condition and duration of exercise training (food composition, swimming speed).

IV. SUMMARY

The main muscle fiber types of fish axial muscle, namely red, white, intermediate, or pink, so-called tonic, red muscle rim, transitional zone, and scattered dorsal and ventral fibers, are described with respect to their location, origin, fine structure, and biochemical and functional characteristics. Ontogenic changes in structural and functional features are discussed, as well as interspecific variations. Red muscle fibers, the most aerobic fiber type, are characterized by high amounts of oxidative enzymes, mitochondria, lipid droplets, and capillaries. In white muscle fibers, metabolic economy is less important than maximum power output, therefore their characteristics emphasize their recruitment for burst-swimming activities. Intermediate muscle fibers, as their name suggests, are between red and white muscle fibers in a variety of their characteristics. The variety of these and other fiber types is described. A further focus of attention is the plasticity of muscle phenotype. Particularly, the factors that lead to adaptational changes in fish axial muscle (season, temperature, oxygen tension, and exercise training) are discussed. Cold acclimation/acclimatization, hypoxic conditions, and endurance training demonstrate a general trend for transition to more aerobic fiber types and a more aerobic type of muscle.

ACKNOWLEDGEMENTS

We are very grateful to John R. Haslett and Paul T. Loughna, and three anonymous reviewers for their constructive criticisms of earlier versions of this chapter. Their comments and corrections are much appreciated. Andreas Zankl helped in preparing the digitalized illustrations. Some of the authors' own work was supported by FWF grant number P12190-BIO to A. M. S.

REFERENCES

Akster, H. A. (1983). A comparative study of fibre type characteristics and terminal innervation in head and axial muscle of the carp (*Cyprinus carpio* L.): a histochemical and electronmicroscopical study. *Neth. J. Zool.* **33**, 164–188.

Akster, H. A. (1985). Morphometry of muscle fibre types in the carp (*Cyprinus carpio* L.). *Cell Tissue Res.* **241**, 193–201.

Alexander, R. McN. (1969). The orientation of muscle fibres in the myomeres of fishes. *J. Mar. Biol. Assoc. U.K.* **49**, 263–290.

Alsop, D. H., and Wood, C. M. (1997). The interactive effects of feeding and exercise on oxygen consumption, swimming performance and protein usage in juvenile rainbow trout (*Oncorhynchus mykiss*). *J. Exp. Biol.* **200**, 2337–2346.

Altringham, J. D., and Johnston, I. A. (1990). Modelling muscle power output in a swimming fish. *J. Exp. Biol.* **148**, 395–402.

Archer, S. D., and Johnston, I. A. (1989). Kinematics of labriform and subcarangiform swimming in the Antarctic fish *Notothenia neglecta*. *J. Exp. Biol.* **143**, 195–210.

Arthur, P. G., West, T. G., Brill, R. W., Schulte, P. M., and Hochachka, P. W. (1992). Recovery metabolism of skipjack tuna (*Katsuwonus pelamis*) white muscle: rapid and parallel changes in lactate and phosphocreatine after exercise. *Can. J. Zool.* **70**, 1230–1239.

Arthur, P. G., West, T. G., and Hochachka, P. W. (1993). Guides to improving white muscle performance: the roles of glycolysis, phosphagens, lactate and intracellular buffers in tuna. *In* "Hypoxia and Molecular Medicine," Vol. VI, pp. 67–71, Queen City Print, Lake Louise.

Ball, D., and Johnston, I. A. (1996). Molecular mechanisms underlying the plasticity of muscle contractile properties with temperature acclimation in the marine fish *Myoxocephalus scorpius*. *J. Exp. Biol.* **199**, 1363–1373.

Barets, A. (1961). Contribution a l'etude des systemes moteur lent et rapide du muscle lateral des teleosteens. *Arch. Anat. Morph. Exp.* **50**, 91–187.

Bennett, A. F. (1978). Activity metabolism of the lower vertebrates. *Annu. Rev. Physiol.* **40**, 447–469.
Bennett, A. F. (1990). Thermal dependence of locomotor capacity. *Am. J. Physiol.* **259**, R253–R258.
Billeter, R., Weber, H., Lutz, Howald, H., Eppenberger, H., and Jenny, E. (1980). Myosin types in human skeletal muscle fibres. *Histochemistry* **65**, 249–259.
Billeter, R., Heizmann, C. W., Howald, H., and Jenny, E. (1981). Analysis of myosin light and heavy chain types in single human skeletal muscle fibres. *Eur. J. Biochem.* **116**, 389–395.
Bischoff, R. (1990). Interaction between satellite cells and skeletal muscle fibers. *Development* **109**, 943–952.
Block, B. A. (1986). Structure of the brain and eye heater tissues in marlins, sailfish, and spearfish. *J. Morphol.* **190**, 169–189.
Boddeke, R., Slijper, E. J., and Van der Stelt, A. (1959). Histological characteristics of the body musculature in fishes in connection with their mode of life. *Proc. K. Ned. Akad. Wet. Ser.* **C 62**, 576–588.
Bone, Q. (1964). Patterns of muscular innervation in the lower chordates. *Int. Rev. Neurobiol.* **6**, 99–147.
Bone, Q. (1966). On the function of the two types of myotomal muscle fibres in elasmobranch fish. *J. Mar. Biol. Assoc. U.K.* **46**, 321–349.
Bone, Q. (1975). Muscular and energetic aspects of fish swimming. *Swimming Flying Nature* **2**, 493–528.
Bone, Q. (1978). Locomotor muscle. *In* "Fish Physiology" (Hoar, W. S., and Randall, D. J., eds.), Vol. VII, pp. 361–424. Academic Press, New York, San Francisco, London.
Bone, Q., Kicenuik, J., and Jones, D. R. (1978). On the role of the different fibre types in fish myotomes at intermediate speeds. *Fish Bull.* **76**, 691–699.
Bone, Q., Johnston, I. A., Pulsford, A., and Ryan, K. P. (1986). Contractile properties and ultrastructure of three types of muscle fibres in the dogfish myotome. *J. Muscle Res. Cell Motil.* **7**, 47–56.
Bottinelli, R., Schiaffino, S., and Reggiani, C. (1991). Force-velocity relations and myosin heavy chain isoform compositions of skinned fibres from rat skeletal muscle. *J. Physiol. Lond.* **437**, 655–672.
Brett, J. R., and Groves, T. D. D. (1979). Physiological energetics. *In* "Fish Physiology" (Hoar, W. S., Randall, D. J., and Brett, J. R., eds.), Vol. VIII, pp. 279–352. Academic Press, New York.
Brill, R. W. (1996). Selective advantages conferred by the high performance physiology of tunas, billfishes, and dolphin fish. *Comp. Biochem. Physiol.* **113A**, 3–15.
Brooks, S., and Johnston, I. A. (1993). Influence of development and rearing temperature on the distribution, ultrastructure and myosin-subunit composition of myotomal muscle fibre types in the plaice, *Pleuronectes platessa. Mar. Biol.* **117**, 501–513.
Brooks, S., Vieira, V. L. A., Johnston, I. A., and Macheru, P. (1995). Muscle development in larvae of fast growing tropical freshwater fish, the curimata-pacu. *J. Fish Biol.* **47**, 1026–1037.
Broughton, N. M., Goldspink, G., and Jones, N. V. (1981). Histological differences in the lateral musculature of 0-group roach, *Rutilus rutilus* (L.) from different habitats. *J. Fish Biol.* **18**, 117–122.
Calvo, J., and Johnston, I. A. (1992). Influence of rearing temperature on the distribution of muscle fibre types in the *Scophthalmus maximus* at metamorphosis. *J. Exp. Mar. Biol. Ecol.* **161**, 45–55.
Carey, F. G. (1982). A brain heater in the swordfish. *Science* **216**, 1327–1329.
Carpene, E., and Veggetti, A. (1981). Increase in muscle fibres in the lateralis muscle (white portion) of Mugilidae (Pisces, Teleostei). *Experientia* **37**, 191–193.
Chayen, N. E., Rowlerson, A. M., and Squire, J. M. (1993). Fish muscle structure: fibre types in flatfish and mullet fin muscles using histochemistry and antimyosin antibody labelling. *J. Muscle Res. Cell Motil.* **14**, 533–542.
Childress, J. J., and Seibel, B. A. (1998). Life at stable low oxygen levels: adaptations of animals to oceanic oxygen minimum layers. *J. Exp. Biol.* **201**, 1223–1232.
Coughlin, D. J., and Rome, L. C. (1996). The roles of pink and red muscle in powering steady swimming in scup, *Stenotomus chrysops. Am. Zool.* **36**, 666–677.

Coughlin, D. J., Zhang, G., and Rome, L. C. (1996). Contraction dynamics and power production of pink muscle of the scup (*Stenotomus chrysops*). *J. Exp. Biol.* **199,** 2703–2712.
Crockford, T., and Johnston, I. A. (1990). Temperature acclimation and the expression of contractile protein isoforms in the skeletal muscles of the common carp (*Cyprinus carpio* L.). *J. Comp. Physiol.* **160,** 23–30.
Curtin, N. A., and Davies, R. E. (1975). Very high tension with very little ATP breakdown by active skeletal muscle. *J. Mechanochem. Cell Motil.* **5,** 147–154.
Curtin, N. A., and Woledge, R. C. (1988). Power output and force-velocity relationship of live fibres from white myotomal muscle of the dogfish, *Scyliorhinus canicula. J. Exp. Biol.* **140,** 187–197.
Curtin, N. A., and Woledge, R. C. (1993). Efficiency of energy conversion during sinusoidal movement of red muscle fibres from the dogfish, *Scyliorhinus canicula. J. Exp. Biol.* **185,** 195–206.
Davie, P. S., Wells, R. M. G., and Tetens, V. (1986). Effects of sustained swimming on rainbow trout muscle structure, blood oxygen transport, and lactate dehydrogenase isozymes: evidence for increased aerobic capacity of white muscle. *J. Exp. Zool.* **237,** 159–171.
Davison, W. (1983). Changes in muscle cell ultrastructure following exercise in *Salmo trutta. Experientia* **39,** 1017–1018.
Davison, W. (1994). Exercise training in the banded wrasse *Notolabrus fucicola* affects muscle fibre diameter but not muscle mitochondrial morphology. *N. Z. Nat. Sci.* **21,** 11–16.
Davison, W. (1997). The effects of exercise training on teleost fish, a review of recent literature. *Comp. Biochem. Physiol.* **117A,** 67–75.
Davison, W., and Goldspink, G. (1977). The effect of prolonged exercise on the lateral musculature of the brown trout (*Salmo trutta). J. Exp. Biol.* **70,** 1–12.
Davison, W., and Goldspink, G. (1978). The effect of training on the swimming muscles of the goldfish (*Carassius auratus). J. Exp. Biol.* **74,** 115–122.
Davison, W., Goldspink, G., and Johnston, I. A. (1976). The division of labour between fish myotomal muscles during swimming. *J. Physiol. Lond.* **263,** 185–186.
Davison, W., Forster, M. E., Franklin, C. E., and Taylor, H. H. (1988). Recovery from exhausting exercise in an Antarctic fish, *Pagothenia borchgrevinki. Polar Biol.* **8,** 167–171.
De Graaf, F., van Raamsdonk, W., Hasselbaink, H., Diegenbach, P. C., Mos, W., Smit-Onel, M. J., van Asselt, E., and Heuts, B. (1990). Enzyme histochemistry of the spinal chord and the myotomal musculature in the teleost fish *Brachydanio rerio*. Effects of endurance training and prolonged reduced locomotory activity. *Z. Mikrosk. Anat. Forsch.* **104,** 593–606.
Devoto, S. H., Melancon, E., Eisen, J. S., and Westerfield, M. (1996). Identification of separate slow and fast muscle precursor cells *in vivo,* prior to somite-formation. *Development* **122,** 3371–3380.
Dickson, K. A. (1996). Locomotor muscle of high-performance fishes: What do comparisons of tunas with ectothermic sister taxa reveal? *Comp. Biochem. Physiol. A* **113,** 39–49.
Driedzic, W. R., and Hochachka, P. W. (1976). Control of energy metabolism in fish white muscle. *Am. J. Physiol.* **230,** 579–582.
Du, S. J., Devoto, S. H., Westerfield, M., and Moon, R. T. (1997). Positive and negative regulation of muscle cell identity by members of the *hedgehog* and *TGF*-beta gene families. *J. Cell Biol.* **139,** 145–156.
Dunn, J. F., and Johnston, I. A. (1986). Metabolic constraints on burst-swimming in the Antarctic teleost *Notothenia neglecta. Mar. Biol.* **91,** 433–440.
Dunn, J. F., Archer, S. D., and Johnston, I. A. (1989). Muscle fibre types and metabolism in post-larval and adult stages of nototheniod fish. *Polar Biol.* **9,** 213–223.
Eaton, R. C., Bombardieri, R. A., and Meyer, O. H. (1977). The Mauthner-initiated startle response in teleost fish. *J. Exp. Biol.* **66,** 65–81.
Egginton, S., and Cordiner, S. (1997). Cold-induced angiogenesis in seasonally acclimatized rainbow trout (*Oncorhynchus mykiss). J. Exp. Biol.* **200,** 2263–2268.
Egginton, S., and Johnston, I. A. (1982a). Muscle fibre differentiation and vascularisation in the juvenile European eel (*Anguilla anguilla* L.). *Cell Tissue Res.* **222,** 563–577.

Egginton, S., and Johnston, I. A. (1982b). A morphometric analysis of regional differences in myotomal muscle ultrastructure in the juvenile eel (*Anguilla anguilla* L.). *Cell Tissue Res.* **222,** 579–596.

Egginton, S., and Sidell, B. D. (1989). Thermal acclimation induces adaptive changes in subcellular structure of fish skeletal muscle. *Am. J. Physiol.* **256,** R1–R9.

El-Fiky, N., Hinterleitner, S., and Wieser, W. (1987). Differentiation of swimming muscles and gills, and development of anaerobic power in the larvae of cyprinid fish (Pisces, Teleostei). *Zoomorphology* **107,** 126–132.

Farrell, A. P., Johansen, J. A., Steffensen, J. F., Moyes, C. D., West, T. G., and Suarez, R. K. (1990). Effects of exercise training and coronary artery ablation on swimming performance, heart size, and cardiac enzymes in rainbow trout, *Oncorhynchus mykiss. Can. J. Zool.* **68,** 1174–1179.

Farrell, A. P., Johansen, J. A., and Suarez, R. K. (1991). Effects of exercise-training on cardiac performance and muscle enzymes in rainbow trout, *Oncorhynchus mykiss. Fish Physiol. Biochem.* **9,** 303–312.

Felsenfeld, A. L., and Curry, M. (1988). A nonmotile zebrafish mutant affecting the earliest stages of myofibrillogenesis. *J. Cell. Biochem.* **12C** (Supplement), 337.

Fleming, J. R., Crockford, T., Altringham, J. D., and Johnston, I. A. (1990). Effects of temperature acclimation on muscle relaxation in the carp: a mechanical, biochemical, and ultrastructural study. *J. Exp. Zool.* **255,** 286–295.

Focant, B., and Huriaux, F. (1976). Light chains of carp and pike skeletal muscle myosins. Isolation and characterization of the most anodic light chain on alkaline pH electrophoresis. *FEBS Lett.* **65,** 16–19.

Focant, B., Huriaux, F., and Johnston, I. A. (1976). Subunit composition of fish myofibrils: the light chains of myosin. *Int. J. Biochem.* **7,** 129–133.

Focant, B., Melot, F., Collin, S., Chikou, A., Vandewalle, P., and Huriaux, F. (1999). Muscle parvalbumin isoforms of *Clarias gariepinus, Heterobranchus longifilis* and *Chrysichthys auratus:* isolation, characterization and expression during development. *J. Fish Biol.* **54,** 832–851.

Franklin, C. E., and Johnston, I. A. (1997). Muscle power output during escape responses in an Antarctic fish. *J. Exp. Biol.* **200,** 703–712.

Freadman, M. A. (1979). Role of partitioning of swimming musculature of striped Bass, *Morone saxatilis* Walbaum and Bluefish, *Pomatomus saltatrix* L. *J. Fish Biol.* **15,** 417–423.

Fry, F. E. J., and Hart, J. S. (1948). Cruising speed of goldfish in relation to water temperature. *J. Fish Res. Bd. Can.* **7,** 175–199.

Gamperl, A. K., and Stevens, E. D. (1991). Sprint-training effects on trout (*Oncorhynchus mykiss*) white muscle structure. *Can. J. Zool.* **69,** 2786–2790.

Gamperl, A. K., Bryant, J., and Stevens, E. D. (1988). Effect of sprint training protocol on growth rate, conversion efficiency, food consumption and body composition of rainbow trout, *Salmo gairdneri* Richardson. *J. Fish Biol.* **33,** 861–870.

Gerday, C. (1982). Soluble calcium-binding proteins from fish and invertebrate muscle. *Mol. Physiol.* **2,** 63–87.

Gerlach, G. F., Turay, L., Malik, K. T. A., Lida, J., Scutt, A., and Goldspink, G. (1990). Mechanisms of temperature acclimation in the carp: a molecular biological approach. *Am. J. Physiol.* **259,** R237–R244.

Gill, H. S., Weatherley, A. H., Lee, R., and Legere, D. (1989). Histochemical characterization of myotomal muscle of five teleost species. *J. Fish Biol.* **34,** 375–386.

Gillis, G. B. (1998). Neuromuscular control of anguilliform locomotion: patterns of red and white muscle activity during swimming in the American eel *Anguilla rostrata. J. Exp. Biol.* **201,** 3245–3256.

Goldspink, G. (1975). Biochemical energetics of fast and slow muscles. *In* "Comparative Physiology—Functional Aspects of Structural Materials" (Bolis, L., Mandrell, H. P., and Schmidt-Nielson, K., eds.), pp. 173–185. Elsevier, Amsterdam.

Goldspink, G. (1985). Malleability of the motor system: a comparative approach. *J. Exp. Biol.* **115,** 375–391.
Goldspink, G., Howells, K. F., and Ward, P. S. (1976). Effects of exercise on muscle fibre size. In "Medicine Sport **9.** Advances in Exercise Physiology" (Jokl, E., ed.), pp. 103–113. Karger, Basel.
Goldspink, G., Turay, L., Hansen, E., Ennion, S., and Gerlach, G. (1992). Switches in fish myosin genes induced by environment temperature in muscle of the carp. *Symp. Soc. Exp. Biol.* **46,** 139–149.
Granzier, H. L. M., Wiersma, J., Akster, H. A., and Osse, J. W. M. (1983). Contractile properties of a white- and a red-fibre type of the m. hyohyoideus of the carp (*Cyprinus carpio* L.). *J. Comp. Physiol.* **149,** 441–449.
Greaser, M. L., Moss, R. L., and Reiser, P. J. (1988). Variations in contractile properties of rabbit single muscle fibres in relation to troponin T isoforms and myosin light chains. *J. Physiol.* **406,** 85–98.
Greene, C. W., and Greene, C. H. (1913). The skeletal musculature of the king salmon. *Bull. U. S. Bur. Fish.* **33,** 25–59.
Greer-Walker, M. (1970). Growth and development of the skeletal muscle fibres of the cod (*Gadus morhua* L.). *J. Cons. Int. Explor. Mer.* **33,** 228–244.
Greer-Walker, M. (1971). Effect of starvation and exercise on the skeletal muscle fibres of the cod (*Gadus morhua* L.) and the coalfish (*Gadus virens* L.), respectively. *J. Cons. Int. Explor. Mer.* **33,** 421–427.
Greer-Walker, M., and Emerson, L. (1978). Sustained swimming speeds and myotomal muscle function in the trout, *Salmo gairdneri. J. Fish Biol.* **13,** 475–481.
Greer-Walker, M., and Pull, G. A. (1973). Skeletal muscle function and sustained swimming speeds in the coalfish *Gadus virens* L. *Comp. Biochem. Physiol.* **44,** 495–501.
Greer-Walker, M., and Pull, G. A. (1975). A survey of red and white muscle in marine fish. *J. Fish Biol.* **7,** 295–300.
Guderley, H. (1990). Functional significance of metabolic responses to thermal acclimation in fish muscle. *Am. J. Physiol.* **259,** R245–R252.
Halver, J. E. (1972). "Fish Nutrition." Academic Press, New York.
Hamoir, G., Focant, B., and Disteche, M. (1972). Proteinic criteria of differentiation of white, cardiac and various red muscles in carp. *Comp. Biochem. Physiol.* **41,** 665–674.
Hanel, R., Karjalainen, J., and Wieser, W. (1996). Growth of swimming muscles and its metabolic cost in larvae of whitefish at different temperatures. *J. Fish Biol.* **48,** 937–951.
Hardewig, I., Van Dijk, P. L. M., and Poertner, O. H. (1998). High-energy turnover at low temperatures: recovery from exhaustive exercise in Antarctic and temperate eelpouts. *Am. J. Physiol.* **274,** R1789–R1796.
Hazel, J. R., and Prosser, C. L. (1974). Molecular mechanisms of temperature compensation in poikilotherms. *Physiol. Rev.* **54,** 620–677.
Heap, S. P., and Goldspink, G. (1986). Alterations to the swimming performance of carp, *Cyprinus carpio*, as a result of temperature acclimation. *J. Fish Biol.* **29,** 747–753.
Heap, S. P., Watt, P. W., and Goldspink, G. (1985). Consequences of thermal change on the myofibrillar ATPase of 5 freshwater teleosts. *J. Fish Biol.* **26,** 733–738.
Hinterleitner, S., Huber, M., Lackner, R., and Wieser, W. (1992). Systemic and enzymatic responses to endurance training in two cyprinid species with different life styles (Teleostei: Cyprinidae). *Can. J. Fish. Aquat. Sci.* **49,** 110–115.
Holeton, G. F., and Randall, D. J. (1967). Changes in blood pressure in the rainbow trout during hypoxia. *J. Exp. Biol.* **46,** 297–305.
Holloszy, J. O., and Booth, F. W. (1976). Biochemical adaptations to endurance exercise in muscle. *Annu. Rev. Physiol.* **38,** 273–291.

Holopainen, I. J., and Hyvärinen, H. (1985). Ecology and physiology of crucian carp (*Carassius carassius* L.) in small Finnish ponds with anoxic conditions in winter. *Verh. Int. Ver. Limnol.* **22**, 2566–2570.

Hoppeler, H. (1986). Exercise-induced ultrastructural changes in skeletal muscle. *Int. J. Sports Med.* **7**, 187–204.

Hoppeler, H., Howald, H., Conley, K. E., Lindstedt, S. L., Claassen, H., Vock, P., and Weibel, E. R. (1985). Endurance training in humans: aerobic capacity and structure of skeletal muscle. *J. Appl. Physiol.* **59**, 320–327.

Houlihan, D. F., and Laurent, P. (1987). Effects of exercise training on the performance, growth, and protein turnover of rainbow trout (*Salmo gairdneri*). *Can. J. Fish. Aquat. Sci.* **44**, 1614–1621.

Hubley, M. J., Locke, B. R., and Moerland, T. S. (1997). Reaction-diffusion analysis of the effects of temperature on high-energy phosphate dynamics in goldfish skeletal muscle. *J. Exp. Biol.* **200**, 975–988.

Hudson, R. C. L. (1973). On the function of the white muscles in teleosts at intermediate swimming speeds. *J. Exp. Biol.* **58**, 509–522.

Hughes, G. M. (1973). Respiratory responses to hypoxia in fish. *Am. Zool.* **13**, 475–489.

Huriaux, F., and Focant, B. (1985). Electrophoretic and immunological study of myosin light chains from fresh water teleost fishes. *Comp. Biochem. Physiol.* **82**, 737–743.

Huriaux, F., Melot, F., Vandewalle, P., Collin, S., and Focant, B. (1996). Parvalbumin isotypes in white muscle from three teleost fish. Characterization and their expression during development. *Comp. Biochem. Physiol. B* **113**, 475–484.

Hwang, G. C., Watabe S., and Hashimoto, K. (1990). Changes in carp myosin ATPase induced by temperature acclimation. *J. Comp. Physiol.* **160**, 233–239.

Jayne, B. C., and Lauder, G. V. (1994). How swimming fish use slow and fast muscle fibers: implications for models of vertebrate muscle recruitment. *J. Comp. Physiol. A* **175**, 123–131.

Jayne, B. C., and Lauder, G. V. (1995). Red muscle motor patterns during steady swimming in largemouth bass: effects of speed and correlations with axial kinematics. *J. Exp. Biol.* **198**, 1575–1587.

Jobling, M. (1994). "Fish Bioenergetics." Chapman & Hall, New York.

Johnson, T. P., and Bennett, A. F. (1995). The thermal acclimation of burst escape performance in fish: an integrated study of molecular and cellular physiology and organismal performance. *J. Exp. Biol.* **198**, 2165–2175.

Johnson, T. P., and Johnston, I. A. (1991). Power output of fish muscle fibres performing oscillatory work: effects of acute and seasonal temperature change. *J. Exp. Biol.* **157**, 409–423.

Johnson, T. P., Bennett, A. F., and McLister, J. D. (1996). Thermal dependence and acclimation of fast start locomotion and its physiological basis in rainbow trout (*Oncorhynchus mykiss*). *Physiol. Zool.* **69**, 276–292.

Johnston, I. A. (1980). Specializations of fish muscle. *In* "Development and Specializations of Muscle. Soc. Exp. Biol. Sem. Ser. Symp. 7" (Goldspink, D. F., ed.), pp. 123–148. Cambridge University Press, Cambridge.

Johnston, I. A. (1981). Structure and function of fish muscles. *Symp. Zool. Soc. Lond.* **48**, 71–113.

Johnston, I. A. (1982a). Biochemistry of fish myosins and contractile properties of fish skeletal muscle. *J. Mol. Physiol.* **2**, 15–29.

Johnston, I. A. (1982b). Capillarization, oxygen diffusion distances and mitochondrial content of carp muscles following acclimation to summer and winter temperatures. *Cell Tissue Res.* **222**, 325–337.

Johnston, I. A. (1982c). Quantitative analyses of ultrastructure and vascularization of the slow muscle fibres of the anchovy. *Tissue Cell* **14**, 319–328.

Johnston, I. A. (1983). On the design of fish myotomal muscles. *Mar. Behav. Physiol.* **9**, 83–98.

Johnston, I. A. (1987). Respiratory characteristics of muscle fibres in a fish (*Chaenocephalus aceratus*) that lacks haem pigments. *J. Exp. Biol.* **133**, 415–428.

Johnston, I. A. (1993a). Temperature influences muscle differentiation and the relative timing of organogenesis in herring (*Clupea harengus*) larvae. *Mar. Biol.* **116,** 363–379.
Johnston, I. A. (1993b). Phenotypic plasticity of fish muscle to temperature change. *In* "Fish Ecophysiology" (Rankin, J. C., and Jensen, F. B., eds.), pp. 322–340. Chapman & Hall, London.
Johnston, I. A. (1994). Developmental aspects of temperature adaptation in fish muscle. *Bas. Appl. Myol.* **4,** 353–368.
Johnston, I. A., and Bernard, L. M. (1982a). Routine oxygen consumption and characteristics of the myotomal muscle in tench: effects of long-term acclimation to hypoxia. *Cell Tissue Res.* **227,** 161–177.
Johnston, I. A., and Bernard, L. M. (1982b). Ultrastructure and metabolism of skeletal muscle fibres in the tench: effects of long-term acclimation to hypoxia. *Cell Tissue Res.* **227,** 179–199.
Johnston, I. A., and Bernard, L. M. (1983). Utilization of the ethanol pathway in carp following exposure to anoxia. *J. Exp. Biol.* **104,** 73–78.
Johnston, I. A., and Bernard, L. M. (1984). Quantitative study of capillary supply to the skeletal muscles of crucian carp *Carassius carassius* L.: effects of hypoxic acclimation. *Physiol. Zool.* **57,** 9–18.
Johnston, I. A., and Goldspink, G. (1973). Quantitative studies of muscle glycogen utilization during sustained swimming in crucian carp (*Carassius carassius* L.). *J. Exp. Biol.* **59,** 607–615.
Johnston, I. A., and Herring, P. J. (1985). The transformation of muscle into bioluminescent tissue in the fish *Benthalbella infans* Zugmayer. *Proc. R. Soc. Lond. B* **225,** 213–218.
Johnston, I. A., and Horne, Z. (1994). Immunocytochemical investigations of muscle differentiation in the Atlantic herring (*Clupea harengus*: Teleostei). *J. Mar. Biol. Assoc. U. K.* **74,** 79–91.
Johnston, I. A., and Lucking, M. (1978). Temperature induced variation in the distribution of different types of muscle fibre in the goldfish (*Carassius auratus*). *J. Comp. Physiol.* **124,** 111–116.
Johnston, I. A., and Maitland, B. (1980). Temperature acclimation in crucian carp (*Carassius carassius* L.); morphometric analyses of muscle fibre ultrastructure. *J. Fish Biol.* **17,** 113–125.
Johnston, I. A., and Moon, T. W. (1980a). Endurance exercise training in the fast and slow muscles of a teleost fish (*Pollachius virens*). *J. Comp. Physiol.* **135,** 147–156.
Johnston, I. A., and Moon, T. W. (1980b). Exercise training in skeletal muscle of brook trout (*Salvelinus fontinalis*). *J. Exp. Biol.* **87,** 177–194.
Johnston, I. A., and Moon, T. W. (1981). Fine structure and metabolism of multiply innervated fast muscle fibres in teleost fish. *Cell Tissue Res.* **219,** 93–109.
Johnston, I. A., Patterson, S., Ward, P. S., and Goldspink, G. (1974). The histochemical demonstration of myofibrillar adenosine triphosphatase activity in fish muscle. *Can. J. Zool.* **52,** 871–877.
Johnston, I. A., Davison, W., and Goldspink, G. (1975a). Adaptations in Mg^{2+}-activated myofibrillar ATPase activity induced by temperature acclimation. *FEBS Lett.* **50,** 293–295.
Johnston, I. A., Ward, P. S., and Goldspink, G. (1975b). Studies on the swimming musculature of the rainbow trout. I. Fibre types. *J. Fish Biol.* **7,** 451–458.
Johnston, I. A., Davison, W., and Goldspink, G. (1977). Energy metabolism of carp swimming muscles. *J. Comp. Physiol.* **114,** 203–216.
Johnston, I. A., Bernard, L. M., and Maloiy, G. M. (1983). Aquatic and aerial respiration rates, muscle capillary supply and mitochondrial volume density in the air breathing catfish (*Clarias mossambicus*), acclimated to either aerated or hypoxic water. *J. Exp. Biol.* **105,** 317–338.
Johnston, I. A., Sidell, B. D., and Driedzic, W. R. (1985). Force-velocity characteristics and metabolism of carp muscle fibres following temperature acclimation. *J. Exp. Biol.* **119,** 239–249.
Johnston, I. A., Camm, J.-P., and White, M. (1988). Specialisations of swimming muscles in the pelagic Antarctic fish *Pleuragramma antarcticum*. *Mar. Biol.* **100,** 3–12.
Johnston, I. A., Fleming, J. D., and Crockford, T. (1990). Thermal acclimation and muscle contractile properties in cyprinid fish. *Am. J. Physiol.* **259,** R231–R236.
Johnston, I. A., Van Leeuwen, J. L., Davies, M. L. F., and Beddow, T. (1995). How fish power predation fast-starts. *J. Exp. Biol.* **198,** 1851–1861.

Johnston, I. A., Calvo, J., Guderley, H., Fernandez, D., and Palmer, L. (1998). Latitudinal variation in the abundance and oxidative capacities of muscle mitochondria in perciform fishes. *J. Exp. Biol.* **201**, 1–12.
Jones, P. L., and Sidell, B. D. (1982). Metabolic responses of striped Bass (*Morone saxatilis*) to temperature acclimation. II. Alterations in metabolic carbon sources and distributions of fiber types in locomotory muscle. *J. Exp. Zool.* **219**, 163–171.
Josephson, R. K. (1985). Mechanical power output from striated muscle during cyclic contractions. *J. Exp. Biol.* **114**, 493–512.
Karasinski, J., Zawadowska, B., and Supikova, I. (1994). Myosin isoforms in selected muscle fibre types of the pond loach *Misgurnus fossilis* L. *Comp. Biochem. Physiol.* **107B**, 249–253.
Katz, S. L., and Shadwick, R. E. (1998). Curvature of swimming fish midlines as an index of muscle strain suggests swimming muscle produces net positive work. *J. Theor. Biol.* **193**, 243–256.
Kaufmann, R., and Wieser, W. (1992). Influence of temperature and ambient oxygen on the swimming energetics of cyprinid larvae and juveniles. *Environ. Biol. Fishes* **33**, 87–95.
Kilarski, W. (1990). Histochemical characterization of myotomal muscle in the roach, *Rutilus rutilus* (L.). *J. Fish Biol.* **36**, 353–362.
Kilarski, W., and Kozlowska, M. (1983). Ultrastructural characteristic of the teleostean muscle fibres and their nerve endings. The Stickleback (*Gasterosteus aculeatus* L.). *Z. Mikrosk. Anat. Forsch.* **97**, 1022–1036.
Kilarski, W., and Kozlowska, M. (1985). Histochemical and electron microscopical analysis of muscle fiber in myotomes of teleost fish. *Gegenbaurs Morphol. Jahrb.* **131**, 55–72.
Kilarski, W., and Kozlowska, M. (1987). Comparison of ultrastructure and morphometrical analysis of tonic, white and red muscle fibers in the myotome of teleost fish (*Noemacheilus barbatulus* L.). *Z. Mikrosk. Anat. Forsch.* **101**, 636–648.
Kilarski, W., and Zawadowska, B. (1985). Distribution of tonic muscle fibers in myotomes of some teleosts. Proc. 14th Eur. Con. on Muscle and Motility, Ulm, 31B.
Kilarski, W. M., Romek, M., Kozlowska, M., and Görlich, A. (1996). Short-term thermal acclimation induces adaptive changes in the inner mitochondrial membranes of fish skeletal muscle. *J. Fish Biol.* **49**, 1280–1290.
Korneliussen, H., Dahl, H. A., and Paulsen, J. E. (1978). Histochemical definition of muscle fibre types in the trunk musculature of a teleost fish (Cod, *Gadus morhua* L.). *Histochemistry* **55**, 1–6.
Koumans, J. T. M., and Akster, H. A. (1995). Myogenic cells in development and growth of fish. *Comp. Biochem. Physiol.* **110A**, 3–20.
Koumans, J. T. M., Akster, H. A., Dulos, G. J., and Osse, J. W. M. (1990). Myosatellite cells of *Cyprinus carpio* (Teleostei) in vitro: isolation, recognition and differentiation. *Cell Tissue Res.* **261**, 173–181.
Kryvi, H., Flood, P. R., and Gulyaer, D. (1980). The ultrastructure and vascular supply of the fibre types in the sturgeon *Acipenser stellatus*. *Cell Tissue Res.* **212**, 114–126.
Kryvi, H., Flatmark, T., and Totland, G. H. (1981). The myoglobin content in red, intermediate and white fibres of the swimming muscles in three species of shark—a comparative study using high-performance liquid chromatography. *J. Fish Biol.* **18**, 331–338.
Lackner, R., Wieser, W., Huber, M., and Dalla Via, J. (1988). Responses of intermediary metabolism to acute handling stress and recovery in untrained and trained *Leuciscus cephalus* (Cyprinidae, Teleostei). *J. Exp. Biol.* **140**, 393–404.
Lännergren, J. (1987). Contractile properties and myosin isoenzymes of various kinds of *Xenopus* twitch muscle fibres. *J. Muscle Res. Cell Motil.* **8**, 260–273.
Langfeld, K. S., Crockford, T., and Johnston, I. A. (1991). Temperature acclimation in the common carp: force-velocity characteristics and myosin subunit composition of slow muscle fibres. *J. Exp. Biol.* **155**, 291–304.
Lauff, R. F., and Wood, C. M. (1996). Respiratory gas exchange, nitrogenous waste excretion and fuel usage during aerobic swimming in juvenile rainbow trout. *J. Comp. Physiol. B* **166**, 501–509.

Liu, D. W. C., and Westerfield, M. (1990). The formation of terminal fields in the absence of competitive interactions among primary motoneurons in the zebrafish. *J. Neurosci.* **10,** 3947–3959.

Lomholt, J. P., and Johansen, K. (1979). Hypoxia acclimation in carp: how it affects O_2 uptake, ventilation, and O_2 extraction from water. *Physiol. Zool.* **52,** 38–49.

Londraville, R. L., and Sidell, B. D. (1990). Ultrastructure of aerobic muscle in Antarctic fishes may contribute to maintenance of diffusive fluxes. *J. Exp. Biol.* **150,** 205–220.

Love, R. M., Munro, L. J., and Robertson, I. (1977). Adaptation of the dark muscle of cod to swimming activity. *J. Fish Biol.* **11,** 431–436.

Lowey, S., Waller, G., and Trybus, K. M. (1993). Skeletal muscle myosin light chains are essential for physiological speeds of shortening. *Nature* **365,** 454–456.

Luther, P. K., Munro P. M. G., and Squire, J. M. (1995). Muscle ultrastructure in the teleost fish. *Micron* **26,** 431–459.

McArdle, H. J., and Johnston, I. A. (1981). Ca^{2+}-uptake by tissue sections and biochemical characteristics of sarcoplasmic reticulum isolated from fish fast and slow muscles. *Eur. J. Cell Biol.* **25,** 103–107.

Martinez, I., Olsen, R. L., Ofstad, R., Janmot, C., and d'Albis, A. (1989). Myosin isoforms in mackerel (*Scomber scombrus*) red and white muscles. *FEBS Lett.* **252,** 69–72.

Mascarello, F., Romanello, M. G., and Scapolo, P. A. (1986). Histochemical and immunohistochemical profile of pink muscle fibres in some teleosts. *Histochemistry* **84,** 251–255.

Matsuoka, M., and Iwai, T. (1984). Development of the myotomal musculature in the Red Sea bream. *Bull. Jpn. Soc. Fish.* **50,** 29–35.

Meyer-Rochow, V. B., and Ingram, J. R. (1993). Red-white muscle distribution and fibre growth dynamics: a comparison between lacustrine and riverine populations of the Southern smelt *Retropinna retropinna* Richardson. *Proc. R. Soc. Lond. B Biol. Sci.* **252,** 85–92.

Milligan, C. L., and McDonald, D. G. (1988). In vivo lactate kinetics at rest and during recovery from exhaustive exercise in coho salmon (*Oncorhynchus kisutch*) and starry flounder (*Platichthys stellatus*). *J. Exp. Biol.* **135,** 119–131.

Morley, S. A., and Batty, R. S. (1996). The effects of temperature on "S-strike" feeding of larval herring *Clupea harengus* L. *Mar. Fresh. Behav. Physiol.* **28,** 123–136.

Mosse, P. R. L., and Hudson, R. C. L. (1977). Muscle fibre types identified in the myotomes of marine teleosts: a behavioural, anatomical and histochemical study. *J. Fish Biol.* **11,** 417–430.

Mourik, J., Raeven, P., Steur, K., and Addink, A. O. T. (1982). Anaerobic metabolism of red skeletal muscle of goldfish, *Carassius auratus* (L.). Mitochondrial produced acetaldehyde as anaerobic electron acceptor. *FEBS Lett.* **137,** 111–114.

Moyes, C. D., Buck, L. T., Hochachka, P. W., and Suarez, R. K. (1989). Oxidative properties of carp red and white muscle. *J. Exp. Biol.* **143,** 321–331.

Moyes, C. D., Buck, L. T., and Hochachka, P. W. (1990). Mitochondrial and peroxisomal fatty acid oxidation in elasmobranchs. *Am. J. Physiol.* **258,** R756–R762.

Moyes, C. D., Mathieu-Costello, O. A., Brill, R. W., and Hochachka, P. W. (1992). Mitochondrial metabolism of cardiac and skeletal muscles from a fast (*Katsuwonus pelamis*) and a slow (*Cyprinus carpio*) fish. *Can. J. Zool.* **70,** 1246–1253.

Nag, A. C. (1972). Ultrastructure and adenosine triphosphatase activity of red and white muscle fibers of the caudal region of a fish, *Salmo gairdneri*. *J. Cell Biol.* **55,** 42–57.

Nag, A. C., and Nursall, J. R. (1972). Histogenesis of white and red muscle fibres of trunk muscles of a fish *Salmo gairdneri*. *Cytobios* **6,** 227–246.

O'Connell, C. P. (1981). Development of organ systems in the northern anchovy, *Engraulis mordax*, and other teleosts. *Am. Zool.* **21,** 429–446.

Ono, R. D. (1983). Dual motor innervation in the axial musculature of fishes. *J. Fish Biol.* **22,** 395–408.

Palzenberger, M., and Pohla, H. (1989). Estimates of respiratory areas of fish gills: a critical view. *Progr. Zool.* **35,** 569–572.

Patterson, J. M., and Zakon, H. H. (1997). Transdifferentiation of muscle to electric organ: regulation of muscle-specific proteins is independent of patterned nerve activity. *Dev. Biol.* **186,** 115–126.

Patterson, S., and Goldspink, G. (1976). Mechanisms of myofibril growth and proliferation in fish muscle. *J. Cell Sci.* **22,** 607–616.

Patterson, S., Johnston, I. A., and Goldspink, G. (1975). A histochemical study of the lateral muscles of five teleost species. *J. Fish Biol.* **7,** 159–166.

Pearson, M. P., Spriet, L. L., and Stevens, E. D. (1990). Effect of sprint training on swim performance and white muscle metabolism during exercise and recovery in rainbow trout (*Salmo gairdneri*). *J. Exp. Biol.* **149,** 45–60.

Penney, R. K., and Goldspink, G. (1980). Temperature adaptation of sarcoplasmic reticulum of fish muscle. *J. Therm. Biol.* **5,** 63–68.

Penney, R. K., and Goldspink, G. (1981a). Compensation limits of fish muscle myofibrillar ATPase enzyme to environmental temperature. *J. Therm. Biol.* **4,** 269–272.

Penney, R. K., and Goldspink, G. (1981b). Regulatory proteins and thermostability of myofibrillar ATPase in acclimated goldfish. *Comp. Biochem. J.* **178,** 373–379.

Pierobon Bormioli, S., Sartore, S., Vitadello, M., and Schiaffino, S. (1980). "Slow" myosins in vertebrate skeletal muscle. An immunofluorescence study. *J. Cell Biol.* **85,** 672–681.

Ramirez-Zarzoza, G., Gil, F., Moreno, F., Vazquez, J. M., and Latorre, R. (1991). Estudio histoquimico de las fibras musculares de algunos teleosteos marinos y de agua dulce. *Anat. Histol. Embryol.* **20,** 169–179.

Ramirez-Zarzoza, G., Gil, F., Latorre, R., Ortega, A., Garcia-Alcaraz, A., Abellan, E., Vazquez, J. M., Lopez-Albors, O., and Moreno, F. (1995). The larval development of lateral musculature in gilthead sea bream *Sparus auratus* and sea bass *Dicentrarchus labrax*. *Cell Tissue Res.* **280,** 217–224.

Rodnick, K. J., and Sidell, B. D. (1995). Effects of body size and thermal acclimation on parvalbumin concentration in white muscle of striped bass. *J. Exp. Zool.* **272,** 266–274.

Rome, L. C. (1990). Influence of temperature on muscle recruitment and muscle function in vivo. *Am. J. Physiol.* **259,** R210–R222.

Rome, L. C. (1994). The mechanical design of the fish muscular system. In "Mechanics and Physiology of Animal Swimming" (Maddock, L., Bone, Q., and Rayner, J. M. V., eds.), pp. 75–97. Cambridge University Press, Cambridge.

Rome, L. C., and Sosnicki, A. A. (1990). The influence of temperature on mechanics of red muscle in carp. *J. Physiol. Lond.* **427,** 151–169.

Rome, L. C., and Swank, D. (1992). The influence of temperature on power output and scup red muscle during cyclical length changes. *J. Exp. Biol.* **171,** 261–281.

Rome, L. C., Loughna, P. T., and Goldspink, G. (1984). Muscle fibre activity in carp as a function of swim speed and muscle temperature. *Am. J. Physiol.* **247,** R272–278.

Rome, L. C., Loughna, P. T., and Goldspink, G. (1985). Temperature acclimation: improved sustained swimming performance in carp at low temperatures. *Science* **228,** 194–196.

Rome, L. C., Funke, R. P., Alexander, R. McN., Lutz, G., Aldridge, H., Scott, F., and Freadman, M. (1988). Why animals have different muscle fibre types. *Nature* **335,** 824–827.

Rome, L. C., Sosnicki, A. J., and Choi, I.-H. (1992a). The influence of temperature on muscle function in the fast swimming scup. II. The mechanics of red muscle. *J. Exp. Biol.* **163,** 281–295.

Rome, L. C., Choi, I.-H., Lutz, G., and Sosnicki, A. J. (1992b). The influence of temperature on muscle function in the fast swimming scup. I. Shortening velocity and muscle recruitment during swimming. *J. Exp. Biol.* **163,** 259–279.

Rome, L. C., Swank, D., and Corda, D. (1993). How fish power swimming. *Science* **261,** 340–343.

Rowlerson, A., and Spurway, N. C. (1988). Histochemical and immunohistochemical properties of skeletal muscle fibres from *Rana* and *Xenopus*. *Histochemistry* **20,** 657–673.

Rowlerson, A., Scapolo, P. A., Mascarello, F., Carpene, E., and Veggetti, A. (1985). Comparative study of myosins present in the lateral muscle of some fish; species variations in myosin isoforms and their distribution in red, pink and white muscle. *J. Muscle Res. Cell Motil.* **6**, 601–640.

Sänger, A. M. (1992a). Quantitative fine structural diversification of red and white muscle fibres in cyprinids. *Environ. Biol. Fishes* **33**, 97–104.

Sänger, A. M. (1992b). Effects of training on axial muscle of two cyprinid species: *Chondrostoma nasus* (L.) and *Leuciscus cephalus* (L.). *J. Fish Biol.* **40**, 637–646.

Sänger, A. M. (1993). Limits to the acclimation of fish muscle. *Rev. Fish Biol. Fish.* **3**, 1–15.

Sänger, A. M. (1997). The so-called tonic muscle fibre type in cyprinid axial muscle: their morphology and response to endurance exercise training. *J. Fish Biol.* **50**, 487–497.

Sänger, A. M. (1999). Morphometric analysis of axial muscle in cyprinid fish provides support that intracellular lipid plays a role in slow fibre oxygen supply. *J. Fish Biol.* **54**, 1029–1037.

Sänger, A. M., and Lackner, R. (1991). Training effects on the fine structure of fish axial muscle. *J. Muscle Res. Cell Motil.* **12**, 83.

Sänger, A. M., Claassen, H., and Adam, H. (1988). The arrangement of muscle fiber types in the axial muscle of various cyprinids (Teleostei). *Zool. Anz.* **221**, 44–49.

Sänger, A. M., Goldschmid, A., and Adam, H. (1989). Muscle fiber type distribution of various cyprinids. *Progr. Zool.* **35**, 561–563.

Sänger, A. M., Kim, Z. S., and Adam, H. (1990). The fine structure of muscle fibres of roach, *Rutilus rutilus* (L.), and chub, *Leuciscus cephalus* (L.), Cyprinidae, Teleostei: interspecific differences and effects of habitat and season. *J. Fish Biol.* **36**, 205–213.

Scapolo, P. A., and Rowlerson, A. (1987). Pink lateral muscle in the carp (*Cyprinus carpio* L.): histochemical properties and myosin composition. *Experientia* **43**, 384–386.

Scapolo, P. A., Veggetti, A., Mascarello, F., and Romanello, M. G. (1988). Developmental transitions of myosin isoforms and organisation of the lateral muscle in the teleost *Dicentrarchus labrax* (L.). *Anat. Embryol.* **178**, 287–295.

Schulte, P. M., Moyes, C. D., and Hochachka, P. W. (1992). Integrating metabolic pathways in post-exercise recovery of white muscle. *J. Exp. Biol.* **166**, 181–195.

Shadwick, R. E., Steffensen, J. F., Katz, S. L., and Knower, T. (1998). Muscle dynamics in fish during steady swimming. *Am. Zool.* **38**, 755–770.

Shoubridge, E. A., and Hochachka, P. M. (1980). Ethanol: novel end-product of vertebrate anaerobic metabolism. *Science* **209**, 308–309.

Sidell, B. D. (1980). Response of goldfish (*Carassius auratus* L.) muscle to acclimation temperature: alterations in biochemistry and proportions of different fibre types. *Physiol. Zool.* **53**, 98–107.

Sidell, B. D. (1998). Intracellular oxygen diffusion: the roles of myoglobin and lipid at cold body temperature. *J. Exp. Biol.* **201**, 1118–1127.

Sidell, B. D., and Johnston, I. A. (1985). Thermal sensitivity of contractile function in chain pickerel *Esox niger*. *Can. J. Zool.* **63**, 811–816.

Sisson, J. E., III., and Sidell, B. D. (1987). Effect of thermal acclimation on muscle fiber recruitment of swimming striped bass (*Morone saxatilis*). *Physiol. Zool.* **60**, 310–320.

Spierts, I. L. Y., and Van Leeuwen, J. L. (1999). Kinematics and muscle dynamics of C- and S-starts of carp (*Cyprinus carpio* L.). *J. Exp. Biol.* **202**, 393–406.

Spierts, I. L. Y., Akster, H. A., and Granzier, H. L. (1997). Expression of titin isoforms in red and white muscle fibres of carp (*Cyprinus carpio* L.) exposed to different sarcomere strains during swimming. *J. Comp. Physiol. B* **167**, 543–551.

Staron, R. S., and Pette, D. (1986). Correlation between myofibrillar ATPase activity and myosin heavy chain composition in rabbit muscle fibres. *Histochemistry* **86**, 19–23.

Steffensen, J. F. (1997). Metabolic cold adaptation in Antarctic fish? Ann. Meeting SEB.

Stoiber, W. (1991). Differenzierung der Rumpfmuskulatur larvaler Cypriniden: Feinstruktur und Morphometrie. Diplomarbeit, Salzburg, pp. 239.

Stoiber, W. (1996). Ontogenesis of axial muscle in teleost fish: an investigation into the source of new muscle fibres and the temperature dependence of growth dynamics. Ph.D. thesis, Salzburg, pp. 373.

Stoiber, W., Haslett, J. R., Goldschmid, A., and Sänger, A. M. (1998). Patterns of superficial fibre formation in the European pearlfish (*Rutilus frisii meidingeri*) provide a general template for slow muscle development in teleost fish. *Anat. Embryol.* **197,** 485–496.

St. Pierre, J., Charest, P.-M., and Guderley, H. (1998). Relative contribution of quantitative and qualitative changes in mitochondria to metabolic compensation during seasonal acclimatisation of rainbow trout *Oncorhynchus mykiss. J. Exp. Biol.* **201,** 2961–2970.

Tatarczuch, L., and Kilarski, W. (1982). Histochemical analysis of muscle fibers in myotome of teleost fishes (*Carassius auratus gibelio*). *Fol. Histochem. Cytochem.* **20,** 143–170.

Taylor, S. E., Egginton, S., and Taylor, E. W. (1996). Seasonal temperature acclimatisation of rainbow trout: cardiovascular and morphometric influences on maximal sustainable exercise level. *J. Exp. Biol.* **199,** 835–845.

te Kronnie, G., Tatarczuch, L., van Raamsdonk, W., and Kilarski, W. (1983). Muscle fibre types in the myotome of stickleback, *Gasterosteus aculeatus* L.: a histochemical, immunohistochemical and ultrastructural study. *J. Fish Biol.* **22,** 303–316.

Temple, G. K., and Johnston, I. A. (1997). The thermal dependence of fast-start performance in fish. *J. Therm. Biol.* **22,** 391–401.

Temple, G. K., and Johnston, I. A. (1998). Testing hypotheses concerning the phenotypic plasticity of escape performance in fish of the family Cottidae. *J. Exp. Biol.* **201,** 317–331.

Thys, T. (1997). Spatial variation in epaxial muscle activity during prey strike in largemouth bass (*Micropterus salmoides*). *J. Exp. Biol.* **200,** 3021–3031.

Timmons, B. A., Araujo, J., and Thomas, T. R. (1985). Fat utilization enhanced by exercise in a cold environment. *Med. Sci. Sports Exerc.* **17,** 673–678.

Totland, G. K., Kryvi, H., Bone, Q., and Flood, P. R. (1981). Vascularization of the lateral muscle of some elasmobranchiomorph fishes. *J. Fish Biol.* **18,** 223–234.

Totland, G. K., Kryvi, H., Jodestol, K. A., Christiansen, E. N., Tangeras, A., and Slinde, E. (1987). Growth and composition of the swimming muscle of adult Atlantic salmon (*Salmo salar* L.) during long-term sustained swimming. *Aquaculture* **66,** 299–313.

Tullis, A., and Block. B. A. (1997). Histochemical and immunohistochemical studies on the origin of the blue marlin heater cell phenotype. *Tissue Cell* **29,** 627–642.

Turner, J. D., Wood, C. M., and Clark, D. (1983a). Lactate and protein dynamics in the rainbow trout (*Salmo gairdneri*). *J. Exp. Biol.* **104,** 247–268.

Turner, J. D., Wood, C. M., and Hobe, H. (1983b). Physiological consequences of severe exercise in the inactive benthic flathead sole (*Hippoglossoides elassodon*): a comparison with the active pelagic rainbow trout (*Salmo gairdneri*). *J. Exp. Biol.* **104,** 269–288.

Tyler, S., and Sidell, B. D. (1984). Changes in mitochondrial distribution and diffusion distances in muscle of goldfish (*Carassius auratus*) upon acclimation to warm and cold temperatures. *J. Exp. Zool.* **232,** 1–10.

Unguez, G. A., and Zakon, H. H. (1998). Phenotypic conversion of distinct muscle fiber populations to electrocytes in a weakly electric fish. *J. Comp. Neurol.* **399,** 20–34.

Usher, M. L., Stickland, N. C., and Thorpe, J. E. (1994). Muscle development in Atlantic salmon (*Salmo salar*) embryos and the effect of temperature on muscle cellularity. *J. Fish Biol.* **44,** 953–964.

van den Thillart, G. (1982). Adaptations of fish energy metabolism to hypoxia and anoxia. *Mol. Physiol.* **2,** 49–61.

van den Thillart, G. (1986). Energy metabolism of swimming trout (*Salmo gairdneri*). *J. Comp. Physiol. B* **156,** 511–520.

van den Thillart, G., Kesbeke, F., and van Waarde, A. (1980). Anaerobic energy metabolism of gold-

fish, *Carassius auratus* (L.): influence of hypoxia and anoxia on phosphorylated compounds and glycogen. *J. Comp. Physiol.* **136,** 45–52.
van den Thillart, G., van Waarde, A., Muller H. J., Erkelens, C., Addink, A., and Lugtenburg, J. (1989). Fish muscle energy metabolism measured by in vivo ^{31}P-NMR during anoxia and recovery. *Am. J. Physiol.* **256,** R922–R929.
van Raamsdonk, W., Pool, C. W., and te Kronnie, G. (1978). Differentiation of muscle fiber types in the teleost *Brachydanio rerio. Anat. Embryol.* **153,** 137–155.
van Raamsdonk, W., te Kronnie, G., Pool, C. W., and van de Laarse, W. (1980). An immune histochemical and enzymic characterization of the muscle fibres in myotomal muscle of the teleost *Brachydanio rerio,* Hamilton-Buchanan. *Acta histochem.* **67,** 200–216.
van Raamsdonk, W., van't Veer, L., Veeken, K., Heyting, C., and Pool, C. W. (1982). Differentiation of muscle fiber types in the teleost *Brachydanio rerio,* the zebrafish. *Anat. Embryol.* **164,** 51–62.
van Waarde, A. (1983). Aerobic and anaerobic ammonia production by fish. *Comp. Biochem. Physiol.* **74B,** 675–684.
van Waarde, A., and De Wilde-van Berg Henegouwen, M. (1982). Nitrogen metabolism in goldfish, *Carassius auratus* (L.). Pathways of aerobic and anaerobic glutamate oxidation in goldfish liver and muscle mitochondria. *Comp. Biochem. Physiol.* **72B,** 133–136.
Veggetti, A., Mascarello, F., Scapolo, P., and Rowlerson, A. (1990). Hyperplastic and hypertrophic growth of lateral muscle in *Dicentrarchus labrax* (L.). *Anat. Embryol.* **182,** 1–10.
Veggetti, A., Mascarello, F., Scapolo, P. A., Rowlerson, A., and Candia Carnevali, M. D. (1993). Muscle growth and myosin isoform transitions during development of a small teleost fish, *Poecilia reticulata* (Peters) (Atheriniformes, Poeciliidae): a histochemical, immunohistochemical, ultrastructural and morphometric study. *Anat. Embryol.* **187,** 353–361.
Videler, J. J. (1993). "Fish Swimming." Chapman & Hall, London.
Vieira, V. L. A., and Johnston, I. A. (1992). Influence of temperature on muscle fibre development in larvae of the herring *Clupea harengus. Mar. Biol.* **112,** 333–341.
Vieira, V. L. A., and Johnston, I. A. (1996). Muscle development in the tambaqui, an important Amazonian food fish. *J. Fish Biol.* **49,** 842–853.
Wachtel, A. W. (1964). The ultrastructural relationships of electric organs and muscle. I. Filamentous systems. *J. Morphol.* **114,** 325–360.
Wakeling, J. M., and Johnston, I. A. (1998). Muscle power output limits fast-start performance in fish. *J. Exp. Biol.* **201,** 1505–1526.
Waterman, R. E. (1969). Development of the lateral musculature in the teleost, *Brachydanio rerio:* A fine structural study. *Am. J. Anat.* **125,** 457–494 (1969).
Weatherley, A. H., and Gill, H. S. (1981). Characteristics of mosaic muscle growth in the rainbow trout *Salmo gairdneri. Experientia* **37,** 1102–1103.
Webb, P. W. (1978). Fast-start performance and body form in seven species of teleost fish. *J. Exp. Biol.* **74,** 211–226.
Weber, J.-M., and Haman, F. (1996). Pathways for metabolic fuels and oxygen in high performance fish. *Comp. Biochem. Physiol.* **113A,** 33–38.
Wells, R. M. G., Grigg, G. C., Beard, L. A., and Summers, G. (1989). Hypoxic responses in a fish from a stable environment: blood oxygen transport in the antarctic fish *Pagothenia borchgrevinki. J. Exp. Biol.* **141,** 97–111.
Wieser, W. (1995). Energetics of fish larvae, the smallest vertebrates. *Acta Physiol. Scand.* **154,** 279–290.
Wieser, W., and Kaufmann, R. (1998). A note on interactions between temperature, viscosity, body size, and swimming energetics in fish larvae. *J. Exp. Biol.* **201,** 1369–1372.
Willemse, J. J., and van den Berg, P. G. (1978). Growth of striated muscle fibres in the m. lateralis of the European eel *Anguilla anguilla* L. (Pisces, Teleostei). *J. Anat.* **125,** 447–460.

Winterbottom, R. (1974). A descriptive synonymy of the striated muscles of the teleostei. *Proc. Acad. Natl. Sci. Philadelphia* **125,** 225–317.
Wissing, J., and Zebe, E. (1988). The anaerobic metabolism of the bitterling *Rhodeus amarus* (Cyprinidae, Teleostei). *Comp. Biochem. Physiol.* **89,** 299–303.
Woledge, R. C. (1992). Relaxation as a determinant of the locomotory role of muscle. Ann. Meeting SEB.
Wood, S. C., and Johansen, K. (1972). Adaptations to hypoxia by increased HbO_2 affinity and decreased red cell ATP concentration. *Nature New Biol.* **237,** 278–279.
Young, P. S., and Cech, J. J. (1993). Improved growth, swimming performance, and musculare development in exercised-conditioned young-of-the-year striped bass (*Morone saxatilis*). *Can. J. Fish. Aquat. Sci.* **50,** 703–707.
Zammit, V. A., and Newsholme, E. A. (1979). Activities of enzymes of fat and ketone-body metabolism and effects of starvation on blood concentration of glucose and fat fuels in teleost and elasmobranch fish. *Biochem. J.* **184,** 313–322.
Zawadowska, B., and Karasinski, J. (1988). Parvalbumins, isoforms of myosin heavy chain in histochemically defined fish skeletal muscles. *In* "Sarcomeric and Non-Sarcomeric Muscles: Basic and Applied Research Prospects for the 90's" (Carraro, U., ed.), pp. 549–554. Unipress, Padova.
Zawadowska, B., and Kilarski, W. (1984). Histochemical characterization of the muscle fiber types of the teleost (*Esox lucius* L.). *Acta Histochem.* **75,** 91–100.
Zhang, G., Swank, D. M., and Rome, L. C. (1996). Quantitative distribution of muscle fiber types in the scup *Stenotomus chrysops*. *J. Morphol.* **229,** 71–81.

8

HORMONAL REGULATION OF MUSCLE GROWTH

THOMAS P. MOMMSEN
THOMAS W. MOON

I. Introduction
II. Growth in Fish
III. Non-Hormonal Regulation of Growth
   A. Localization
   B. Exercise
   C. Specific Dynamic Action
   D. Cell Volume
IV. Hormonal Effects
   A. Growth Hormone
   B. Prolactin
   C. Insulin-Like Growth Factors
   D. Insulin
   E. Thyroid Hormones
   F. Cortisol
   G. Gonadal Steroids
   H. Adrenergic Agonists
V. Future Directions
   Acknowledgments
   References

I. INTRODUCTION

Physiologically, growth is usually defined as an increase in cell number and/or cell size concomitant with a positive change in energetic (caloric) content of the organism under consideration. Therefore, the normal, ongoing turnover of body constituents does not fall within this definition, unless a positive contribution is made to whole body energetics; i.e., rates of synthesis of body constituents exceed those for simple replacement rates. Operationally, different definitions are used to describe fish growth. Most common is the use of the "condition factor" (k), relating weight and length of a fish by the equation k = weight · [length]$^{-3}$ · 100. While this definition may be useful on the fish farm, it clearly lacks the important energetic component that needs to be considered when true growth is assessed.

Partitioning of resources into growth in vertebrates is controlled by numerous biotic and abiotic factors, with any factor altering the energetics impacting growth. Generally, these factors range from temperature and photoperiod, through food quality and availability and animal density to circannual variability, reproductive phases, and pollutants. For fish, one should also consider oxygen concentration, salinity, medium density, water quality, and water flow. Further, inter- and intraspecific competition, social interactions, predation, and disease can impact growth rates, even if, or especially if, all other parameters have been strictly controlled. Finally, genetic factors may be involved at every step in this process and always are an overriding consideration when discussing factors affecting growth.

II. GROWTH IN FISH

And still, for two important reasons, fish growth does not follow the general picture derived for other vertebrates. First, with the exception of some short-lived species, such as annual fishes, fish growth is indeterminate, i.e., growth continues throughout their lives. Although the relative growth rates are decreased with age, the same does not apply to absolute growth rates that have been shown to stay nearly constant over a size range from 45 to 800 g in crucian carp (*Carassius carassius*) (Dutta, 1994). The reasons for the low relative rates of growth in adult fishes are many, but may include the potential limitations of the intestinal system that scales differently than muscle to animal mass and diffusion rates in the white muscle itself.

The specific growth rate (G) scales with body weight (W) according to the allometric function $G = a \cdot W^b$, where *a* is the growth rate of a fish weighing 1 g; *b*, the weight exponent, is a negative number, empirically determined to be between −0.35 and −0.45. However, even under "optimum" conditions, fish growth is not constant, and periods of weight gain may alternate with periods of growth in length (Brown, 1974) or fast growth intervals alternate with cycles of lower growth rates (Noel and Le Bail, 1997). In arctic char (*Salvelinus alpinus*), for instance, cyclical growth rates are correlated with food conversion efficiency (Dabrowski *et al.*, 1992). Superimposed on such internal periodicity can be other phases governed by lunar, reproductive, or circannual cycles (Horseman and Meier, 1978; Farbridge and Leatherland, 1987).

Second, fish muscle grows by hyperplasia (increase in cell number) as well as by hypertrophy (increase in cell size). The post-hatch hyperplasia is a feature that sets the fishes apart from all other vertebrates, where hyperplastic muscle growth occurs only in very early ontogenetic stages and muscle fiber number is supposed to be fixed at birth—subsequent muscle growth is therefore primarily due to hypertrophy of existing muscle fibers. In the rainbow trout (*Oncorhynchus mykiss*), as in other fish species, muscle growth through hyperplasia continues well past the juvenile stages, although the ratio between the hyperplastic and the hypertrophic contributions to muscle growth decreases with age and size. Trout exceeding 65 cm no longer show any hyperplastic growth of muscle, but continue to grow by hypertrophy of existing fibers (Weatherley *et al.*, 1980; Stickland, 1983). The situation in growing carp (*Cyprinus carpio*) is not as clear-cut. Although the number of small (<200 μm^2) fibers tends to decrease with aging in carp, substantial numbers of small diameter fibers can be found throughout the observational growing period, which covered fish from 14 g (average fish weight) to 574 g (Alfei *et al.*, 1994). Periods of hyperplastic growth—assessed by immunocytochemical demonstration of 5-bromo-2'-deoxyuridine incorporation and presence of proliferating cell nuclear antigen (PCNA/cyclin) as a proliferation marker—appear to alternate with periods of muscle growth involving hypertrophy, although the two processes are not exclusive of each other (Alfei *et al.*, 1994).

Finally, the aquatic existence of fish enables them to exploit system buoyancy. Therefore, storing energy as hydrated, relatively low energy protein compared to lipid, does not come with the two penalties inherent to living on land: high weight and requirement of nitrogen detoxification for excretion. On both counts, fishes are at an advantage. First, minimizing weight gain is not crucial in a buoyant medium, and drag-generating increases in volume (scaling at W^{-3}) are at least partially compensated by the acquisition of functional tissue—muscle. Second, excess nitrogen produced during the utilization of the storage protein can be excreted efficiently and cheaply as ammonia or ammonium across gill and skin, except under a few extreme ambient conditions (pH, salinity, ammonia) that prevent direct release of ammonia (Randall *et al.*, 1989; McGeer *et al.*, 1994; Danulat, 1995).

Muscle accounts for more than half of the bulk of our piscine friends. Yet, the potentially overriding importance of this particular tissue to metabolism and growth has been less obvious. The lack of attention is likely due to the presumed sluggish metabolic rate of white muscle, its low fractional rate of protein synthesis, and low abundance of mitochondria and is compounded by the experimental difficulties to deal with muscle in animals lacking obvious muscular appendages that can be manipulated experimentally. However, a closer look should dispel these beliefs of metabolic inertia and kindle our interest into understanding the regulatory principles behind these fascinating "metabolic machines" (Hochachka, 1994). Protein synthesis in the crucian carp accounts for more than half of its total oxygen uptake (Meyer-Burgdorff and Rosenow, 1995), with liver and white muscle demanding the largest individual shares; white muscle tissue due to its bulk, and liver because of its extraordinarily high protein turnover. Even higher

Table I
The Importance of Muscle Protein Synthesis to Overall Protein Synthesis in Two Species of Teleost Fishes

	Fractional rate of protein synthesis	Growth rate efficiency
Atlantic cod (*Gadus morhua*)		
Gill	10.1	10%
Intestine	6.8	15%
Ventricle	4.8	24%
White muscle	1.94	50%
Crucian carp (*Carassius carassius*)		
Brain	0.41 ± 0.9	
Heart	1.24 ± 0.1	
Liver	10.0 ± 0.1	
White muscle	0.25 ± 0.15	
Red muscle	0.76 ± 0.1	

Note: In Atlantic cod, the fractional rate of protein synthesis is dependent on the overall growth rate. Data above are for fish growing at 1% day^{-1} (Houlihan *et al.*, 1989). Growth rates for carp were not stated (Smith *et al.*, 1996).

percentages can be expected for marine species, since carp, a freshwater species, expends a higher proportion of oxygen uptake on osmoregulation than marine species (Sparks *et al.*, 1999). Yet, as clearly seen from the data in Table I, it is the white muscle that makes an overwhelming contribution to overall protein accretion due to the exceptionally high efficiency of protein deposition. Within the white muscle (of rainbow trout), the myofibrillar fraction accounts for almost two thirds of the protein synthetic activity (Fauconneau *et al.*, 1995), implying that it is accretion of myofibrillar proteins that makes the single most important contribution to fish growth.

The many forms and functions for muscle tissues in fishes are covered in great detail elsewhere in this volume and need not be repeated here. However, at the outset of our task, we would like to make the case for increased attention to hormonal regulation of muscle growth in fishes. A simple calculation shall point out the overwhelming potential of muscle not only as a hormonal target, but also as a metabolic sink for nutrients, both in the short term, i.e., post prandially, and in the long term, when real growth can be observed. Fish muscle is the home of many hormone receptors, including those for the two known myogens, insulin and insulin-like growth factor I (IGF-I). The importance of insulin receptors is apparent immediately: the tissue is capable of binding as much insulin as the liver on a per gram tissue wet weight basis (Table IIA), with some differences in receptor abundance between species (Párrizas *et al.*, 1994b; 1995a). However, recalculated for a 100 g carp (*C. carpio*), where muscle accounts for at least 50% of the weight and the liver for 3%, the fish's body contains about 24-fold more insulin receptors

8. HORMONAL REGULATION OF MUSCLE GROWTH

Table II
Insulin Receptors in Fish Muscle: Abundance and Kinetic Characteristics

A. Insulin receptors

	Insulin binding (fmoles (g tissue)$^{-1}$)	Binding constant (K_d − nMol)	Specific binding (% of total)
Carp (*Cyprinus carpio*)			
Muscle	20.6	0.24	12.6
Liver	14.3	0.52	7.5
Brown trout (*Salmo trutta*)			
Muscle	7.2	0.47	2.5
Liver	4.4	0.76	2.4

Note: Data recalculated from Gutiérrez and Plisetskaya, 1994.

B. Insulin and IGF-I receptors in muscle of common carp and brown trout

	Insulin binding (fmoles (g tissue)$^{-1}$)	Binding constant (K_d − nMol)	IGF-I binding (fmoles (g tissue)$^{-1}$)	Binding constant (K_d − nMol)	Ratio (IGF/insulin)
Carp	25.1	0.28	51.6	0.26	2.1
Trout	6.6	0.41	22.1	0.43	3.3

Note: Data recalculated from Párrizas *et al.*, 1995a.

in the muscle than in the liver. If the slight difference in binding constant is taken into consideration, the gap widens even further; yet, the fish liver insulin receptor has attracted most of the scientific attention (Gutiérrez and Plisetskaya, 1991a,b; 1994). As shown in Table IIB for carp and brown trout (*Salmo trutta*), adult fish muscle has a two- to threefold greater ability to bind IGF-I than insulin; in other species of fishes, the ratio ranges between 2 and 6.4 (Párrizas *et al.*, 1995a). This situation is diametrically opposite that in mammals, where IGF-I is supposed to lose most of its importance as a muscle mitogen after birth, and as a result, insulin receptors are three- to fivefold more abundant than IGF-I receptors (Zorzano *et al.*, 1988). This mismatch between other tissues and white muscle is also reflected in the relative perfusion of tissues. Weight-specific blood flow in red muscle is 10 times higher than in white muscle, yet, because of its overall bulk, white muscle accounts for over 50% of total blood flow (Cameron, 1975).

A mixture of large diameter and small diameter fibers makes up fish white muscle, with the small diameter fibers thought to relate to the potential for hyperplastic growth. After hatching, the size of large fibers continues to increase, and the proportion of small diameter fibers decreases. As mentioned already, this mixture of hyperplasia and hypertrophy is unique to the fishes. In the context of hormonal regulation and early life history, it is interesting to note that temperature differentially affects the method of muscle accretion, depending on the species. Although total muscle mass at hatching is independent of temperature, two marine species (herring, *Clupea harengus;* plaice, *Pleuronectes platessa*) show hypertrophic muscle growth at the lower incubation temperatures, while in two species reared in freshwater (Atlantic salmon, *Salmo salar;* whitefish, *Coregonus lavaretus*), hyperplastic muscle growth prevails. At elevated incubation temperatures, opposing results were found; i.e., hyperplasia in the marine fish and hypertrophy in the freshwater species (Blaxter, 1992; Brooks and Johnston, 1993; Hanel *et al.,* 1996). One wonders whether this switch could be related to the differing amounts of free amino acids available in eggs of fresh- and saltwater species, as suggested by Blaxter (1992) who also points out that hypertrophy, which does not require differentiation, should be less energetically demanding than hyperplasia. An alternative explanation could lie with differential expression of insulin and IGF-I during phases of development (Chen *et al.,* 1994) and different temperature sensitivity of gene expression, similar to what was shown for temperature-sensitive myosin heavy chain (MHC) expression in carp (see below). Fish muscle contains satellite cells, myogenic stem cells, located between the sarcolemma and basal lamina of mature fibers. These are small spindle-shaped cells characterized by a heterochromatic nucleus and ribosomes and polysomes, but few other organelles. In mammals, muscle perturbation, such as exercise or denervation, will initiate division of satellite cells and result in the generation of myoblasts. These myoblasts, in turn, are useful as a source of nuclei for fusion and can either form new muscle fibers or fuse with existing fibers. The occurrence of satellite cells had been confirmed for fish muscle almost 30 years ago (Nag and Nursall, 1972) and their role in fish muscle growth has been established (Akster *et al.,* 1995). Fish myosatellite cells have also been used successfully to establish long term cell cultures that will differentiate to yield multinucleated, striated myotubes (Powell *et al.,* 1989, Koumans *et al.,* 1990) and constitute an extremely useful, albeit underutilized, experimental tool (Koumans and Akster, 1995).

The maintenance of MHC transcripts is noted throughout the time from hatching to adulthood in rainbow trout white muscle (Gauvry and Fauconneau, 1996) (see Chapter 5, this volume). This situation is in contrast to birds and mammals, where these transcripts are only present in embryos or during fetal development. As mentioned, the number of fibers does not increase beyond embryonic development, and skeletal muscle growth in adult mammals is due to hypertrophy of existing fibers only. In juvenile and adult fishes, in contrast, myotomal muscles grow by hyperplasia as well as hypertrophy. Both processes involve the myosatellite cells, which develop into the new muscle fiber in the case of hyperplasia and which fuse with existing muscle fibers in the case of hypertrophic growth. Obviously, differentiation is an ongoing process with fish growth, both in red and in white skeletal muscles. Since more than 85% of the cellular RNA is ribosomal, it is generally assumed that the total amount of RNA per tissue can serve as a relatively good proxy for protein synthesis, an assumption that is valid also for fish

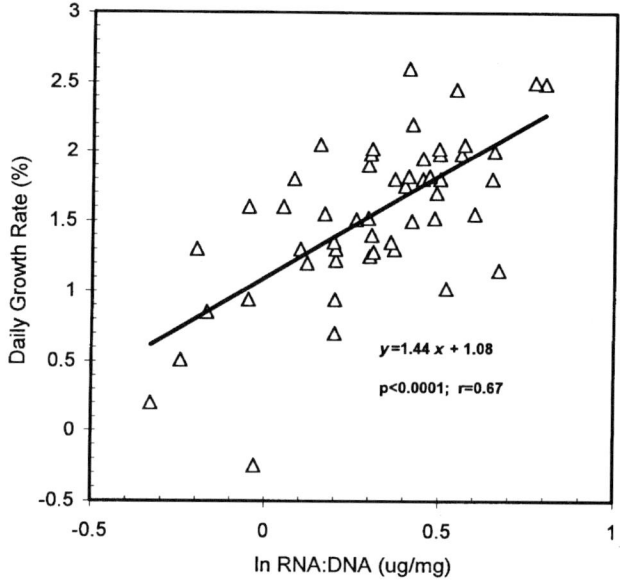

Fig. 1. Correlation of growth and muscular RNA/protein ratio in growing juvenile Atlantic salmon (*Salmo salar*). The specific growth rate (%body weight per day) measured over 10 days is plotted against the natural log of the total RNA to DNA ratio in salmon muscle. Fish weight ranged from 2.5 to 8 g at the start of the experiment. Regression equation is y = 1.44x + 1.08 (r = 0.67, $p < 0.001$). Data recalculated and replotted from Arndt *et al.*, 1994.

muscle (Fig. 1) (Arndt *et al.*, 1994), since overall growth rate of Atlantic salmon scales quite well with the natural logarithm of the ratio of RNA over DNA. A similar linear relationship holds between the ratio of ribosomal RNA to DNA and protein synthesis rates in white skeletal muscle of Atlantic cod (*Gadus morhua*) (Lied and Rosenlund, 1984) and a high RNA-DNA ratio could serve as a suitable indicator of recent growth rates in brown trout muscle (Grant, 1996). The fact that the DNA/protein ratio decreases with age in fishes, but the ratio itself is larger in red muscle than in white muscle of carp (*C. carpio*), implies a longer lasting importance of hyperplastic growth in red muscle than in white (Koumans *et al.*, 1994). In a direct comparison of slow- and fast-growing strains of rainbow trout, the slow-growing strain has significantly lower values for the muscle RNA/protein ratio. The faster growing strain, in contrast, is characterized by higher levels of muscle DNA, suggestive of higher cell number and smaller cell size and the importance of hyperplasia to growth (Valente *et al.*, 1998). The faster growing strain shows a larger contribution of hyperplasia to growth compared with the slower growing strain (Table III). During their first 9 months, as the strains reach body weights of 106 g (slow growth) and 193 g (fast growth), respectively, hyperplasia accounts for the majority of muscle growth. Interestingly, and reiterating a

Table III
Relative Contribution of Hyperplasia and Hypertrophy to Growth in Two Strains of Rainbow Trout (*Oncorhynchus mykiss*) Growing at Different Relative Rates

Age (from hatching)	4–19 days (whole animal)	60–89 days (muscle)	89–117 days (muscle)	117–146 days (muscle)	146–179 days (muscle)	179–243 days (muscle)
Mass (g)						
Slow strain	0.10	1.95	4.41	6.91	25.4	106
Fast strain	0.10	3.07	8.29	16.3	50.9	193
Hypertrophy (% of total growth)						
Slow strain	47	13	12	59	13	25
Fast strain	38	6	12	45	11	29
Hyperplasia (% of total growth)						
Slow strain	53	87	88	41	87	75
Fast strain	62	94	88	55	89	71

Note: Data from Valente *et al.*, 1998. Direct comparison of Cornec (fast-growing) and Mirwart (slow-growing) strains of rainbow trout (*O. mykiss*).

point made several times in this review, the contributions fluctuate between intervals, but not in a predictable manner, contradicting the previous statement about the decreasing importance of hyperplasia in rainbow trout past a critical weight.

In red muscles, indicator enzymes for oxidative metabolism, namely citrate synthase and hydroxyacyl-Coenzyme A dehydrogenase, increase with muscle accretion, while these enzymes decrease in white muscle. Interestingly, the muscle contents of cytochrome C oxidase increase with fish weight throughout life (Kiessling *et al.*, 1991), suggesting that differential expression continues through the life of a muscle cell. The effects of scaling with size in different muscles has been discussed extensively elsewhere (Walsh *et al.*, 1989; Somero and Childress, 1990; Goolish, 1991).

Since under normal conditions the rate of protein turnover in fish muscle is very low (usually below 0.1% per day), a strong correlation exists between muscle protein synthesis and muscle protein accretion rate. Further, because of the bulk of the tissue and again partially due to its low endogenous turnover rate, the rate of muscle protein synthesis is correlated with the accretion rate for whole body protein. As a result, the rate of whole body growth can be delineated from an assessment of muscle protein synthesis. The usefulness of this approach has been demonstrated with *in vitro* systems (Lied and Rosenlund, 1984; Houlihan *et al.*, 1995) as well as *in vivo* (Smith, 1981). Protein synthetic rate, measured using constant infusion of ^{14}C-tyrosine, was 0.38% day^{-1} in feeding rainbow trout (Smith, 1981), and only slightly higher in mullet (*Mugil cephalus*) at 0.54% day^{-1}

(Haschemeyer and Smith, 1979). The overall growth rate of the trout was slightly lower at 0.25%, implying that high protein turnover tissues such as liver (17.4% day^{-1}) or gill (4.7% day^{-1}) make an insignificant contribution to the overall protein turnover of the fish. Also, the observation that muscle, but not liver or gill, protein synthesis drops substantially after a 15-day fast at 12°C, from 0.38% day^{-1} to 0.09% day^{-1} (Smith, 1981), confirms the importance of muscle to overall protein synthesis and can be taken to indicate that muscle protein synthesis is under tight control. This line of reasoning can be taken further; namely to the level of the myofibrillar fraction that accounts for the majority (64%) of protein synthesis in white muscle of the rainbow trout (Fauconneau *et al.*, 1995). Further, MHCs represent a major fraction of the myofibrillar proteins, and hence constitute a useful and representative measure of overall protein accretion (Goll *et al.*, 1992).

Numerous factors are known to control or influence the rate of protein synthesis in muscle. In most vertebrates, these factors include physical parameters such as tension, immobilization, or denervation and chemical parameters, such as nutrients, ionic conditions, and not least hormones. Ultimately, the actual rate of protein synthesis will reflect the concerted actions of these factors, making it rather difficult to assess the importance or role of individual parameters. In the fishes, the unique composition of muscle tissues and their unique growth patterns add still another set of factors into the equation. Unfortunately, to date, little information has been gathered on these points in fish muscle and hence the picture arising is fragmentary at best. Because of the dearth of data for piscine systems, we have to draw on comparative models from other vertebrates, but their general applicability to fishes must be queried in numerous instances. The two fundamental differences between muscle growth in fishes and in other vertebrates are essential in this discussion. Muscle can "grow" by different methods, depending not only on biochemical input, but also on the definition. For instance, muscle in Pacific salmon species during the freshwater phase of their spawning migration will increase the bulk of their muscle by increasing the water content of the fibers, while at the same time, degrading and oxidizing the muscle constituents, largely protein and lipid (Mommsen *et al.*, 1980). To exclude such pseudogrowth, we want to reiterate that we define growth as an increase in caloric value, achieved by hypertrophy of existing cells, hyperplasia, or both. Because of the relative bulk of the muscle in fishes, it is unlikely that simple relocation of body components to the muscle from non-muscle tissues will ever be an important factor contributing to muscle growth. As a result, we can restrict our discussion to non-fasting fish, and to fish not undergoing metabolically unique phases in their life cycles and hence are actively increasing the bulk of their muscle tissue.

In fishes, just as in mammals, protein synthesis and degradation are under coordinated regulation, and the rate of muscle growth is tightly linked to degradation. This relationship certainly holds for a number of fish, including the rainbow trout as an exemplary species (Houlihan et al., 1995). However, protein degradation does not always follow this tenet. Under catabolic situations, brought about by long term starvation, anorexic exercise (as in salmon on their spawning migration) or hormonally by corticosteroids, sex steroids, or high doses of thyroid hormones, muscle proteolysis can function quite independently of protein synthesis.

Skeletal muscle needs to be supported not only from within, with collagen and other tensile and resilient structures, but muscle growth must be correlated, however loosely, with increases in the abundance of matrix structures, including cartilage and bone. After all, although buoyancy plays an important role to fish structure, the skeleton still supports the bulk of the carcass tissue, including muscle, and hence at least in non-piscine vertebrates, a relatively good correlation exists between skeletal size and body weight, especially during hyperplastic growth early in development.

Other, seemingly ancillary, parameters can influence muscle growth and performance. Cold acclimation in striped bass (*Morone saxatilis*), for instance, increases red muscle fatty acid binding protein by almost 40%, an increase that is instrumental in the temperature compensation of fatty acid oxidation (Londraville and Sidell, 1996); these changes coincide with low temperature-dependent proliferation in mitochondria and increases in activities of oxidative enzymes in red muscle, including key enzymes in the oxidation of fatty acids. Unfortunately, the metabolic cost of this increased expression of the binding proteins together with the other compensatory mechanisms remains to be assessed. An analysis of the potential effects on muscle fiber composition and on muscle growth per se could make a valuable contribution to our understanding of muscle adaptability under different physiological conditions, including temperature.

It is already clear that the overall scope of muscle adaptation seems to have been largely overshadowed by the researchers' inherent bias, namely that muscle, although plastic in more than just functional ways, is a metabolically slow and hence relatively unexciting tissue. Quite obviously, we feel that this aversion is utterly misplaced. We hope this review, together with the other chapters in this volume, serves to rekindle the interest in muscle biochemistry and physiology. We predict that some very exciting insights can be gained from these analyses, based upon the many uniquely piscine features covered here.

III. NON-HORMONAL REGULATION OF GROWTH

A. Localization

Depending on the location in the fish body, muscle fibers experience different strains during swimming. Strains are higher in posterior than anterior fibers of the same type and because of their larger distance from the body axis, red muscle

8. HORMONAL REGULATION OF MUSCLE GROWTH

fibers tend to be exposed to larger strain than white fibers (Altringham *et al.*, 1993; van Leeuwen, 1995). Therefore, location can be used to probe muscle differences related to the relative inherent strain-scape. The relative distribution of connectin (titin), an elastic protein, present in striated muscle and stretching between the Z- and the M-lines of the sarcomere provides an interesting case in point. Titin isoforms with high molecular mass are found in red muscle, while isoforms in white muscle tend to be smaller. Further, for both muscle types, the titin molecular masses were larger in posterior than in anterior muscles that experience relatively smaller strains (Spierts *et al.*, 1997). It therefore seems that strain and tension may influence the type and amount of protein synthesized to fit the functionality of individual areas of the locomotory machinery. Such positional differences may also provide a way to probe—in a natural setting—parameters that are accessible in mammals by exercise, inactivity, length restriction, denervation, or similar experimental treatments, but are less applicable to piscine muscle.

Based on studies on locomotion and geometric orientation, regional differences in function, composition and properties can be expected within the white muscle. Indeed, such regional differences have been found in enzyme titers and mitochondrial densities along the fish axis, with oxidative machinery predominating in the white muscle anterior to the caudal peduncle, while anaerobic enzyme activities are preeminent in the bulk (dorsal of the midsection) white muscle in Atlantic cod (*G. morhua;* Martínez *et al.*, 1999) and flying fish (*Hirundichthys affinis*) (T.P. Mommsen and M.D. Mallet, in preparation). It does not come as a surprise, then, that mobilization of protein during starvation is not uniform across all areas of white muscle. In the starving sturgeon (*Acipenser transmontanus*), for instance, protein is mobilized from the dorsal regions of the white muscle, an area that shows the slowest recovery in size during refeeding (Kiessling *et al.*, 1993). Similarly, in starving cod, enzymatic and functional integrity is maintained preferentially in the muscle fibers closest to the tail (Martínez *et al.*, 1999). Starvation also leads to a substantial decrease in the average size of white muscle fibers, while red muscle fibers seem to be barely affected, implying the preferential use of structural protein from white muscle. By analogy, one can expect that muscle growth will be subject to similar differences on different scales, where local strain and sensory and hormonal input will govern not only where muscle fibers will be rebuilt, but also what type of fiber composition will be on the menu. Similar considerations apply to the intracellular muscle constituents, where myofibrillar components appear to turn over at different rates than the rest of the muscle constituents (Velick, 1956; Fauconneau *et al.*, 1995).

There is much evidence in both mammals and fish (Fauconneau *et al.*, 1995) that red and cardiac muscles have higher fractional protein synthetic rates than white muscle. Fauconneau *et al.* (1995) noted using the phenylalanine flooding technique in white muscle of rainbow trout that the mitochondrial and nuclear fractions of white muscle had rates well above those of the total protein or myofibrillar fractions; white muscle was the only muscle type showing this difference. This difference in rates is thought to be scaled with tissue oxidative capacity (e.g., Lowery and Somero, 1990), although contractile activity may also be a significant contributing factor.

B. Exercise

Sustained exercise in fish is correlated with increases in growth and protein turnover and higher plasma levels of cortisol and growth hormone. The increase in growth may be due to reduced aggressiveness in exercising fish, diluting the dominance structure—and with it feeding dominance—and increasing appetite. In some species, exogenous growth hormone increases both dominance and feed uptake (Johnsson and Björnsson, 1994; Johnsson et al., 1996). The concomitant rise in cortisol may increase the rate of muscle protein degradation (for a discussion of the proteolytic actions of cortisol, cf. Mommsen et al., 1999) and thus offset some of the growth-promoting effects of growth hormone, together resulting in the observed increase in protein turnover. In addition, moderate sustained exercise will alter the composition of muscle fibers, reflected in augmented oxidative potential of white fibers and increased relative amounts of red fibers (Davie et al., 1986).

Mammals also demonstrate exercise-induced changes in muscle protein synthesis, although the precise mechanisms responsible have not been clearly demonstrated (Tipton and Wolfe, 1998). Generally, exercise results in increased muscle protein synthesis, probably related to enhanced blood flow and amino acid delivery to the tissues. In humans, IGF-I increases during physical exercise (Eliakim et al., 1998), a specific response declining with age. Chronic exercise will restore IGF-I sensitivity of aged-muscle by increasing receptor protein and binding affinities (Willis et al., 1998). By analogy, similar changes in IGF-I sensitivity may be responsible for decreased relative growth rates seen as fish age.

C. Specific Dynamic Action

Following the infusion with amino acids, simulating a meal, the metabolic activity of fish increases, reflected in higher rates of oxygen uptake and ammonia excretion. This so-called specific dynamic action (SDA) is observed after any proteinaceous meal, and in response to amino acid infusion in channel catfish (*Ictalurus punctatus*), and can be prevented by preceding infusion with the protein synthesis inhibitor cycloheximide (Brown and Cameron, 1991). It appears therefore that SDA is largely correlated with amino acid handling, and especially protein biosynthesis. The cycloheximide effect can also be interpreted to mean that under normal conditions the protein synthetic pathway in fish muscle is limited by the availability of amino acids (as in mammals; see Tipton and Wolfe, 1998).

However, hormonal influences should not be ignored. As pointed out elsewhere, many amino acids are important insulinotropins in fishes, and resulting increased insulin levels could be instrumental in accelerating the amino acid uptake into muscle, and thus in protein synthesis. Simple mass-action effects may not, however, explain everything. It can further be concluded that protein synthesis in muscle is a potent regulator of intracellular and, secondarily plasma, amino acid pools. It is straightforward to estimate that the amino acids and proteins in the meal (most fish meals are high in protein to achieve high growth rates) of a fish will easily overwhelm endogenous pools of amino acids. Yet large fluctuations in the concentrations of amino acids are only found in the plasma and not propagated into any tissues. Again, because of their bulk and local actions of insulin (and IGF-I?), muscle tissues are the most probable candidates for turning wily amino acids into (inert) storage and functional proteins. Since only 20% of the amino acid nitrogen infused in the studies above are recovered as excreted nitrogen, and tissue and plasma amino acid titers quickly assume normal (low) concentrations, the bulk of amino acids must have been actively funneled into protein (Brown and Cameron, 1991). As described above, muscle—and within it myofibrillar components—must account for the bulk of protein accretion. The same conclusion can be reached via an entirely different route. Houlihan and colleagues (1989) analyzed the fractional rates and protein retention efficiencies in Atlantic cod (*G. morhua*) at two different growth rates. When the fish were not growing, the fractional rate of protein synthesis of white muscle was 0.46%, increasing to 1.94% at a growth rate of 1% per day. Comparative values for the gill were 4.4 and 10.1%, respectively, clearly showing that the fold-increase was most pronounced in muscle. At the same time, the growth rate efficiency, calculated as the percentage of protein synthesized that is retained as growth, was 50% for white muscle, compared with around 10% for the cod gill (see Table I). In the fasted channel catfish, rates of fractional protein synthesis are 0.98% for the white muscle and 39.5% for the liver. After the amino acid infusion mentioned above, the rate increases to 2.29% for the white muscle and 78.1% for the liver, respectively. Relative increases are 2.3-fold for the white muscle and twofold for the liver (Brown and Cameron, 1991). Considering the bulk of the muscle compared with the liver and other organs, it becomes clear why muscle protein synthesis can be directly correlated to body growth in fishes and is usually accepted to reflect body growth.

D. Cell Volume

Feeding and exercise alter myofibrillar swelling in mammals, and there is no reason to believe that the situation in fishes should be different. In addition, euryhaline species can undergo drastic changes in plasma osmolarity and it is likely that intracellular ionic conditions are affected as species adapt to altered ambient osmolarities. In rat skeletal muscle, changes in glycogen synthase with altered medium osmotic strength involve interactions of the extracellular matrix with integrin and the cytoskeleton; these effects are specific to glycogen synthase, since the rate of glucose transport remains unaltered through the osmotic insults (Low et al., 1997). Incubation in hypo-osmotic medium will decrease the rate of glutamine and alanine release from skeletal muscle (Parry-Billings et al., 1991) and increase the rate of glutamine uptake by muscle, involving activation of phosphatidylinositol 3-kinase (PI3-kinase) (Low et al., 1997). Interestingly, osmotic stress may activate some of the same transduction routes as insulin and IGF-I, i.e., MAP kinases, Ras, PI3-kinases, and other protein kinases (reviewed in Lang et al., 1998) and some of these have been confirmed for cardiac myocytes (Sadoshima et al., 1996). Overall, cell swelling re-orients cell metabolism toward anabolic pathways, including sustained protein and lipid synthesis and parallel decreases in glycolytic flux and in the rate of protein degradation (Lang et al., 1998). It would be interesting to analyze whether SDA in fishes is associated with alterations in cell volume. It might explain the instant protein synthetic response following infusion of amino acids in the channel catfish. Then again, these effects may be indirect. Insulin is likely to be released in response to the amino acid infusion in the catfish—mediated by its effects on ion transporters, including Na^+/K^+-ATPase (Waldegger et al., 1997)—and insulin is known to increase cell volume in mammalian systems (Lang et al., 1998). The question remains whether this phenomenon applies to muscle and possibly also to IGF-I. In addition, the occurrence of directly osmosensing MAP-kinase cascades has been described (Maeda et al., 1995). PI3-kinase, an intracellular link for growth hormone, insulin, and IGF-I message transduction pathways, is an essential component of muscle cells to regulate amino acid transport induced by volume changes (Low et al., 1997).

Finally, factors controlling other energy-demanding processes will have a bearing on the energy potentially allocated to growth. Without specifically singling out muscle as a target, it has been shown numerous times that decreases in osmoregulatory demands can increase the growth rate of fishes. For instance, the euryhaline Mozambique tilapia (*Oreochromis mossambicus*) grows more quickly, especially at weights below 60 g, and grows to much larger sizes if kept in brackish water or seawater (Sparks et al., 1999). At the same time, the routine metabolic rate of the fish was significantly lower at higher ambient salinity, implying that it is indeed the energy cost of osmoregulation that limits the growth rate of fish (Ron et al., 1995).

It is apparent that numerous non-hormonal factors will affect muscle protein accretion is fishes. Although many studies have addressed a limited selection of such factors in fish, these studies are generally descriptive and limited in scope, and in only a few cases have we a good mechanistic explanation of the events involved.

IV. HORMONAL EFFECTS

A number of endocrine and paracrine factors affect muscle growth in fishes, and some of these principles are listed in Table IV. Unfortunately, the information is quite limited, since often the description of the growth-promoting effect itself and not the mechanisms behind the effect were at the center of attention. A number of studies have implied that hormones exert important effects on the rate of protein synthesis in muscle, but direct evidence is largely lacking. Apart from a few exceptions, the application of an exogenous (usually mammalian) hormone will yield the desired result and increase muscle growth, but such experiments will not deliver much insight into the exact mechanisms of action of the hormone or even on the role of the native, endogenous hormone. In some cases even the site of action will be obscured by the experimental approach. Injection with growth hormone, for instance, could successfully stimulate growth through changes in aggression or stimulation of appetite. As well, a positive response will fail to identify the site of action, since growth hormone could stimulate IGF-I production in the liver, indirectly targeting the muscle, or growth hormone could

Table IV
Hormones with Positive Effects on Muscle Growth in Fishes. See Text for References and Discussion of These Factors

Hormone	Effects
Growth hormone	↑ Protein synthesis; ↑ amino acid uptake and incorporation; ↑ lipolysis; affects reproduction
Growth hormone transgenics	Early maturation
IGF-I	Hypoglycemia; ↑ protein synthesis; ↓ protein degradation; ↑ amino acid uptake; cell proliferation
Insulin	↑ Amino acid uptake; ↑ protein synthesis; antilipolytic
Glucagon	↑ Amino acid uptake
Prolactin	↑ Intestinal amino acid transport; ↑ protein synthesis
Thyroid hormone	Metamorphic; ↑ amino acid uptake; ↑ food conversion efficiency; positive nitrogen balance; ↑ protein synthesis
Steroids	
Estrogen	↓ Myosin synthesis
Androgen	↑ Growth performance; ↑ protein synthesis; ↑ intestinal amino acid uptake
Corticosteroids	↑ Plasma amino acids; ↑ proteolysis
Anti-SRIF	Blocks effects of growth hormone
β-Adrenergic agonists	↑ Protein synthesis; ↑ plasma fatty acids; ↓ muscle lipids

stimulate muscle IGF-I production for paracrine use, or it could alter muscle protein synthesis directly.

Although usually mammalian models (and with them, inherent biases) are applied to fishes, a number of differences between fish and mammalian systems makes the use of such models unworkable. Some of these are rooted in the specifics of muscle growth in fishes, some in the unique fiber composition, and others still in the unique characteristics of fish swimming (see above). For instance, a useful model in mammalian muscle physiology is "inactivity and immobilization" of a limb or muscle. This induces a generally catabolic state associated with muscle atrophy, deterioration of function and de-differentiation that can be exploited to study protein degradation, and downregulation of protein synthesis. Unfortunately, a similar situation cannot be used in fishes, where immobilization is difficult to achieve and because of immersion in a buoyant medium, local immobilization and other experimental tools cannot be exploited as readily.

We will focus, therefore, on a few hormonal effectors where more than just the growth-promoting action in fishes has been identified. A recent review has examined some of these areas, but with the emphasis on whole fish growth (Mommsen, 1998).

A. Growth Hormone

Growth hormone is a 20–22 kDa polypeptide, produced and released by the somatotrophic cells located in the anterior pituitary. Fish growth hormones have been isolated from over 30 species (Bernardi et al., 1993) and the typical version has 187 or 188 residues and is characterized by 4 or 5 cysteines and a single assumed glycosylation site. The structure of the growth hormone gene(s) has been analyzed for a few species, and promoter regions have been sequenced and contain a literal grab bag of hormone responsive elements and cis-acting transcription factors. Reflecting the situation in mammals (Theill and Karin, 1993), the growth hormone promoter in the rainbow trout contains a thyroid hormone responsive element as well as other consensus sequences, potentially regulating growth hormone gene expression by glucocorticoids, retinoic acid, and cAMP (Yowe and Epping, 1995; Wong et al., 1996; Yang et al., 1997). In addition, an estrogen-responsive element (ERE) was noted in one of the two growth hormone genes of the rainbow trout (Yang et al., 1997). At least in the coho salmon (Oncorhynchus kisutch), one of the growth hormone genes also appears to be sex-linked (Forbes et al., 1994). Growth hormone belongs within the growth hormone/prolactin gene family, together with prolactin, somatolactin, and the mammalian placental lactogens; hormones in this family are closely related and are thought to have been derived from a common ancestor (Chen et al., 1994). Although aptly named, growth hormone has important functions beyond the regulation of somatic growth in fishes. For instance, the hormone plays an important role in smoltification and in seawater adaptation in salmonids (McCormick, 1996) and in the hepatic biosynthesis of antifreeze proteins in marine teleosts (Idler *et al.*, 1989). Finally, it regulates a number of metabolic functions that, at first sight, seem to be unrelated to and independent of somatic growth.

In addition to the multitude of different targets, three interdependent dynamic processes contribute to the actions of growth hormone in mammals. The first is the pulsatile release of hormone from the somatotrophs, with target tissues responding to changes in frequency and amplitude of hormone peaks more so than to the actual concentration. The second is contributed by the growth hormone binding protein corresponding to the extracellular domain of the growth hormone receptor. The binding protein protects growth hormone from degradation and constitutes a ready supply of hormone. The third is the growth hormone receptor whose availability and expression are regulated by growth hormone itself. Many of these processes have recently been reviewed (Mommsen, 1998) in fish and will not be discussed further.

While the fish growth hormone receptor has been characterized through hormone-binding studies (Hirano, 1991; Yao *et al.*, 1991; Ng *et al.*, 1992), information on sequence, expression, or the possible relationship to growth hormone binding proteins is not yet available. The absence of such information leaves a rather large gap in our understanding of growth hormone dynamics in fish. The mammalian receptor is part of the large cytokine/erythropoietin receptor gene superfamily, identified by a single transmembrane domain (Waters *et al.*, 1994). In the case of growth hormone binding, homodimerization of the receptor results in activation of tyrosine kinases associated with the receptor, the so-called Janus kinases. These in turn, will eventually lead to the activation of different transcription factors of the stat family. Alternate routes of message transduction can be activated. One involves phospholipase C and protein kinase C, which, together with diacylglycerol, can also tie into the MAP kinases and transcription factors. Another pathway leads to activation of MAP kinases through ras and raf and thus ultimately converges on transcription factors as targets. Finally, increased phosphorylation of insulin receptor substrates (IRS) and activation of PI3-kinase have been implicated in the growth-promoting actions of growth hormone in mammals (Argetsinger *et al.*, 1993; 1995). Whether such interactions occur within fish are yet to be determined. Growth hormone is a powerful stimulator of the expression of IGF-I genes in skeletal muscle, and many of the local physiological functions of growth hormone are thought to be mediated by IGF-I (which ties into similar, if not identical, parts of the transduction system outlined). Evidence for *direct* actions of growth hormone on muscle is not quite as strong. For instance, transcripts of growth hormone receptor are present in mammalian muscles, but direct and specific binding of growth hormone has been more difficult to demonstrate. However, growth hormone may stimulate satellite cell proliferation in the chicken in an IGF-I-independent manner (Hodik *et al.*, 1997).

Irrespective of the mechanism of action, exogenous growth hormone will increase the rate of body weight accretion in feeding fishes. For example, mammalian growth hormone, in the form of an intraperitoneal cholesterol implant, stimulates whole body protein synthesis in trout; the treatment also leads to an increase in the RNA/protein ratio, indicative of increased transcriptional activity (Foster *et al.*, 1991). In this case, little effect was noted on the rate of protein degradation and growth hormone exerted no effect on the efficiency of deposition of the newly synthesized protein. Numerous reports on the effectiveness of growth hormone to accelerate the rate of fish (muscle) growth and fishes are available. These have been reviewed elsewhere (McLean and Donaldson, 1993; Mommsen, 1998) and need not be repeated here. Therefore, we shall focus on a few concepts involved and on changes pertaining specifically to muscle.

Regular injection with bovine or human growth hormone leads to an increased proportion of small diameter fibers in rainbow trout muscle, indicating a key function for growth hormone in the regulation of hyperplastic muscle growth (Fauconneau *et al.*, 1997). At times, but not always, exogenous growth hormone affects the length of the fish. In several studies with salmonids, Fauconneau, Le Bail, and colleagues reported length increases (Foster *et al.*, 1991; Fauconneau *et al.*, 1997) or no effects of growth hormone in trout and Atlantic salmon (Fauconneau *et al.*, 1996). Specific growth rate and muscle fractional protein synthesis rates were, however, increased. In one of these studies where increases in fish length were noted, growth hormone caused a substantial decrease in muscle protein (Foster *et al.*, 1991). Although these fish were growing in the framework of the definition laid out in the introduction, some weight acquired was actually water and the caloric increases were not as large as the values for mass suggest. This line of thought can be taken a little further, because of the relationship of growth hormone to prolactin and its involvement in osmoregulation, at least in some fishes. One wonders whether the muscle cells might not have undergone changes in cell volume, especially swelling, during the growth hormone-dependent accretion processes (water? see above). Using mammalian models, protein synthesis would ensue governed directly by transient cell swelling.

Although fish seem to respond to administration of exogenous growth hormone in many different ways, it is noteworthy that immunoneutralization of growth hormone affects only a single parameter, muscle protein synthesis, which was decreased from $1-0.3\%$ day^{-1} (Fauconneau *et al.*, 1996). Clearly, growth hormone plays an essential role in the maintenance of protein synthesis, and the muscle can be identified as the most important target. Compared with liver, the muscle contains much smaller numbers of specific growth hormone binding sites (Yao *et al.*, 1991; Ikuta *et al.*, 1995), but the bulk of the tissue more than compensates for the shortfall in receptor activity per gram of tissue. Overall, the muscle commands more than 2.5 times the specific binding sites for growth hormone of the liver (Yao *et al.,* 1991). Considering the effects of growth hormone on muscle accretion, it is not surprising that the hormone has a positive influence of the growth of bone and cartilage in fishes (Takagi *et al.,* 1992), again, most likely mediated by local or hepatic production of IGF-I (see below).

Apart from the direct action on muscle and those mediated via local or hepatic IGF-I production, growth hormone can alter the rate of muscle protein synthesis by a number of indirect routes. For instance, the hormone has been shown to increase appetite (Markert *et al.,* 1977; Johnsson and Björnsson, 1994), dominance (Johnsson *et al.,* 1996), nitrogen retention, and amino acid incorporation (Cheema and Matty, 1978). The growth hormone-dependent lipolytic action mentioned below can also be seen to be positive to protein synthesis, in that it will deliver oxidizable substrate to the system and thus divert amino acids away from oxidative pathways. The protein-sparing action of lipolysis finds additional support in the decelerated release of nitrogenous compounds under the influence of exogenous growth hormone (Cheema and Matty, 1978). As shown elsewhere, muscle growth may be limited by the availability of building blocks, it may be a relevant observation that growth hormone accelerates the rate of amino acid uptake from the gut (Sun and Farmanfarmaian, 1992) and from the plasma into the muscle (Cheema and Matty, 1978), adding two paths to the potential routes of growth promotion in fish muscle.

A further confirmation for the role of growth hormone in normal growth processes is seen through the effects of immunoneutralization of the neuropeptide somatostatin-14 (somatotropin release inhibitory factor, SRIF). The fittingly named SRIF antagonizes the actions of growth hormone at a number of different levels: (1) it inhibits the release of growth hormone from the somatotrophs of the anterior pituitary; (2) it inhibits the biosynthesis of growth hormone; (3) it up-regulates the expression of IGF-I-binding proteins, which, *de facto,* lowers the systemic availability of IGF-I; and (4) it compromises the hepatic production of IGF-I in response to growth hormone. Considering these actions, it is not surprising that immunoneutralization of SRIF transiently increases plasma concentrations of growth hormone in coho salmon (*O. kisutch;* Diez *et al.,* 1992) and significantly enhances growth in juvenile chinook salmon (*O. tshawytscha;* Mayer *et al.,* 1994).

Potentially, growth hormone also controls muscle lipid stores, although, again, mammalian models need to be consulted to provide a framework for lipids in piscine systems and data on fish muscle are very rare. An acute increase in growth hormone in mammals will mimic insulin-like, lipogenic effects, while with chronic growth hormone exposure, mammals will experience the lipolytic face of the hormone. Under chronic conditions, growth hormone-dependent protein accretion and carbohydrate sparing are accompanied by decreases in lipid deposits. A similar insulin/anti-insulin dichotomy is noted for juvenile vs. adult mammals and birds. Early in ontogeny, growth hormone exerts lipogenic, insulin-like actions, while in adults, growth hormone activates lipolysis, inhibits fatty acid biosynthesis, and opposes the lipogenic functions of insulin (Goodman, 1993). From the few studies done on fishes, we can conclude that the lipolytic side prevails in juveniles and adults, although one study on channel catfish reported increases in body lipid after application of exogenous growth hormone (Wilson et al., 1988). In salmonids, application of exogenous growth hormone decreases body lipid reserves (Higgs et al., 1975) and in juvenile coho salmon, the hormone is generally lipolytic, activating hepatic triacylglycerol-lipase (Sheridan, 1986). Similarly, in carp (*C. carpio*), porcine growth hormone substantially reduces body lipid (Fine et al., 1993) and in goldfish (*Carassius auratus*) injection with growth hormone increases plasma fatty acid concentrations, again indicative of a lipolytic action. The hormone counteracts the depression of lipase in hypophysectomized fish and restores normal enzyme activity. Growth hormone application causes a decrease in the average size of perivisceral adipocytes in rainbow trout and decreases the overall lipid content around the digestive tract, while white muscle lipid remains unaltered by the treatment (Fauconneau et al., 1997). Surprisingly, isolated fish adipocytes fail to respond to common lipolytic hormones, including growth hormone (Migliorini et al., 1992). Nevertheless, lipolysis can be accelerated *in vitro* by exogenous cAMP and, indirectly, affect the same regulatory site through inhibition of phosphodiesterase that prolongs the endogenous cAMP message (Migliorini et al., 1992). Unfortunately the adipocytes dispersed throughout muscle tissue have to date proven to be a poorly accessible experimental target. In Atlantic salmon and in rainbow trout, the myosepta are major sites for the localization of adipocytes, with minor abundance of adipocytes in connective tissue surrounding the white muscle fibers. There is a steeply decreasing gradient in adipocyte abundance from the belly region to the dorsal region of the white muscle and the mean size of adipose cells increases with body weight and muscle lipid content in rainbow trout (Fauconneau et al., 1997). While not a major storage site for lipids, the red muscle contains lipid droplets in the endomysium around the muscle fibers and smaller, finely dispersed lipid droplets can be located in red, but not white, muscle cells (Zhou et al., 1996).

Little attention has been devoted to possible interactions and to the potentiating effects of growth hormone with other lipolytic hormones. In mammals, for instance, glucocorticoids potentiate the lipolytic actions of epinephrine or glucagon in the presence of growth hormone. In summary, lipolytic actions for growth hormone prevail in fishes, but it is not obvious how many of these functions are mediated directly or locally through the production of IGF-I. Indications in mammals are that growth hormone can regulate adipose metabolism directly, without the involvement of IGF-I.

In addition to its clear involvement in skeletal muscle growth, growth hormone also plays a vital, if as yet less well-defined, role in fish reproduction, with potential ramifications on muscle protein turnover. Surprisingly high concentrations of growth hormone are reported in the plasma of sexually mature specimens of numerous species, and an interesting interaction seems to exist with sex steroid hormones, as already implied by the existence of estrogen-responsive elements in one of the two rainbow trout growth hormone genes (Criswell et al., 1996). Administration of exogenous growth hormone stimulates gonadal steroid production in several species (Le Gac et al., 1993), and early maturation was one of the effects noted in transgenic chinook salmon (O. tshawytscha) carrying extra copies of the growth hormone gene (Devlin et al., 1995). In a type of feed-forward mechanism, the application of gonadal steroids augments plasma growth hormone concentrations (Holloway and Leatherland, 1997).

The obvious role of growth hormone in fish muscle growth makes this hormone an ideal candidate to enhance productivity in aquaculture situations. Two technical problems have, however, blocked the use of this hormone in aquaculture. The first is the availability of large amounts of homologous hormone. A recent report (Ben-Atia et al., 1999) has expressed gilthead sea bream (Sparus aurata) growth hormone cDNA in E. coli and produced large quantities of sea bream growth hormone. This product shows high binding affinity to hepatic microsomal fractions and exhibits growth-stimulating activity when given orally to sea bream larvae or intraperitoneally to juveniles. The second problem is the production of a fish that carries a transgene for growth hormone. Du and co-workers (1992) injected into fertilized, non-activated Atlantic salmon eggs a chimeric gene construct containing an antifreeze protein gene promoter from ocean pout linked to a chinook salmon growth hormone cDNA clone. The transgene was present and these transgenic fish demonstrated variable, but in many cases, exceptionally dramatic increases in growth rates. Similarly, transgenic tilapia lines have been generated using tilapia growth hormone cDNA, and these lines demonstrate enhanced growth (Martínez et al., 1996; Rahman et al., 1998). Growth enhancement may occur as a result of either an increase in food intake or an increase in efficiency of food conversion to tissue mass, especially muscle, but data on the latter are generally inconclusive. There is some evidence that coho salmon fry carrying the transgene for sockeye salmon growth hormone under control by the metallothionein B promoter grow faster but show no evidence for expression of the transgene. In older animals, the transgene is expressed, the fish show much accelerated growth, and plasma growth hormone titers are elevated some 40-fold compared with normal coho salmon (Mori and Devlin, 1999). Studies in transgenic mice have reported an increase in body mass correlated with a 30-fold increase in plasma levels of growth hormone, but a significantly lower proportion of muscle expressed as a percentage of body mass (Sharma et al., 1996). This study concludes by hypothesizing that it is locally produced IGFs that are more critical to muscle growth than circulating growth hormone concentrations. So even though growth hormone transgenes are beginning to appear in aquaculture, more study is needed to better understand their capabilities and whether coinciding changes in

IGF expression are needed to demonstrate consistent growth enhancement in such growth hormone transgenics.

B. Prolactin

Prolactin, considered the most versatile member of the growth hormone family of peptides, possesses mitogenic, somatotropic, and osmoregulatory properties, with the latter restricted to freshwater-adapted fishes. Only one of the two prolactins found in tilapia (*O. mossambicus*)—tPRL$_{177}$ (Specker *et al.*, 1985)—has growth-promoting activity, assessed by thymidine incorporation (DNA synthesis) and sulfate (extracellular matrix) uptake into ceratobranchial cartilage, most likely mediated by hepatic production of IGF-I (Shepherd *et al.*, 1997a). Interestingly, PRL$_{177}$ displays a lower affinity for the single class of PRL receptors described for the Nile tilapia (*O. niloticus*) than the somatotropically inactive PRL$_{188}$ (Sandra *et al.*, 1995), but PRL$_{177}$ displaces growth hormone from its receptor (Shepherd *et al.*, 1997a). Because plasma PRL$_{177}$ concentrations are sufficiently high to displace tGH in freshwater-adapted tilapia only, this growth-promoting effect of PRL may be ancillary to other functions of the hormone. Apart from these IGF-I-mediated actions, PRL may positively influence growth by its acceleration of transport process in the fish intestine (Mainoya, 1982). This is likely to have secondary effects on the efficiency of muscle growth stimulators such as growth hormone, insulin, IGF-I, or T_3 under conditions of amino acid limitations.

Prolactin accelerates the incorporation of labeled glycine into liver protein and to lower the plasma concentration of amino acids, but the hormone fails to exert any positive effects on the growth rate of goldfish. Application of mammalian growth hormone, in contrast, increases growth rate in these fish, without altering the plasma amino acid titers or glycine incorporation into liver protein (Prack *et al.*, 1980). One wonders about the ultimate fate of amino acids with prolactin exposure, since obviously muscle, the largest sink for amino acids, does not constitute a primary target for prolactin. A slightly different picture is obtained for other species. In a study on sockeye salmon (*O. nerka*), PRL exerts a positive effect on growth, but the growth-promoting actions are well below those observed for growth hormone under the same conditions (Clarke *et al.*, 1977). Prolactin shifts the RNA/DNA ratio in catfish (*Heteropneustes fossilis*) muscle in favor of RNA; in addition, the hormone induces and significantly increases the amount of lipid deposition in muscle. Interestingly, none of these parameters are affected in the liver (Singh and Prasad, 1987).

C. Insulin-Like Growth Factors

IGF-I is a single-chain protein with a molecular mass of around 7.5 kDa that belongs to the insulin superfamily of peptides. It is highly conserved across the

8. HORMONAL REGULATION OF MUSCLE GROWTH

entire vertebrate line, including the fishes (Chan *et al.*, 1993, 1994; Duguay *et al.*, 1995). The hormone is similar to insulin and IGF-II in amino acid sequence, receptor-binding, and biological activities. At times, insulin and IGF-I can replace each other on their respective receptor proteins, although the affinity of the receptors tends to be highest for their own (name-giving) ligand. However, actual concentrations of free hormone reaching any particular tissue through the circulatory system or via paracrine and autocrine routes are difficult to assess due to the presence of various IGF-binding proteins (Duan, 1998) that differ in their site of synthesis and physiological roles. Plasma concentrations of IGF-I fluctuate widely with the nutritional state of the fish and are highly dependent on the concentrations of other hormones. In starvation, plasma IGF-I concentrations are depressed, as is the amount of free IGF-I, while under the influence of growth hormone, both total IGF-I and the free form of the hormone are increased. The total amount of IGF in fish plasma is normally about 25 ng mL^{-1}. Since the amount of free IGF-I is in the range of 40 pg mL^{-1}, little more than 0.1% of the total IGF is present in the biologically active form (Plisetskaya, 1998). A similar calculation for mammals produced numbers in the 0.5–2% range (Murphy, 1998). Even at this low concentration, there is sufficient free IGF available to bind to the receptor, although concentrations may be too low to interfere with the binding of insulin to the insulin receptor.

The importance of IGF-I-binding proteins is most apparent in tissue culture of L-6 myotubes, where an N terminally truncated variant of IGF-I exerts a ninefold higher potency than full length IGF-I, largely because the truncated IGF-I is not bound by the binding protein (Bagley *et al.*, 1989). All fish tissues analyzed to date contain transcripts for IGF-I (Chen *et al.*, 1994) including all developmental stages, and transcription is highly responsive to activation by growth hormone, especially in liver and possibly gill (Duan, 1998). In contrast to the liver, where IGF-I is produced for systemic functions, other tissues, including muscle, express IGF-I largely for autocrine or paracrine actions and may not be regulated by growth hormone. IGF-I is truly a pluripotent hormone, in that it enhances proliferation and differentiation, while also regulating apoptosis, mitosis, chemotaxis, and renal Na$^+$ transport. Other evidence suggests that the hormone suppresses protein degradation, accelerates the rate of amino acid and glucose uptake by tissues, and increases whole body parameters as diverse as aggressiveness, food consumption, growth, and salinity tolerance. IGF-I dependent decreases in plasma amino acid and glucose titers are consequences of accelerated tissue uptake. Although it resembles proinsulin most closely in structure, its mitogenic ability measured in a salmonid test system *in vitro* is much higher than that for proinsulin (Urbinati *et al.*, 1994).

At the muscle level, the hormone is confirmed to increase the rate of DNA and protein synthesis. Unfortunately, little information is available on other functions of IGF-I or on the mechanism(s) of any of these observed alterations in fishes. IGF-I exerts its physiological actions largely through specific IGF-I receptors, heterotetrameric membrane receptors with substantial similarity to insulin receptors. Some of the actions of IGF-I can potentially also be mediated by insulin receptors, while affinity of the IGF-II/mannose 6-phosphate receptor for IGF-I is very low. Transcripts of the IGF-I receptor are detectable in muscle, gill, and liver. Processing of the receptor apparently is under complex post-translational regulation that changes with developmental status. In early developmental stages of turbot (*Scophthalmus maximus*), the receptor mRNA is poly-adenylated, while after hatching, only a de-adenylated form of message is detectable (Elies *et al.*, 1996; 1998).

IGFs stimulate a multitude of anabolic responses in myoblasts and enhance both their differentiation and proliferation. Usually mitogenesis and myogenesis are considered to be mutually exclusive. However, more recent studies have shed light on this enigma and have presented evidence that the two functions are utilizing different post-receptor signaling pathways (Florini *et al.*, 1996). In addition, myoblasts themselves produce at least four different types of IGF-binding proteins that may regulate the functions of IGF-I locally. Finally, mammalian myoblasts express large amounts of IGF-II, which can stimulate myogenesis in an autocrine fashion. One of the IGF-I inducible gene products involved in muscle differentiation is myoD, encoding a transactivating factor for a number of muscle genes, including those for MHCs (Florini and Ewton, 1992). The terminal differentiation in myoblasts via IGF-I is controlled by myogenin, belonging to a family of closely related muscle determination genes. Interestingly, the use of antisense codons to the myogenin mRNA interferes with the stimulation of terminal differentiation of L6A1 myoblasts, but not with other IGF-I-mediated functions, including cell proliferation and incorporation of amino acids or nucleotides (Florini and Ewton, 1990). At any rate, the presence of inhibitory binding proteins and members of the myogenin and myoD families of differentiation factors (Hughes *et al.*, 1993; Florini *et al.*, 1996), together with the recent identification of a muscle-specific growth factor derived from differential processing of the IGF gene (Yang *et al.*, 1996), add further complexity to the control of muscle differentiation and myogenesis.

This discussion should supply a challenging framework for work on fishes. At present, we can only provide a few glimpses into the piscine situation. Clearly, the unique features of fish muscle with its separation into distinct muscle fibers, occurrence of temperature-sensitive myosin genes, temperature-dependent synthesis of fatty acids binding proteins, and the often indeterminate growth, could be developed into an interesting general model for hormonal control of hyperplasia and hypertrophy in vertebrates. It is likely that analysis of fish muscle will also furnish useful insights into autocrine, paracrine, and endocrine control of these processes, superimposed on intrinsic control mechanisms, especially for the multifaceted IGFs.

IGF expression in muscle is controlled by a number of hormonal factors, not least insulin and IGF-I, leading to an autocrine feed-forward activation of IGF-I gene expression and implying independence of the growth hormone system. IGFs function as powerful stimulators of many anabolic processes that result in muscle growth. Among these processes are amino acid uptake, cell proliferation, protein synthesis, glucose transport, and differentiation to form post-mitotic myotubes. IGFs tip the balance toward muscle growth even further by curtailing the rate of protein degradation. c-Myc, linked to cyclins and cyclin-dependent protein kinases, appears to be one of the key players in cellular decision making (Helbing *et al.*, 1998) because of its extraordinary duality in function. Its expression is not only stimulated when quiescent cells enter the cell cycle, but also in apoptosis of growth arrested cells. The increase in myc gene expression is counteracted by the IGF-I receptor that has an anti-apoptotic function (O'Connor *et al.*, 1997). In mammalian muscle, hypertrophy is correlated with expression of the c-myc gene and expression of the gene declines with age, implying a role for c-myc expression in skeletal muscle growth (Whitelaw and Hesketh, 1992). The presence of c-myc has been confirmed for the fishes (Van Beneden *et al.*, 1986), but the dynamics of expression in muscle are yet to be analyzed. In Atlantic salmon the rate of myc gene expression during early life stages is relatively higher at low ambient temperatures (Matschak and Stickland, 1996) and correlates with nuclear numbers at hatching (Usher *et al.*, 1994).

Growth hormone stimulates the expression of IGF-I genes in muscle, but a number of cases have been reported where stimulation of IGF-I expression is independent of growth hormone. Generally IGF-I levels rise in teleosts after injection of growth hormone, although these two hormones can become dissociated depending upon the level of protein feeding. This suggests some negative feedback system involving IGF-I on growth hormone secretion (see Melamed *et al.*, 1998). Both IGF-I and IGF-II are thought to mediate their actions through a common receptor, the so-called IGF-I receptor. The IGF-I receptor regulates cell proliferation by a number of routes controlled by different regions of the receptor. One route is transformation, a signal transmitted most likely by the C terminal sections of the receptor. The two other control routes, most likely more important to growing muscle, are inhibition of apoptosis, also localized to the C terminal region (Liu *et al.*, 1998), but activating separate intracellular pathways, and finally, activation of mitogenesis (Resnicoff and Baserga, 1998). The latter is likely the most relevant function to hyperplastic muscle growth in fishes. Because of the abundance of both insulin and IGF-I receptors on muscle and appreciable overlap in their ligand specificity (Li *et al.*, 1998; Plisetskaya, 1998; Navarro *et al.*, 1999) which includes the idea that IGF-II may transduce its mitogenic actions through the insulin receptor (Louvi *et al.*, 1997), it is difficult to discern—if deemed necessary—the relative contributions and mechanisms of insulin and IGF-I to growth, differentiation, and even metabolism. Perhaps in the future, the availability of hormone or receptor knockout fish will help resolve this enigma. At any rate, in adult or aging mammals, insulin receptors prevail over IGF-I receptors, while fish muscle contains considerably higher titers of IGF-I receptors than insulin receptors at all ages (Párrizas et al., 1995b). Perhaps it is the inhibitory action of IGF-I on apoptosis (Párrizas et al., 1997) and its profound role in mitogenesis that control indeterminate muscle growth in fishes, although, at this point, the role of apoptosis in fish muscle development is not clear. It would be interesting to compare receptor abundance and key steps in message transduction for annual fish, such as members of the genus *Cynolebias* (Cyprinodontiformes: Rivulidae) or other determinate species (Okuda et al., 1998).

Unfortunately, analysis of the mechanisms of action for IGF-I (or insulin) in fishes is still in its infancy. Of the many common components involved in the postreceptor processes for these two peptides in mammals, e.g., insulin receptor substrates, ras, Raf, the MEK cascade, PI3-kinase, Janus kinases, protein kinase B, initiation factors, etc. (cf. Butler et al., 1998; Mommsen, 1998), only protein tyrosine kinase has been analyzed in any detail (Gutiérrez et al., 1995; Párrizas et al., 1995a). This analysis shows, not really surprisingly, that activation of this initial part of the cascade is similar in all vertebrates.

When subjected to activity, mammalian striated muscle expresses a highly specific growth factor, termed Mechano Growth Factor (MGF). The structure of this MGF indicates that it is derived from the IGF-I gene by alternative splicing. The peptide possesses different exons than the liver type IGF-I and lacks the glycosylation found in liver IGF-I (Yang et al., 1996). Fish muscle is known to express IGFs at a much lower rate than most other tissues, at about 3% of the liver. Yet, as mentioned above, fish muscle contains much higher titers of IGF-I receptors than insulin receptors, a situation that is unique among adult vertebrates. In contrast to mammals, adult rainbow trout express substantially more IGF-II than IGF-I, with only one of the four IGF-I mRNA isoforms detectable (Chen et al., 1994). This makes one wonder whether the presence of IGF-II may play a role in the indeterminate muscle growth observed in many fish species.

IGF-I affects both proteoglycan biosynthesis and cell proliferation; in fact, the ability of IGF-I to stimulate incorporation of sulfate into branchial cartilage is used as a bioassay of IGF-I presence and activity (Duan, 1998). Interestingly, these two important stimulatory actions are independent of each other, as shown for mammalian and avian bone and chondrocytes (Kemp et al., 1988). IGF-I increases the rate of collagen type I biosynthesis by increasing the rate of transcription (McCarthy et al., 1989).

Interleukin-15 (IL-15), a protein containing a four-helix bundle reminiscent of that found in other cytokines and also in growth hormone (Horseman and Yu-Lee, 1994), has been shown to stimulate differentiation and accretion of MHCs in mammalian muscle, without affecting muscle cell proliferation. Interestingly, this effect is as powerful as the hypertrophic actions of IGF-I, but is independent of IGF-I and additive to those of IGF-I (Quinn et al., 1995).

8. HORMONAL REGULATION OF MUSCLE GROWTH

In summary, IGFs are some of the main players controlling muscle growth in fishes, be it as hormones in their own right or as handmaidens of other endocrine principles, or through cross-talk at receptor or post-receptor levels, or even at the level of transcription factors. Unfortunately, at this point, the information is still too patchy to delineate an overall picture of the mechanisms or specific sites of action for IGFs. Such a picture will have to await more information on IGF-I expression, processing, secretion and the role of IGF-binding proteins in fish muscle, and until methods are at hand to differentiate between the role of endocrine, paracrine, and autocrine functions of IGF-I in muscle. Clearly, early development and post-hatching dynamics in muscle composition and growth of fishes are ideal systems to probe the many facets of IGFs.

D. Insulin

Apart from the well-known metabolic actions on carbohydrate and lipid metabolism, the protein anabolic effects of insulin in mammals have been known for a long time. In this, the hormone stimulates protein synthesis and curtails protein degradation, resulting in muscle protein accretion. Part of the action of insulin is to regulate, i.e., speed up, the rate of amino acid uptake into muscle, rapidly decreasing the pool size in plasma and subsequently also in muscle due to its strong effects on protein synthesis. Similar actions have been reported for the fishes, but an overall appreciation of the importance of endogenous insulin to amino acid flux is yet to be published. The fact that exogenous insulin, including heterologous insulins from mammals, activates the amino acid uptake routes in fishes should not be surprising, considering the conservation of insulin receptors and the conservative nature of pathways involving amino acid transport.

In contrast to the many potential production sites of IGF-I and its intricate relationship with the plethora of binding proteins, the insulin-complex appears much simpler. Insulin is a 51 amino acid residue peptide, almost exclusively produced in the β cells of the Brockmann bodies. Although plasma concentrations tend to be much higher than in mammals (Mommsen and Plisetskaya, 1991), they are still smaller than those for total IGF-I present in fish plasma. However, since there are no known binding proteins for insulin, its effective concentration in plasma could be a few orders of magnitude higher than that for the biologically active form of IGF-I. There have been suggestions that not all of the insulin measured in radioimmunoassays is biologically active, but may include the C peptide fragment of insulin that lacks *in vivo* activity (Plisetskaya, 1998). In the course of vertebrate evolution, insulin itself appears to be less conserved in sequence than IGF-I with which it shares a number of similarities, not least in receptor-binding and intracellular targets. In mammals, more clearly so than in fishes, two different, and at times separated, areas of insulin functions can be distinguished. On one side are the metabolic actions that in mammals are heavily biased toward regulation of glucose homeostasis at the experimental expense of lipid metabolism (McGarry, 1992). Many of these actions are linked to insulin-dependent changes in phosphorylation of pivotal enzymes involving serine/threonine kinases and phosphatases, but the intracellular events precipitated by insulin binding to the extracellular portion of its receptor are still not entirely clear (Saltiel, 1996). On the other side are the growth-promoting actions, and these are thought to be controlled through a number of transcription factors including the MEKK/MEK and MAP-kinase pathways (Mader and Sonenberg, 1995; Saltiel, 1996; Sonenberg and Gingras, 1998).

Ince and Thorpe (1978) present evidence that the kidney, the liver, and the gastrointestinal tract are the major sites for accumulation of ^{125}I-insulin after injection into the rainbow trout, with muscle making a minor contribution only. Alas, again, this accumulation is based on unit tissue weight and does not take the white muscle bulk in the fish into account. Given the differences in mass, skeletal muscle will have a higher total number of insulin receptors than any other tissue (Párrizas et al., 1995b; cf. Table II). The number of muscle insulin receptors increases in mammals and in fish maintained on a high carbohydrate diet that results in increases in circulating insulin levels (Párrizas et al., 1994a; Baños et al., 1998). These binding differences, however, disappear quickly if the fish do not continue to feed (Gutiérrez et al., 1991). In this same study, downregulation of insulin binding to liver was found, suggesting that the role of insulin in these two tissues is quite different. Trout fed a high carbohydrate diet for an extended period also show higher insulin binding to skeletal muscle and tyrosine kinase activities (Párrizas et al., 1994b). Hyperinsulinemia induced by arginine injection is correlated with short term increases in specific binding of insulin to skeletal muscle. In carp, specific binding increases from 5.8–9.6%, but only from 0.8–1.5% in brown trout (Párrizas et al., 1994a). Both basal binding and hyperinsulinemia-dependent increases in carp are similar in muscle and in liver, if normalized to protein content. The increases in binding are not associated with increases in the specific activity of receptor tyrosine kinase (RTK). In fact, insulin-dependent increases in RTK activity assessed in vitro are indistinguishable for both trout and carp even though receptor numbers are higher in carp. Insulin and IGF-I binding to red skeletal muscle of trout and carp are higher than values reported in white muscle and arginine-stimulated insulin release resulted in a decrease in binding within 4 hr (Baños et al., 1997). Unfortunately, none of the above experiments revealed any correlation between insulin and/or IGF-I binding and any physiological parameters in these muscles. No putative mammalian GLUT-4 transporter has been demonstrated in any tissue of tilapia (O. niloticus; see Wright et al., 1998), an observation that does not exclude future identification of piscine-specific GLUT-4. Our recent studies with membrane vesicles isolated from rainbow trout, American eel (Anguilla rostrata), and black bullhead (Ameiurus melas) do, however, demonstrate a saturable, D-glucose-specific, transporter; alas, at this early

8. HORMONAL REGULATION OF MUSCLE GROWTH

point, we have no indication from mRNA or protein analyses that this transporter has similarities with mammalian GLUT-1 or -4 (N. Legate, A. Bonen, and T.W. Moon, submitted). Furthermore, even though the fish may represent an important tool to examine the role of insulin on transcriptional activation in the possible absence of metabolic activation, our knowledge of this system beyond the tyrosine kinase is nonexistent.

In non-muscle tissues, hyperinsulinemia does not lead to the expected downregulation of insulin receptors as in mammals, but instead to a slight increase in the abundance of insulin receptors in brain and liver. These conclusions are not entirely clear-cut as in some cases downregulation is seen at least in liver (Gutiérrez and Plisetskaya, 1991a; Plisetskaya and Duguay, 1993). Overall binding of insulin to muscle receptors is more specific and insulin receptor numbers are greater in omnivorous than carnivorous fish species (Párrizas et al., 1994b; cf. Table II).

A fairly common thread to the growth-promoting hormones listed here is that they tend to increase the rate of amino acid transport at different levels, be it from the gut into the plasma or into the liver or into the muscle. Even the increased activity of intestinal proteolytic enzymes after administration of some hormones should be mentioned in this context (Lone and Matty, 1981). Without taking away from the importance of transcription factors, it is conceivable that protein synthesis is simply a slave to mass action and that ultimately transporters and availability of precursors *in situ* decide on the direction and velocity of flux, provided the molecular machinery is in place to deal with amino acid abundance. Evidence points toward myofibrillar components as the main "beneficiaries" of the swift removal of amino acids. The apparent surprising stability of intracellular amino acid titers in muscle implies a tight regulation of the "set-point," even under conditions of amino acid flooding or cell volume changes due to amino acid flux, without giving away any mechanisms regulating the set-point.

Insulin brings about a strong anti-lipolytic action on fish adipocytes and antagonizes the activation of lipase and fatty acid release by glucagon (see Mommsen and Plisetskaya, 1991). The hormone is likely to oppose the lipolytic actions of any hormone employing the adenylyl cyclase/cAMP route of message transduction. As in mammals, insulin probably works by interfering with phosphorylation of the responsible enzymes or directly by causing dephosphorylation of these enzymes by activating selective phosphatases (Saltiel, 1996). In rainbow trout, insulin directly stimulates lipogenesis and enhances the flux of glucose and amino acid carbon into muscle lipids (Tashima and Cahill, 1968; Ablett et al., 1981). Unfortunately, the research into lipid metabolism seems to have been focused on fish liver and little attention has been devoted to muscle. This is in spite of the fact that muscle may contain substantial amounts of lipids, and adipocytes are found in association with red muscle cells and often in close proximity to muscle cells (see above discussion). Nevertheless, some indirect observations offer a glimpse of the potential importance of lipid to muscle metabolism and growth in fishes. Lipoprotein lipase is a regulatory enzyme of the endothelial cells and facilitates the uptake of lipids by non-hepatic tissues by hydrolyzing very low density lipoproteins (VLDLs). The activity of lipoprotein lipase in rainbow trout is positively correlated with plasma insulin concentrations; i.e., it is high in actively feeding fish and severely depressed in starving fish (Black and Skinner, 1986). However, any direct alteration in enzyme activity by insulin as in mammals, needs to be confirmed for the fishes. Also, insulin is instrumental in reducing the plasma concentrations of fatty acids and muscle lipogenesis appears as an obvious sink, the connection of this decrease to lipid biosynthesis and the nature of flux regulation remain to be established.

In chinook salmon (*O. tshawytscha*) fed diets with different lipid contents, plasma insulin is positively correlated with dietary fat, while total plasma IGF-I scale positively with body fat levels. Also, there is an implied negative feedback of adiposity on food intake and possibly growth, suggesting that dietary and body lipid can potentially offset the anabolic actions of insulin and IGF-I (Shearer *et al.*, 1997). Applying a mammalian paradigm, the effects of IGF-I on lipolysis could be indirect through IGF-I-dependent depression of insulin titers, since IGF-I receptors are absent from adipocytes (Hussain *et al.*, 1995). The fact that insulin fails to control glucose transport in fish adipocytes (Christiansen *et al.*, 1985) is surprising and in total opposition to the situation in mammals. It is, however, in general agreement with the observation that GLUT-4, the insulin-sensitive mammalian glucose transporter, is not expressed in fish tissues (Wright *et al.*, 1998). The observation simply confirms the suspicion that the main actions of insulin in fishes are related to amino acid transport, especially in muscle where it counts the most, and to a much smaller degree to glucose homeostasis.

In mammals, insulin, in concert with other anabolic hormones, promotes the proliferation of satellite cells during the (limited) postnatal muscle development and one wonders whether the chronically elevated (by mammalian standards) titers of insulin may play a role in the hyperplastic muscle growth found in the fishes. However, this does not mean to imply that insulin is the sole endocrine principle behind the muscle hyperplasia. In fact, a growth hormone responsive stat5-binding element has been located in the rat insulin 1 gene (Galsgaard *et al.*, 1996) and likely cross-talk with other hormones will be of utmost importance. The observation that in Atlantic salmon plasma insulin concentrations scale positively with fish weight (Sundby *et al.*, 1991), but growth by hyperplasia decreases in another salmonid (rainbow trout), cannot be reconciled with our suggestions. However, as implied by the specific growth pattern of carp, substantial species differences in the nature of muscle growth exist. In carp, a substantial and variable contribution of hyperplasia is noted even in relatively large fish (Alfei *et al.*, 1994; Koumans *et al.*, 1994; cf. Table III).

Just like IGF, insulin stimulates the rate of sulfate incorporation into gill cartilage, a phenomenon likely to be extrapolated to other cartilaginous or even ossified structures within the fish body and in agreement with general mitotic actions of the hormone.

We surmise that insulin is involved in fish muscle accretion, principally through its effects on amino acid transport and flooding the cell with exogenous amino acids. The other roles of insulin are less clear in fish. The fact that there is an overabundance of IGF compared with insulin receptors on skeletal muscle of all fishes examined to date, may imply a secondary or potentiating role for insulin in this tissue. Unfortunately, none of the data examining insulin receptors on fish muscle looked at a physiological endpoint for this hormone, and certainly such data are necessary before our understanding of this hormone is complete.

E. Thyroid Hormones

The growth-promoting actions of any hormone are economically most important later in the lives of fishes. However, the best experimental model to probe the importance of thyroid hormones to muscle development and growth is early development, where thyroid hormones are implicated in the control of fin differentiation and, especially in flatfishes, metamorphosis. Metamorphosing larvae of flatfish undergo endogenous surges in thyroid hormones and with them profound alterations in body anatomy and muscle distribution and possibly function. It is already known that developing fish larvae possess much larger proportions of red muscle than white muscle and it would be interesting to see what role triiodothyronine (T_3) might play during early growth in directing muscle development and in control of muscle composition. Thyroid receptor $\alpha 1$ (TRα1) is present and functional very early in zebrafish (*Danio rerio*) development (Brown, 1997), acting as a repressor of retinoic acid signaling, at a time when endogenous thyroid hormones are not available in the embryo (Essner *et al.*, 1997), although a latent role of maternally contributed T_3 should not be ignored (Specker and Sullivan, 1995). To provide a comparative framework, in mammals, all genes coding for the different types of muscle MHC are regulated by thyroid hormones, with the modes and direction of response depending on the individual muscle tissue (Izumo *et al.*, 1986). Thyroid hormones interact with other factors to control MHC expression (Lee *et al.*, 1997). In the carp, 28 different genes encoding MHCs have been characterized, but apart from a temperature-sensitive isoform (Gerlach *et al.*, 1990), their regulatory control of expression needs to be determined.

In mammalian skeletal muscles, disuse or lack of weightbearing as achieved by artificially maintaining the muscle in a shortened position, is generally associated with atrophy, deterioration of function, and de-differentiation, involving decreases in protein synthesis and stepped-up protein degradation. Similarly, denervation causes de-differentiation of muscle, associated with increases in the concentrations of the mRNAs coding for two myogenic regulatory factors;

namely, myoD and myogenin that are normally expressed in immature muscle. Other factors controlling expression of these myogenic factors are thyroid hormone and electrical stimulation. Increased myoD expression is associated with fast MHC expression, while high levels of myogenin expression are found in slow fibers (Hughes et al., 1993). Both myogenic factors can be regulated independently, since the β_2-adrenergic agonist clenbuterol leads to an almost fourfold increase in the presence of myogenin message without altering the rate of expression of myoD (Delday and Maltin, 1997). However, it should be kept in mind that increases in mRNA do not necessarily translate into similar increases in protein (Ulbright and Snodgrass, 1993; Ding et al., 1994).

Thyroid hormones also regulate switches in muscle gene expression and muscle phenotype. In the developing rat, for instance, the prenatal replacement of embryonic to adult fast MHCs is regulated by T_3, as is the induction of cardiac muscle α-chain and downregulation of β-chain expression in adult mammals (Lee et al., 1997). Unfortunately, similar lines of inquiry are yet to be pursued for fish.

There is general agreement, but precious little direct evidence, that thyroid hormones can exert anabolic actions in fish tissues, including muscle (Higgs et al., 1982). These anabolic actions are noticed at low concentrations of the hormone only and may lead to better growth performance, while at higher concentrations, catabolic actions appear to prevail (Matty and Lone, 1985). Of course, T_3, the active form of thyroid hormone, may exert some of its action through central release of growth hormone (Luo and McKeown, 1991). Growth hormone, in turn in a feed-forward mechanism, may increase the availability of T_3 through altering the kinetic behavior of the hepatocyte deiodinase in fish (MacLatchy et al., 1992). In isolated systems, T_3 stimulates the incorporation of labeled leucine into muscle homogenates of brown trout (S. trutta) and tilapia (O. mossambicus), but not in all species. Injection of T_3 into Fundulus heteroclitus does not significantly increase the rate of leucine incorporation into muscle protein, in contrast to insulin that stimulates the rate by 60% (Jackim and LaRoche, 1973) over sham-injected control fish. In sea bream (Chrysophrys major), T_3 added to the feed increases growth rate as well as food conversion efficiency (Woo et al., 1991) and in rainbow trout, the thyroid hormones lead to reduced nitrogen excretion and the same positive nitrogen balance (Smith and Thorpe, 1977) noticed in mammals. In addition, some interesting interactions between thyroid hormone, nitrogen excretion, and temperature are noted in a snakehead (Ophiocephalus punctatus) (Ray and Medda, 1976). Better growth performance in fish treated with thyroid hormones is accompanied by enhanced appetite and increased titers of intestinal enzymes. Further, a common observation is that plasma amino acids increase during treatment, especially in fasted fish, and hence the growth promotion may be rooted in the insulinotropic actions of some of these amino acids. However, it should be pointed out that even direct arginine stimulation (the most powerful insulinotropic amino acid in fishes; Ronner and Scarpa, 1987) results only in small and transient improvements in growth performance (Plisetskaya et al., 1991) and is effective in juvenile fish only. Further, the concurrent substantial hyperglycemia during the treatment with T_3 argues against the involvement of insulin. There is sufficient evidence for fish showing that increased appetite and plasma T_3 will stimulate release of growth hormone from the pituitary to implicate growth hormone in the T_3-dependent increase in growth rate. Again, the evidence for involvement of growth hormone is thin, since T_3 treatment fails to alter lipid metabolism—part and parcel of growth hormone action: plasma unesterified fatty acids, cholesterol, and triacylglycerol remain unaltered (Woo et al., 1991).

Muscle protein and RNA syntheses are stimulated in a dose-dependent manner after injection of tilapia with T_3 or T_4 (Matty et al., 1982). Conversely, the use of the goitrogen thiourea decreases the level of muscle protein in an air-breathing catfish (H. fossilis; Singh, 1979). Although MHCs have been recently cloned and partially sequenced in fish (Gauvry and Fauconneau, 1996; Gauvry et al., 1996; Ojima et al., 1998; Watabe, S. GenBank accession #AB016626–54) (see also Chapter 4, this volume), control of their expression by thyroid (and other) hormones needs to be assessed. Thyroid hormone receptors, products of cellular erbAα1 and c-erbAβ1 genes, possess a duality in function in that they have the ability to activate or repress transcription of specific genes. For instance, in zebrafish thyroid receptor α1 (THRα) can act as a repressor of early development and also interfere with retinoic acid-signaling (Essner et al., 1997), while in mammals thyroid hormones regulate skeletal muscle differentiation and increase transcription of the myoD gene and of fast MHCs. The T_3-response element in the mouse myoD promoter region has been characterized (Muscat et al., 1994). Fish muscle also shows developmentally regulated expression of two factors, related to mammalian Id factors, proteins that inhibit DNA binding/differentiation (cf. Chapter 3). These factors contain a helix-loop-helix (HLH) motif, but lack the basic amino acid domain essential to bind DNA, as found, inter alia, in the mitogenic factors myogenin and myoD, and function by interfering with the DNA binding of the dimerized activating HLH proteins. The interaction between tissue specific expression of such inhibitory and activating HLHs regulates muscle differentiation and progression through the cell cycle. The specific expression pattern for rainbow trout Id's in embryonic muscle and their expression in slow oxidative fibers, but not fast glycolytic fibers, of juveniles and adults, implies roles for these proteins in early myogenesis and in regulating muscle phenotype later in life (Rescan, 1997).

While specific thyroid hormone receptors have been functionally identified in many fish tissues (Bres et al., 1994), data for muscle seem to be unavailable, although expression of different receptors has been described for metamorphosing flatfish (Yamano and Miwa, 1998). A number of hormone genes are known to contain thyroid hormone response elements, including growth hormone, steroid hormones, and retinoic acid; all of these can potentially alter the rate of protein synthesis in muscle. Responsiveness of fish muscle protein synthesis to growth hormone and steroids has been confirmed many times and is detailed elsewhere in this review. Thyroid hormone receptors are best characterized in flatfish due to their key involvement in metamorphosis. The Japanese flounder (*Paralichthys olivaceus*) expresses the two expected types of thyroid hormone receptor—THRα and THRβ—including two subtypes of each receptor (Yamano et al., 1994a; Yamano and Inui, 1995), with considerable sequence homologies to mammalian thyroid hormone receptors containing the hallmarks of other nuclear steroid-binding proteins. In developing flounder larvae, THRα is the predominant type of TH receptor expressed in skeletal muscle and its abundance decreases post-climax. During metamorphosis, THRα reaches a peak in unison with thyroid hormone, while THRβ peaks slightly later and is most abundant in cartilage and surrounding tissues, with some presence in myosepta between the myotomes (Yamano and Miwa, 1998). It appears that both types of receptors undergo tissue-specific alterations in expression during metamorphosis and control different developmental processes.

Reminiscent of the their role in amphibian metamorphosis, thyroid hormones play an overwhelming role during critical stages of fish early development, most pronounced in the metamorphosis of flatfishes. Here, a surge in thyroxine triggers metamorphic changes, and many of the changes can be induced precociously by administration of exogenous thyroid hormones or prevented by the application of goitrogens such as thiourea. One of the prime targets for thyroid hormone actions is the skeletal muscle, and within it, the myosin isoforms. However, in surprising contrast to the situation in mammals and amphibians, where thyroid hormones alter the composition of MHCs (Izumo et al., 1986), in the flatfishes, myosin *light* chains are affected most dramatically, as the larval types are replaced by adult types. At the same time, T_3 prompts production of adult-type troponin T at the expense of the larval troponin. No changes are noted for the composition of the MHCs in the flounder (Yamano et al., 1991; 1994b). In the metamorphosing bullfrog, thyroxine induces apoptosis, accompanied by increases in ubiquitin titers, processes involving protein kinase C (Phillips and Platt, 1994). It appears that the thyroid hormone-driven reorganization of fish muscle, fin differentiation, or fin ray resorption (De Jesus et al., 1990) would be ideally suited to probe apoptosis or the cell cycle in a piscine model. Conversely, the involvement of thyroid hormones in smoltification of salmonid fish (Dickhoff, 1993) could be used as a comparative system, where muscle does not seem to constitute a primary target.

Just like other anabolic hormones discussed here (growth hormone, insulin, IGF-I), T_3 administration will significantly increase the rate of skeletal growth in fishes, measured as an increase in sulfate uptake by gill cartilage (Barber and Barrington, 1972), a result that was recently confirmed histologically (Takagi et al., 1994). In fact, anabolic concentrations of T_3 result in larger increases in fish length than in weight (Higgs *et al.*, 1982). Mechanistically, it appears that some of the skeletal effects are mediated via growth hormone dependent increases in tissue IGF-I, but some effects are independent of growth hormone and IGF-I (Takagi *et al.*, 1994), leaving one to wonder whether THRβ may be involved in this action and that internal fine tuning between anabolic peptides and thyroid hormone may drive the different phases of linear and three-dimensional muscle growth described for trout a quarter century ago (Brown, 1974).

F. Cortisol

Cortisol is the principal glucocorticoid of fish. It has numerous functions, but is most prominent with respect to the stress response (Wendelaar Bonga, 1997). The physiological and metabolic roles of this hormone have recently been reviewed in fish (Mommsen *et al.*, 1999). A cortisol-binding protein has been identified in salmonid skeletal muscle by both competitive binding (Chakraborti *et al.*, 1987) and by mRNA transcript analysis using Northern blotting (Ducouret *et al.*, 1995). Evidence supports its identification and classification as a glucocorticoid receptor (GR) rather than a mineralocorticoid receptor, since no true mineralocorticoid receptor activity has been identified in any fish species to date. Little is known of the structure of this receptor, although circumstantial data suggest it is similar to the mammalian GR (see Mommsen *et al.*, 1999). The number of muscle receptors in brook trout is small compared with either liver or gills (Chakraborti *et al.*, 1987). Again, as noted throughout this chapter, the sheer bulk of muscle relative to other tissue mass means that a significant fraction of all GRs are localized to muscle.

Glucocorticoid receptors are known to act as transcription factors in mammals. Although the DNA-binding domain of the rainbow trout GR is slightly altered relative to that in mammals (Ducouret *et al.*, 1995), there is no reason not to assume that the GR in fish also acts as a transcription factor. What the role of GRs in muscle is or how it might change with developmental stage, age, or environmental conditions, has to our knowledge not been investigated; neither has the regulation of the GR in any tissue of fish, although promoter studies have been undertaken for other fish hormones. Antibodies have been produced against the rainbow trout GR (Tujague *et al.*, 1998), so studies can now be initiated to examine the role of this receptor in tissue function.

Cortisol is implicated in peripheral (muscle) proteolysis in fish. This has been shown during exercise (Milligan, 1997), smoltification (Specker and Schreck, 1982), and spawning migration (Ueda *et al.*, 1984). Cortisol is also known to decrease growth rates in feeding fish, but whether this is a result of increased proteolysis or reduced rates of protein synthesis is generally not clear (see Mommsen *et al.*, 1999). Given the mass of muscle, significant changes in the availability of amino acids will occur with small changes in mass, and it does appear that cortisol may be involved in the partitioning of amino acid carbon into non-protein pathways.

Cortisol interacts with many other metabolic and growth-related hormones. Glucocorticoids and insulin are antagonistic regulators of energy metabolism in mammals (see Strack *et al.*, 1995), but it is unlikely that such a scenario occurs in fish given the major differences seen in the actions of insulin between mammals where insulin functions as a glucostatic hormone and fish where insulin's glucostatic actions are secondary to growth-promoting functions at best (see Mommsen and Plisetskaya, 1991). However, insulin does increase both amino acid uptake and protein synthesis in fish (see Section IV.D), so in this regard the two hormones may be counter-regulatory. Glucocorticoids also affect the bioavailability of IGF-I in mammals by increasing IGF-I receptor phosphorylation and receptor signaling in muscle (Georgino and Smith, 1995) and decreasing IGF-I-binding protein-3 (IGFBP-3), the major IGF-binding protein found in mammalian plasma (Villafuerte *et al.*, 1995). This would increase the circulating availability of IGF-I and possibly its effects on muscle. These binding proteins are thought to be present in fish plasma, where they distort the measurements of biologically active IGF-I (Duan, 1998) and cortisol may play a similar role in fish. A sequence similar to a mammalian glucocorticoid-responses element (GRE) has been localized to the promoter of the barramundi (*Lates calcarifer*) growth hormone gene, although whether it is functional has yet to be determined (Yowe and Epping, 1995).

Unfortunately, we have no clear idea of the exact role of cortisol on fish muscle and this area represents a major future research direction. It is apparent that the role of cortisol should not simply be viewed as being involved in peripheral proteolysis. Undoubtedly, there are important interactions with insulin, IGF, and growth hormone that involve determining compositional and performance changes in fish muscle.

G. Gonadal Steroids

The sex steroids, androgens and estrogens, have been implicated in seasonally dependent changes in muscle protein synthesis in a number of fish species. However, the precise mechanism by which such changes actually occur is not clear. In addition, growth hormone in many fish species modulates gonadal steroidogenesis (see section IV.A) and this important interaction may trigger seasonal growth cycles. Certainly the effects of gonadal steroids on growth hormone is confused by the significant differences noted between species.

Anabolic steroids, especially androgens, are powerful growth promoters in fishes, especially in juveniles, and the literature has been reviewed extensively (Higgs *et al.*, 1982; Zohar, 1989). Unfortunately, the number of descriptive studies on growth activation by steroids, mostly in salmonid fishes, is not matched by a similar abundance of studies addressing mechanisms of actions. In fact, very little insight into concepts can be gained from the studies using anabolic steroids in fishes.

Application of exogenous anabolic steroids results in increased growth performance, but such effects may be restricted to juveniles and cannot be found in adults (Thorarensen et al., 1996). Improved growth performance is linked to increased rates of protein synthesis in muscle (Lone and Ince, 1983), accelerated rates of amino acid uptake from the intestine (Habibi and Ince, 1984) and stepped-up proteolytic capacity in the intestinal tract (Lone and Matty, 1981). In juvenile cherry salmon (*O. masou*), 17α-methyltestosterone-dependent increases in growth are accompanied by increases in nitrogen retention and decreases in ammonia excretion (Fig. 2) (Santandreu and Díaz, 1994) without alterations in urea release. Similar phenomena in mammalian muscle are interpreted to signify increased re-utilization of amino acids in the muscle, concomitant with increased net protein synthesis (Ferrando et al., 1998). Although reduced nitrogen release during androgen treatment is often implied as indicating reduced rates of protein turnover due to decreased proteolysis (Santandreu and Díaz, 1994), at least in

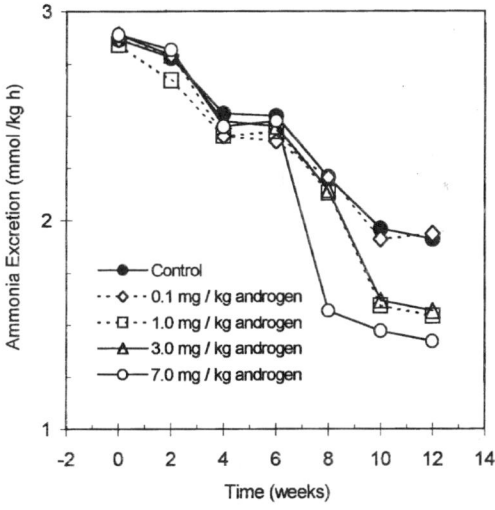

Fig. 2. Nitrogen-sparing effect of 17α-methyltestosterone in masu salmon (*Oncorhynchus masou*; Santandreu and Díaz, 1994). Juvenile masu salmon (initial weight 0.7 g) were fed pellets treated with 0 (control) to 7 mg 17α-methyltestosterone per kilogram of diet. Ammonia excretion was measured twice weekly. Note that during the course of the experiment, nitrogen intake decreases in all groups, and treatment groups take in significantly less protein than control fish. Positive androgen effects on weight become significant after 8 weeks (7 mg androgen kg^{-1}) and 10 weeks (other androgen groups), respectively. Data recalculated and replotted from Santandreu and Díaz, 1994. Excretion rate for urea remained constant throughout the experiments.

mammals, this notion is not supported by the experimental evidence (Hayden et al., 1992). Nevertheless, at times, androgens are thought to function merely as anti-catabolic hormones by preventing the actions of glucocorticoids. In fishes, androgens suppress the activity of the interrenal cells (Pottinger et al., 1996; Young et al., 1996) and thus may increase muscle protein synthesis by reducing the alleged cortisol-driven proteolysis in muscle.

Androgen receptors have been characterized in muscle tissue of fish (Fitzpatrick et al., 1994), and these receptors have been implicated in the circannual and androgen-dependent fluctuations in heart ventricular mass (Davie and Thorarensen, 1997) and in red muscle cross-sectional area (Thorarensen et al., 1996). However, the fluctuations in androgen levels between the sexes are insufficient to isolate androgens as the sole instigators of these observed alterations (Thorarensen et al., 1996). Consequently, and in the absence of similar studies on androgen receptor dynamics for white skeletal muscle, it is quite possible that some of the apparent growth effects of anabolic steroids are mediated by other hormones. For instance, exogenous androgens are able to stimulate growth hormone and prolactin release from the pituitary (Shepherd et al., 1997b) and thyroid hormone delivery (Hunt and Eales, 1979) and generally enhance the activity of the endocrine pancreas (β cells?) (McBride and van Overbeeke, 1971).

Androgen-dependent increases in muscle mass imply growth by hypertrophy, although usually this occurs together with changes in muscle composition. Such changes include the tendency of muscle to augment the oxidative capacity of muscle fibers, a greater abundance of mitochondria and higher density of blood capillaries (Egginton, 1987). This, in turn, implies the involvement of concerted mechanisms, over and above a simple increase in fiber mass through hypertrophic growth. Unfortunately, the situation in the fishes is not yet entirely clear. On the one hand are the many attempts to increase fish weight and growth with application to anabolic steroids, geared toward juvenile fish where mass accretion is accompanied by hyperplasia. On the other hand are the many observations of sexual dimorphism in fish weight and shape, at times with muscle as the most obvious defining characteristic.

Just as in the case of thyroid hormones, a dualism in actions appears to exist for anabolic steroids. At lower exogenous concentrations, anabolic actions prevail, while at higher concentrations, catabolic actions predominate. However, the existing endocrine situation and potential interactions with exogenous hormones may alter the apparent response. In chum salmon (O. keta) during their anorexic spawning migration, for instance, treatment with exogenous androgen potentiates the already increased proteolytic activity noticed in muscle at this particular life stage (Ando et al., 1986b), possibly accentuating the cortisol-driven muscle protein breakdown (Ueda et al., 1984) and leaving open the question of the importance of endogenous androgens that are elevated during the prespawning period. The proteolytic machinery involved in fish muscle protein breakdown has been partially characterized and involves several members of the cathepsin family of proteases (Mommsen et al., 1980; Ando et al., 1986a; Yamashita and Konagaya, 1992; Toyohara et al., 1998) and calpain activation (Watson et al., 1992), while the role of proteasome and ubiquitination (cf. Solomon and Goldberg, 1996; Solomon et al., 1998) remain to be analyzed for fish muscle.

Sex steroids are also known to modulate growth hormone levels in a variety of fish species, including white sucker (*Catostomus commersoni*), rainbow trout, goldfish, coho, Atlantic and chum salmon (see references in Holloway and Leatherland, 1997). Holloway and Leatherland (1997) report that implants of both testosterone and 17β-estradiol increase plasma growth hormone titers, although the effect is greater with 17β-estradiol. The non-aromatizable androgen 5α-dihydrotestosterone failed to affect plasma growth hormone levels in this study, suggesting that only those androgens convertible to estradiol would impact upon circulating growth hormone levels; this may not be true of all fish species. An estrogen-responsive element is located within the rainbow trout growth hormone gene 2, but not gene 1 (Yang et al., 1997) although some authors believe that growth hormone activation may occur through a promiscuous thyroid responsive element (Melamed et al., 1998). 17β-Estradiol may act directly on the pituitary to increase growth hormone content (Zou et al., 1997). These apparent changes in growth hormone should result in enhanced growth, as noted above for growth hormone. The limited amount of data available, however, does not support this happening.

The treatment of mature, male Atlantic salmon with 17β-estradiol (10 mg kg^{-1} once per week) for three weeks significantly increases vitellogenin synthesis by the liver, while significantly decreasing skeletal muscle myosin synthesis (Olin and von der Decken, 1987). In 18-month-old salmon pre-smolts of 65 g, the same treatment significantly decreases MHC protein synthesis (Nazar et al., 1991). These studies imply a re-partitioning of energy with 17β-estradiol, such that growth is reduced to ensure reproductive success. Although growth hormones levels were not measured in these studies, presumably plasma growth hormone levels were elevated by the 17β-estradiol treatment. This provides circumstantial, but significant, support for the role of hormones other than growth hormone acting at the level of the muscle to enhance growth. The most obvious hormone would be IGF (see Section IV.C).

The sex steroids clearly play a role in muscle growth, either directly or through their effects on growth hormone. It should be remembered, however, that growth hormone itself modifies gonadal steroidogenesis, and this feedback relationship may be extremely important for the observed seasonal changes in growth in fish.

H. Adrenergic Agonists

Selective β_2-agonists repartition nutrient use to encourage muscle growth (protein accretion) and reduce lipid disposition (Beerman, 1993). Application of clenbuterol to laboratory rats and farm animals effectively prevents the deterioration of muscle by a transient stimulation of protein synthesis and a long-lasting reduction in degradation. It is this latter effect that has made clenbuterol and other β_2-agonists such popular drugs to increase terrestrial animal meat production. Unfortunately, the causative association between β_2-agonist binding and receptor-specific transduction has not been reported (Beerman, 1993). Long term exposure to clenbuterol not only affects the mass of muscle, but also alters the fiber composition. With rat soleus as a model, the drug increases the proportion of type II (fast-contracting) fibers, and type IIdx MHCs at the expense of type I (slow-contracting, oxidative) MHCs (Criswell et al., 1996). Obviously, the use of β_2-agonist provides a useful model to probe the plasticity of muscles and their components.

Similar trends to introduce β_2-agonists to fish farming, not necessarily as a basic model system, are anticipated and methods to determine β_2-agonist turnover and residues in fish tissues have already been established for clenbuterol (Brambilla et al., 1994). Initial trials with the β_2-agonists ractopamine and L644,969 show increased length (but not weight) and decreased muscle lipid in channel catfish (I. punctatus; Mustin and Lovell, 1993) and increased muscle protein and diminished muscle lipid content in blue catfish (I. furcatus; Webster et al., 1995). In the rainbow trout, ractopamine leads to increases in growth hormone release and higher titers of plasma fatty acids (Vandenbergh and Moccia, 1998; Vandenbergh et al., 1998). This is possibly indicative of lipolytic actions of the agonist (or of growth hormone?), supporting the decreases in muscle lipids noted in the two species of catfish. Otherwise, the effects of this re-partitioning agent in the rainbow trout were modest and spotty. Overall, the response of fish muscle to these β_2-agonists seems to be much smaller, if real at all, than those reported in mammalian or avian muscles.

The apparent lower responsiveness of fish muscle to ractopamine might indicate that the routes of β-adrenoceptor response are generally at a higher basal level in fish than in fowl or mammals, leaving less room for increases and resulting in much muted growth, if any, in fish. Alternatively, it is possible that appropriate receptors are less abundant in fish muscle, or that agonist-dependent downregulation is more rapid in the fishes than in mammals.

The adrenoceptor (AR) complex in vertebrate muscle is not entirely clear, with three different types of β-type receptors having been described. In skeletal muscles of rats, classical β_2-ARs predominate (Plourde et al., 1993) over a small percentage (7–10%) of β_1-type binding proteins (Jensen et al., 1995). The β_2-ARs are upregulated by exercise and about twofold more abundant in type I than type II muscles. Some studies have described so-called "atypical" ARs in rat muscles with characteristics closer to β_3-ARs—found predominantly in adipose tissue—than either β_1- or β_2-ARs (Roberts and Summers, 1998). Binding of agonist to the β_3-AR present in L6 myocytes increases glucose uptake, similar to the actions of insulin, but without employing any parts of the insulin message transduction pathways or targeting the insulin-dependent glucose transporter GLUT-4

(Tanishita *et al.*, 1997). The other two β-ARs are generally related through the c-AMP-protein kinase A signaling pathway (Roberts and Summers, 1998), but it is quite possible that nuclear cAMP-binding could act to enhance transcription and thus account for the longer term changes noted in protein synthesis. In addition to increasing the rate of muscle acquisition in mammals and birds through activation of β_2-AR, clenbuterol negatively affects muscle perfusion and down-regulates the abundance of β_2-ARs (Sillence *et al.*, 1995). Thus, the emerging picture for β-adrenergic actions on vertebrate muscle are clearly multifaceted and ultimately the mechanisms of the positive effects on muscle bulk still need to be elucidated.

No similar studies have to date been undertaken in fish muscle, although one study does indicate that adrenaline decreases the half-time of twitch relaxation in slow muscle fibers presumably through an α_1-AR mechanism as it is blocked by phentolamine (Johnson *et al.*, 1991). With the increasing availability of molecular probes for adrenoceptors in fishes (Svensson *et al.*, 1993; Yasuoka *et al.*, 1996), this entire area of adrenoceptor dynamics in fish muscle is ripe for investigation, especially given the high proportion of fish body weight represented by fast-twitch fibers. The presumed absence of GLUT-4 in fish (Wright *et al.*, 1998) means that muscle lacks one of the key ingredients for a growth-promoting responses to β-type andrenergic agonists.

Myostatin, a member of the transforming growth factor β superfamily of proteins, is significantly expressed in vertebrate skeletal muscle, where it functions as a negative regulator of muscle mass. Null mutations in the mouse myostatin gene, myostatin sequence deletions, and residue substitution in cattle all result in substantial increases in muscle mass (Grobet *et al.*, 1997) together with decreases in intramuscular lipids. In fact, it appears that increased muscle protein growth is often accompanied by decreases in muscular lipid, somewhat independent of the route chosen to increase muscle mass—be it growth hormone, β_2-agonists or genetically determined decreases in myostatin availability—making one wonder what common theme might underlie these phenomena. Unfortunately, the abundance and regulation of myostatin are yet to be analyzed for fishes, although zebrafish muscle myostatin has been cloned and sequenced. Compared with the almost invariant mammalian myostatins, the protein from zebrafish shows sequence homology of only 68% overall, and 88% in the presumed functional C terminal region (McPherron and Lee, 1997).

V. FUTURE DIRECTIONS

This review is not exhaustive in its coverage of hormones and fish muscle growth. Where hormones are discussed, much of the information is correlative at best; examining mechanisms of action has simply not yet happened. With the development of molecular probes for important myogenic factors, a better

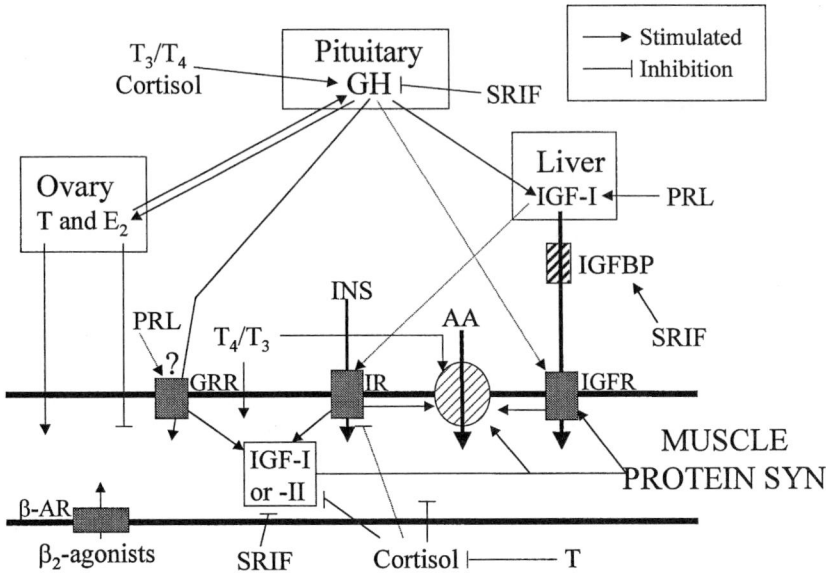

Fig. 3. Integration of factors affecting growth in fish muscle. Dotted lines and ? indicate possible interactions; shaded rectangles are specific hormone receptors (GH = growth hormone; I = insulin; IGF = insulin-like growth factor; β-AR = β-adrenoceptor; R = receptor), patterned rectangle is a binding protein, and the patterned oval represents the amino acid transporter. The thicker lines represent the more important pathways. T = testosterone; E_2 = 17β-estradiol; T_3/T_4 = thyroid hormones; GH = growth hormone; SRIF = somatotropin releasing inhibitory factor; PRL = prolactin; INS = insulin; and AA = amino acids.

understanding of how hormones affect myogenesis will finally be possible. Among these factors would be myogenin, myoD, and c-myc, all factors known to be involved in mammalian muscle growth. It is clear that many factors interact to result in the observed pattern of protein accretion, and certainly the hormones discussed in this review interact in many ways, some of which we have not identified to date either in mammals or fish (see Fig. 3). We would like to conclude our discourse on regulation of muscle growth by posing a few questions that should be asked once such probes become available.

1. Is fish muscle metabolically inert? As noted throughout this review, the relative mass of muscle in fish is very high, meaning that this tissue retains the majority of hormone receptors relative to other body tissues. Yet dogma suggests this tissue is not particularly metabolically demanding and that mitochondrial density is small, contradicted by the fact that protein synthetic rates are substantial. This view needs to be re-examined and possibly new ways of relating metabolic capacity need to be identified for tissues

such as white muscle. Possibly relating metabolic rate to an index of mitochondrial content or of active gene expression beyond RNA/DNA or RNA/protein ratios would provide a much more realistic picture of the metabolism of this tissue.
2. Even though fish grow primarily indeterminantly, some are determinate growers. Could these not represent excellent growth models to separate both the physical and hormonal factors that modulate growth? Even among the indeterminant growers, there are major species differences with respect to growth rates and patterns. Yet, at this point, we do not know the bases for such differences. If we did, it may be possible to augment growth in a particularly prized species. Are there any piscine equivalents, genetically or functionally, to the double muscled strains of cattle? Is it really the availability of proteins/amino acids that limit growth, or is it a particular myogenic factor? What about the linkage between growth hormone and IGF-I or even -II? Even with indeterminant growth, fish growth is seasonally dependent and seasonal heterotrophy could be an ideal model to follow the various key myogenic factors responsible for activation of this process.
3. What is the role of growth hormone? Some evidence suggests that growth hormone can directly act on muscle to stimulate growth. Yet the majority of evidence points to growth hormone affecting IGFs that, in turn, directly stimulate muscle growth. In fact, IGF-I may act independently of growth hormone in some cases, but such conditions and at what point these two systems interact have yet to be identified. As growth hormone possesses important gonadal function as well as growth function, which function predominates and how is this balance established? Is the growth hormone surge at reproduction necessary simply to "prime" the system for growth or to trigger IGF activities?
4. β-Adrenergic agonists are used in the meat industry, yet our understanding of how they work or how they impact upon other bodily functions is very rudimentary as are the ecological and pharmokinetic implications. The characterization of β-receptors has been reported in mammals, but not in any fish, and in no case is there a mechanistic explanation for how agonists specifically alter muscle protein accretion. Still, we continue to use these at least in livestock and they are beginning to be used in aquaculture without a good mechanistic handle.
5. There are potential models that have not been examined here, some of these cover transcription factors and the many additional growth factors identified in other vertebrates or factors that tie into the insulin, IGF-I, or growth hormone message transduction systems. Others models are fish specific, and it is here where work on fishes could make valuable contributions to our conceptual understanding of muscle growth in general. The sonic muscles of vocalizing fish are seasonally sensitive to anabolic steroids and

may represent one such alternate model. Reef fish that are sexually dimorphic and can undergo sex changes are yet another. And nothing has been stated here regarding how toxicants may affect muscle growth, or stress in general. Certainly high density aquaculture is a chronic stress situation and the impact this has on growth and development of fish is poorly understood. It is possible that myogenic factors could play an important role in defining these low, but chronic levels of stress under such conditions.

In conclusion, it is painfully obvious that fish muscle growth needs much further attention and we have barely begun to understand and appreciate the piscine systems for what they really are—ideal models to probe the many aspects of muscle growth that are unique to our finned friends.

ACKNOWLEDGMENTS

We thank Drs. Erika Plisetskaya, Vance Trudeau, and Mathilakath Vijayan for helpful discussions. Research in the authors' laboratories is supported by NSERC Canada.

REFERENCES

Ablett, R. F., Sinnhuber, R. O., and Selivonchick, D. P. (1981). The effect of bovine insulin on [^{14}C]glucose and [^{3}H]leucine incorporation in fed and fasted rainbow trout (*Salmo gairdneri*). *Gen. Comp. Endocrinol.* **44,** 418–427.

Akster, H. A., Koumans, J. T. M., Cuelenaere, J., and Osse, J. W. M. (1995). Uptake of tritiated thymidine in muscle of juvenile carp. *J. Fish Biol.* **47,** 165–167.

Alfei, L., Onali, A., Spano, L., Colombari, P. T., Altavista, P. L., and De Vita, R. (1994). PCNA/cyclin expression and BrdU uptake define proliferating myosatellite cells during hyperplastic muscle growth of fish (*Cyprinus carpio* L.). *Eur. J. Histochem.* **38,** 151–162.

Altringham, J. D., Wardle, C. S., and Smith, C. I. (1993). Myotomal muscle function at different location in the body of a swimming fish. *J. Exp. Biol.* **182,** 191–206.

Ando, S., Hatano, M., and Zama, K. (1986a). Protein degradation and protease activity of chum salmon (*Oncorhynchus keta*) muscle during spawning migration. *Fish Physiol. Biochem.* **1,** 17–26.

Ando, S., Yamazaki, F., Hatano, M., and Zama, K. (1986b). Deterioration of chum salmon (*Oncorhynchus keta*) muscle during spawning migration—III. Changes in protein composition and protease activity of juvenile chum salmon muscle upon treatment with sex steroids. *Comp. Biochem. Physiol.* **83B,** 325–330.

Argetsinger, L. S., Campbell, G. S., Yang, X., Witthuhn, B. A., Silvennoinen, O., Ihle, J. M., and Carter-Su, C. (1993). Identification of JAK2 as a growth hormone receptor-associated tyrosine kinase. *Cell* **74,** 237–244.

Argetsinger, L. S., Hsu, G. W., Myers, M. G., Jr., Billestrup, N., White, M. F., and Carter-Su, C. (1995). Growth hormone, interferon-gamma, and leukemia inhibitory factor promoted tyrosyl phosphorylation of insulin receptor substrate-1. *J. Biol. Chem.* **270,** 14685–14692.

Arndt, S. K. A., Benfey, T. J., and Cunjak, R. A. (1994). A comparison of RNA concentrations and ornithine decarbeoxylase activity in Atlantic salmon (*Salmo salar*) muscle tissue, with respect to specific growth rates and diel variation. *Fish Physiol. Biochem.* **13**, 463–471.

Bagley, C. J., May, B. L., Szabo, L., McNamara, P. J., Ross, M., Francis, G. L., Ballard, F. J., and Wallace, J. C. (1989). A key functional role for the insulin-like growth factor 1 N-terminal pentapeptide. *Biochem. J.* **259**, 665–671.

Baños, N., Moon, T. W., Castejón, C., Gutiérrez, J., and Navarro, I. (1997). Insulin and insulin-like growth factor-I (IGF-I) binding in fish red muscle: regulation by high insulin levels. *Regulat. Pept.* **68**, 181–187.

Baños, N., Baro, J., Castejn, C., Navarro, I., and Gutiérrez, J. (1998). Influence of high-carbohydrate enriched diets on plasma insulin levels and insulin and IGF-I receptors in trout. *Regulat. Pept.* **77**, 55–62.

Barber, S., and Barrington, E. J. W. (1972). Dynamics of uptake and binding of 35S-sulphate by the cartilage of rainbow trout (*Salmo gairdneri*) and the influence of thyroxine. *J. Zool. Lond.* **168**, 107–117.

Beerman, D. H. (1993). β-Adrenergic agonists and growth. In "The Endocrinology of Growth, Development, and Metabolism in Vertebrates" (Schreibman, M. P., Scanes, C. G., and Pang, P. K. T., eds.), pp. 327–366. Academic Press, San Diego.

Ben-Atia, I., Fine, M., Tandler, A., Funkenstein, B., Maurice, S., Cavari, B., and Gertler, A. (1999). Preparation of recombinant gilthead seabream (*Sparus aurata*) growth hormone and its use for stimulation of larvae growth by oral administration. *Gen. Comp. Endocrinol.* **113**, 155–164.

Bernardi, G., D'Onofrio, G., and Caccio, S. (1993). Molecular phylogeny of bony fishes, based on the amino acid sequence of the growth hormone. *J. Mol. Evol.* **37**, 644–649.

Black, D., and Skinner, E. R. (1986). Features of the lipid transport system of fish as demonstrated by studies on starvation in the rainbow trout. *J. Comp. Physiol. B* **156**, 497–502.

Blaxter, J. H. S. (1992). The effect of temperature on larval fishes. *Neth. J. Zool.* **42**, 336–357.

Brambilla, G., Bocca, A., Delisle, M., and Guandalini, E. (1994). Residues of clenbuterol in tissues of the rainbow trout (*Oncorhynchus mykiss*). *Vet. Res. Commun.* **18**, 37–42.

Bres, O., MacLatchy, D. L., and Eales, J. G. (1994). Thyroid hormone: deiodinase and receptor assay. In "Biochemistry and Molecular Biology of Fishes" (Hochachka, P. W., and Mommsen, T. P., eds.), Vol. 3, Analytical Techniques, pp. 447–456. Elseviere Science, Amsterdam.

Brooks, S., and Johnston, I. A. (1993). Influence of development and rearing temperature on the distribution, ultrastructure and myosin sub-unit composition of myotomal muscle-fibre types in the plaice *Pleuronectes platessa*. *Mar. Biol.* **117**, 501–513.

Brown, C. R., and Cameron, J. N. (1991). The relatioship between specific dynamic action (SDA) and protein synthesis rates in the channel catfish. *Physiol. Zool.* **64**, 298–309.

Brown, D. D. (1997). The role of thyroid hormone in zebrafish and axolotl development. *Proc. Natl. Acad. Sci. U. S. A.* **94**, 13011–13016.

Brown, M. E. (1974). Experimental studies on growth. In "Fish Physiology" (Brown, M. E., ed.), pp. 361–371. Little Brown, Philadelphia.

Butler, A. A., Yakar, S., Gewolb, I. H., Karas, M., Okubo, Y., and LeRoith, D. (1998). Insulin-like growth factor-I receptor signal transduction: at the interface between physiology and cell biology. *Comp. Biochem. Physiol.* **121B**, 19–26.

Cameron, J. N. (1975). Blood flow distribution as indicated by tracer microspheres in resting and hypoxic Arctic grayling (*Thymallus arcticus*). *Comp. Biochem. Physiol.* **52A**, 441–444.

Chakraborti, P. K., Weisbart, M., and Chakraborti, A. (1987). The presence of corticosteroid receptor activity in the gills of the brook trout, *Salvelinus fontinalis*. *Gen. Comp. Endocrinol.* **66**, 323–332.

Chan, S. J., Cao, Q.-P., Nagamatsu, S., and Steiner, D. F. (1993). Insulin and insulin-like growth factor genes in fishes and other primitive chordates. In "Biochemistry and Molecular Biology of Fishes" (Hochachka, P. W., and Mommsen, T. P., eds.), Vol. 2, Molecular Biology Frontiers, pp. 407–417. Elsevier Science, Amsterdam.

Cheema, I. R., and Matty, A. J. (1978). Increased uptake of L-leucine-^{14}C in the skeletal muscle of rainbow trout, *Salmo gairdneri*, after administration of growth hormone. *Pak. J. Zool.* **10,** 119–123.

Chen, T. T., Marsh, A., Shamblott, M. J., Chan, K.-M., Tang, Y.-L., Cheng, C. M., and Yang, B.-Y. (1994). Structure and evolution of fish growth hormone and insulinlike growth factor genes. In "Fish Physiology" (Sherwood, N. M., and Hew, C. L.), Vol. XIII—Molecular Endocrinology of Fish, pp. 179–209. Academic Press, San Diego.

Christiansen, D. C., Skarstein, L., and Klungsoyr, L. (1985). Uptake studies in adipocytes isolated from rainbow trout (*Salmo gairdnerii*). A comparison with adipocytes from rat and cat. *Comp. Biochem. Physiol.* **82A,** 201–205.

Clarke, W. C., Farmer, S. W., and Hartwell, K. M. (1977). Effect of teleost pituitary hormone on growth of *Tilapia mossambicus* and on growth and seawater adaptation of sockeye salmon (*Oncorhynchus nerka*). *Gen. Comp. Endocrinol.* **33,** 174–178.

Criswell, D. S., Powers, S. K., and Herb, R. A. (1996). Clenbuterol-induced fiber transition in the soleus of adult rats. *Eur. J. Appl. Physiol.* **74,** 391–396.

Dabrowski, K., Krumschnabel, G., Paukku, L., and Labanowski, J. (1992). Cyclic growth and activity of pancreatic enzymes in alevins of Arctic charr (*Salvelinus alpinus* L.). *J. Fish Biol.* **40,** 511–521.

Danulat, E. (1995). Biochemical-physiological adaptations of teleosts to highly alkaline, saline lakes. In "Biochemistry and Molecular Biology of Fishes" (Hochachka, P. W., and Mommsen, T. P., eds.), Vol. 5—Environmental and Ecological Biochemistry, pp. 229–249. Elsevier Science, Amsterdam.

Davie, P. S., and Thorarensen, H. (1997). Heart growth in rainbow trout in reponse to exogenous testosterone and 17-alpha methyltestosterone. *Comp. Biochem. Physiol.* **117A,** 227–230.

Davie, P. S., Wells, R. M. G., and Tetens, V. (1986). Effects of sustained swimming on rainbow trout muscle structure, blood oxygen transport, and lactate dehydrogenase isozymes: evidence for increased aerobic capacity of white muscle. *J. Exp. Zool.* **237,** 159–171.

De Jesus, E. G., Inui, Y., and Hirano, T. (1990). Cortisol enhances the stimulating action of thyroid hormones on dorsal fin-ray resorption of flounder larvae *in vitro*. *Gen. Comp. Endocrinol.* **79,** 167–173.

Delday, M. I., and Maltin, C. A. (1997). Clenbuterol increases the expression of myogenin but not myoD in immobilized rat muscles. *Am. J. Physiol.* **272,** E941–E944.

Devlin, R. H., Yesaki, T. Y., Donaldson, E. M., Du, S. J., and Hew, C. L. (1995). Production of germline transgenic Pacific salmonids with dramatically increased growth performance. *Can. J. Fish. Aquat. Sci.* **52,** 1376–1384.

Dickhoff, W. W. (1993). Hormones, metamorphosis, and smolting. In "The Endocrinology of Growth, Development, and Metabolism in Vertebrates" (Schreibman, M. P., Scanes, C. G., and Pang, P. K. T., eds.), pp. 519–540. Academic Press, San Diego.

Diez, J. M., Giannico, G., McLean, E., and Donaldson, E. M. (1992). The effect of somatostatins (SRIF-14, 25 and 28), galanin and anti-SRIF on plasma growth hormone levels in coho salmon (*Oncorhynchus kisutch* Walbaum). *J. Fish Biol.* **40,** 887–893.

Ding, J. L., Lim, E. H., and Lam, T. J. (1994). Cortisol-induced hepatic vitellogenin mRNA in *Oreochromis aureus* (*Steindachner*). *Gen. Comp. Endocrinol.* **96,** 276–287.

Du, S. J., Gong, Z. Y., Fletcher, G. L., Shears, M. A., King, M. J., Idler, D. R., and Hew, C. L. (1992). Growth enhancement in transgenic Atlantic salmon by the use of an all fish chimeric growth hormone gene construct. *BioTechnology* **10,** 176–181.

Duan, C. (1998). Nutritional and developmental regulation of insulin-like growth factors in fish. *J. Nutr.* **128,** 306S–314S.

Ducouret, B., Tujague, M., Ashraf, J., Mouchel, N., Servel, N., Valotaire, Y., and Thompson, E. B. (1995). Cloning of a teleost fish glucocorticoid receptor shows that it contains a deoxyribonucleic acid-binding domain different from that of mammals. *Endocrinology* **136,** 3774–3783.

Duguay, S. J., Chan, S. J., Mommsen, T. P., and Steiner, D. F. (1995). Divergence of insulin-like growth factors I and II in the elasmobranch, *Squalus acanthias. FEBS Lett.* **371,** 69–72.

Dutta, H. (1994). Growth in fishes. *Gerontology* **40,** 97–112.

Egginton, S. (1987). Effects of anabolic hormone on aerobic capacity of rat striated muscle. *Pflügers Arch.* **410,** 356–361.

Eliakim, A., Brasel, J. A., Mohan, S., Wong, W. L. T., and Cooper, D. M. (1998). Increased physical activity and the growth hormone IGF-I axis in adolescent males. *Am. J. Physiol.* **275,** R308–R314.

Elies, G., Groigno, L., Wolff, J., Boeuf, G., and Boujard, D. (1996). Characterization of the insulin-like growth factor type 1 receptor messenger in two teleost species. *Mol. Cell. Endocrinol.* **124,** 131–140.

Elies, G., Groigno, L., Wolff, J., Boeuf, G., and Boujard, D. (1998). Cloning of the IGF-1 receptor cDNA of turbot (*Scophthalmus maximus*)—Messenger expression and polyadenylation status. *Ann. N. Y. Acad. Sci.* **839,** 513–514.

Essner, J. J., Breuer, J. J., Essner, R. D., Fahrenkrug, S. C., and Hackett, P. B., Jr. (1997). The zebrafish thyroid hormone receptor α1 is expressed during early embryogenesis and can function in transcriptional repression. *Differentiation* **62,** 107–117.

Farbridge, K. J., and Leatherland, J. F. (1987). Lunar cycles of coho salmon, *Oncorhynchus kisutch* I. Growth and feeding. *J. Exp. Biol.* **129,** 165–178.

Fauconneau, B., Andre, S., Chmaitilly, J., Le Bail, P.-Y., Krieg, F., and Kaushik, S. J. (1997). Control of skeletal muscle fibres and adipose cells size in the flesh of rainbow trout. *J. Fish Biol.* **50,** 296–314.

Fauconneau, B., Gray, C., and Houlihan, D. F. (1995). Assessment of individual protein turnover in three muscle types of rainbow trout. *Comp. Biochem. Physiol.* **111B,** 45–51.

Fauconneau, B., Mady, M. P., and Le Bail, P.-Y. (1996). Effect of growth hormone on muscle protein synthesis in rainbow trout (*Oncorhynchus mykiss*) and Atlantic salmon (*Salmo salar*). *Fish Physiol. Biochem.* **15,** 49–56.

Ferrando, A. A., Tipton, K. D., Doyle, D., Phillips, S. M., Cortiella, J., and Wolfe, R. R. (1998). Testosterone injection stimulates net protein synthesis but not tissue amino acid transport. *Am. J. Physiol.* **275,** E864–E871.

Fine, M., Sakal, E., Vashdi, D., Daniel, V., Levanon, A., Lipshitz, O., and Gertler, A. (1993). Recombinant carp (*Cyprinus carpio*) growth hormone: expression, purification, and determination of biological activity *in vitro* and *in vivo. Gen. Comp. Endocrinol.* **89,** 51–61.

Fitzpatrick, M. S., Gale, W. L., and Schreck, C. B. (1994). Binding characteristics of an androgen receptor in the ovaries of coho salmon, *Oncorhynchus kisutch. Gen. Comp. Endocrinol.* **95,** 399–408.

Florini, J. R., and Ewton, D. Z. (1990). Highly specific inhibition of IGF-I-stimulated differentiation by an antisense oligodeoxyribonucleotide to myogenin mRNA. No effects on other actions of IGF-I. *J. Biol. Chem.* **265,** 13435–13437.

Florini, J. R., and Ewton, D. Z. (1992). Induction of gene expression in muscle by IGFs. *Growth Regulat.* **2** , 23–29.

Florini, J. R., Ewton, D. Z., and Coolican, S. A. (1996). Growth hormone and the insulin-like growth hormone system in myogenesis. *Endocr. Rev.* **17,** 481–517.

Forbes, S. H., Knudsen, K. L., North, T. W., and Allendorf, F. W. (1994). One of two growth hormone genes in coho salmon is sex-linked. *Proc. Natl. Acad. Sci. U. S. A.* **91,** 1628–1631.

Foster, A. R., Houlihan, D. F., Gray, C., Medale, F., Fauconneau, B., Kaushik, S. J., and Le Bail, P. Y. (1991). The effects of ovine growth hormone on protein turnover in rainbow trout. *Gen. Comp. Endocrinol.* **82,** 111–120.

Galsgaard, E. D., Gouilleux, F., Groner, B., Serup, P., Nielsen, J. H., and Billestrup, N. (1996). Identification of a growth hormone-responsive Stat5-binding element in the rat insulin 1 gene. *Mol. Endocrinol.* **10,** 652–660.

Gauvry, L., Ennion, S., Hansen, E., Butterworth, P., and Goldspink, G. (1996). The characterisation of the 5' regulatory region of a temperature-induced myosin-heavy-chain gene associated with myotomal muscle growth in the carp. *Eur. J. Biochem.* **236,** 887–894.

Gauvry, L., and Fauconneau, B. (1996). Cloning of a trout fast skeletal myosin heavy chain expressed both in embryo and adult muscles and in myotubes neoformed *in vitro*. *Comp. Biochem. Physiol.* **115B,** 183–190.

Georgino, F., and Smith, R. J. (1995). Dexamethasone enhances insulin-like growth factor-I effects on skeletal muscle cell proliferation. Role of specific intracellular signaling pathways. *J. Clin. Invest.* **96,** 1473–1483.

Gerlach, G. F., Turay, L., Malik, K. T., Lida, J., Scutt, A., and Goldspink, G. (1990). Mechanisms of temperature acclimation in the carp: a molecular biology approach. *Am. J. Physiol.* **259,** R237–R244.

Goll, D. E., Thompson, V. F., Taylor, R. G., and Christiansen, J. A. (1992). Role of calpain system in muscle growth. *Biochimie* **74,** 225–237.

Goodman, H. M. (1993). Growth hormone and metabolism. In "The Endocrinology of Growth, Development, and Metabolism in Vertebrates" (Schreibman, M. P., Scanes, C. G., and Pang, P. K. T., eds.), pp. 93–115. Academic Press, San Diego.

Goolish, E. M. (1991). Aerobic and anaerobic scaling in fish. *Biol. Rev.* **66,** 33–56.

Grant, G. C. (1996). RNA-DNA ratios in white muscle tissue biopsies reflect recent growth rates of adult brown trout. *J. Fish Biol.* **48,** 1223–1230.

Grobet, L., Royo Martin, L. J., Poncelet, D., Pirottin, D., Brouwers, B., Riquet, J., Schoeberlein, S., Menissier, F., Massabanda, J., Fries, R., Hanset, R., and Georges, M. (1997). A deletion in the bovine myostatin gene causes the double-muscled phenotype in cattle. *Nature Genet.* **17,** 71–74.

Gutiérrez, J., Åsgård, T., Fabbri, E., and Plisetskaya, E. M. (1991). Insulin binding in skeletal muscle of trout. *Fish Physiol. Biochem.* **9,** 351–360.

Gutiérrez, J., Párrizas, M., Maestro, M. A., Navarro, I., and Plisetskaya, E. M. (1995). Insulin and IGF-I binding and tyrosine kinase activity in fish heart. *J. Endocrinol.* **146,** 35–44.

Gutiérrez, J., and Plisetskaya, E. M. (1991a). Insulin binding to liver plasma membranes in salmonids with modified plasma insulin levels. *Can. J. Zool.* **69,** 2745–2750.

Gutiérrez, J., and Plisetskaya, E. M. (1991b). Insulin binding to the liver plasma membranes of coho salmon during smoltification. *Gen. Comp. Endocrinol.* **82,** 466–475.

Gutiérrez, J., and Plisetskaya, E. M. (1994). Peptide receptor assays: insulin receptor. In "Biochemistry and Molecular Biology of Fishes" (Hochachka, P. W., and Mommsen, T. P., eds.), Vol. 3— Analytical Techniques, pp. 431–446. Elsevier Science, Amsterdam.

Habibi, H. R., and Ince, B. W. (1984). A study of androgen-stimulated L-leucine transport by the intestine of rainbow trout (*Salmo gairdneri* Richardson) *in vitro*. *Comp. Biochem. Physiol.* **79A,** 143–149.

Hanel, R., Karjalainen, J., and Wieser, W. (1996). Growth of swimming muscle and its metabolic cost in larvae of whitefish at different temperatures. *J. Fish Biol.* **48,** 937–951.

Haschemeyer, A. E. V., and Smith, M. A. K. (1979). Protein synthesis in liver, muscle and gill of mullet (*Mugil cephalus* L.) *in vivo*. *Biol. Bull.* **156,** 93–102.

Hayden, J. M., Bergen, W. G., and Merkel, R. A. (1992). Skeletal muscle protein metabolism and serum growth hormone, insulin, and cortisol concentrations in growing steers implanted with

estradiol-17β, trenbolone acetate, or estradiol-17β plus trenbolone acetate. *J. Anim. Sci.* **70,** 2109–2119.
Helbing, C. C., Wellington, C. L., Gogela-Spehar, M., Cheng, T., Pinchbeck, G. G., and Johnston, R. N. (1998). Quiescence versus apoptosis: Myc abundance determines pathway of exit from the cell cycle. *Oncogene* **17,** 1491–1501.
Higgs, D. A., Donaldson, E. M., Dye, H. M., and McBride, J. R. (1975). A preliminary investigation of the effect of bovine growth hormone on growth and muscle composition of coho salmon (*Oncorhynchus Risutch*). *Gen. Comp. Endocrinol.* **27,** 240–253.
Higgs, D. A., Fagerlund, U. H. M., Eales, J. G., and McBride, J. R. (1982). Application of thyroid and steroid hormones as anabolic agents in fish culture. *Comp. Biochem. Physiol.* **73B,** 143–176.
Hirano, T. (1991). Hepatic receptors for homologous growth hormone in the eel. *Gen. Comp. Endocrinol.* **81,** 383–390.
Hochachka, P. W. (1994). "Muscles as Molecular and Metabolic Machines" pp. 1–158. CRC Press, Boca Raton, FL.
Hodik, V., Mett, A., and Halevy, O. (1997). Mutual effects of growth hormone and growth factors on avian skeletal muscle satellite cells. *Gen. Comp. Endocrinol.* **108,** 161–170.
Holloway, A. C., and Leatherland, J. F. (1997). Effect of gonadal steroid hormones on plasma growth hormone concentrations in sexually immature rainbow trout, *Oncorhynchus mykiss*. *Gen. Comp. Endocrinol.* **105,** 246–254.
Horseman, N. D., and Meier, A. H. (1978). Prostaglandin and the osmoregulatory role of prolactin in a teleost. *Life Sci.* **22,** 1485–1490.
Horseman, N. D., and Yu-Lee, L. Y. (1994). Transcriptional regulation by helix bundle peptide hormones: Growth hormone, prolactin, and hematopoietic cytokines. *Endocr. Rev.* **15,** 627–649.
Houlihan, D. F., Carter, C. G., and McCarthy, I. D. (1995). Protein synthesis in fish. *In* "Biochemistry and Molecular Biology of Fishes" (Hochachka, P. W., and Mommsen, T. P., eds.), Vol. 4—Metabolic Biochemistry, pp. 191–220. Elsevier Science, Amsterdam.
Houlihan, D. F., Hall, S. J., and Gray, C. (1989). Effects of ration on protein turnover in cod. *Aquaculture* **79,** 103–110.
Hughes, S. M., Taylor, J. M., Tapscott, S. J., Gurley, C. M., Carter, W. J., and Peterson, C. A. (1993). Selective accumulation of MyoD and myogenin mRNAs in fast and slow adult skeletal muscle is controlled by innervation and hormones. *Development* **118,** 1137–1147.
Hunt, D. W. C., and Eales, J. G. (1979). The influence of testosterone proprionate on thyroid function of immature rainbow trout, *Salmo gairdneri* Richardson. *Gen. Comp. Endocrinol.* **37,** 115–121.
Hussain, M. A., Schmitz, O., Christiansen, J. S., Zapf, J., and Froesch, E. R. (1995). Metabolic effects of insulin-like growth factor-I: A focus on insulin sensitivity. *Metabolism* **44,** Suppl. 4, 108–112.
Idler, D. R., Fletcher, G. L., Belkhode, S., King, M. J., and Hwang, S. J. (1989). Regulation of antifreeze protein production in winter flounder: A unique role for growth hormone. *Gen. Comp. Endocrinol.* **74,** 327–334.
Ikuta, K., Hirano, T., and Aida, K. (1995). Radioreceptor assay for salmon growth hormone. *Proc., XIth Int. Symp. Comp. Endocrinol.* Malaga, Spain May 14–20, 1989, p.168, Abstract.
Ince, B. W., and Thorpe, A. (1978). Insulin kinetics and distribution in rainbow trout (*Salmo gairdneri*). *Gen. Comp. Endocrinol.* **35,** 1–9.
Izumo, S., Nadal-Ginard, B., and Mahdavi, V. (1986). All members of the MHC multigene family respond to thyroid hormone in a highly tissue-specific manner. *Science* **231,** 597–600.
Jackim, E., and LaRoche, G. (1973). Protein synthesis in *Fundulus heteroclitus* muscle. *Comp. Biochem. Physiol.* **44A,** 851–866.
Jensen, J., Brors, O., and Dahl, H. A. (1995). Different β-adrenergic receptor density in different rat skeletal muscle fibre types. *Pharmacol. Toxicol.* **76,** 380–385.
Johnson, T. P., Moon, T. W., and Johnston, I. A. (1991). Actions of epinephrine on the contractility of fast and slow skeletal muscle fibres in teleosts. *Fish Physiol. Biochem.* **9,** 83–89.

Johnsson, J. I., and Björnsson, B. T. (1994). Growth hormone increases growth rate, appetite and dominance in juvenile rainbow trout, *Oncorhynchus mykiss*. *Anim. Behav.* **48**, 177–186.

Johnsson, J. I., Jönsson, E., and Björnsson, B. T. (1996). Dominance, nutritional state, and growth hormone levels in rainbow trout (*Oncorhynchus mykiss*). *Horm. Behav.* **30**, 13–21.

Kemp, S. F., Kearns, G. L., Smith, W. G., and Elders, M. J. (1988). Effects of IGF-I on the synthesis and processing of glycosaminoglycan in cultured chick chondrocytes. *Acta Endocrinol.* **119**, 245–250.

Kiessling, A., Hung, S. S. O., and Storebakken, T. (1993). Differences in protein mobilization between ventral and dorsal parts of white epaxial muscle from fed, fasted and refed white sturgeon (*Acipenser transmontanus*). *J. Fish Biol.* **43**, 401–408.

Kiessling, A., Kiessling, K.-H., Storebakken, T., and Åsgård, T. (1991). Changes in the structure and function of the epaxial muscle of rainbow trout (*Oncorhynchus mykiss*) in relation to ration and age. II. Activity of key enzymes in energy metabolism. *Aquaculture* **93**, 357–372.

Koumans, J. T. M., and Akster, H. A. (1995). Myogenic cells in development and growth of fish. *Comp. Biochem. Physiol.* **110A**, 3–20.

Koumans, J. T. M., Akster, H. A., Dulos, G. J., and Osse, J. W. M. 1990. Myosatellite cells of *Cyprinus carpio* (Teleostei) *in vitro*: isolation, recognition and differentiation. *Cell Tissue Res.* **261**, 173–181.

Koumans, J. T. M., Akster, H. A., Witkam, A., and Osse, J. W. M. (1994). Numbers of muscle nuclei and myosatellite cell nuclei in red and white axial muscle during growth of the carp (*Cyprinus carpio*). *J. Fish Biol.* **44**, 391–408.

Lang, F., Busch, G. L., Ritter, M., Völkl, H., Waldegger, S., Gulbins, E., and Häussinger, D. (1998). Functional significance of cell volume regulatory mechanisms. *Physiol. Rev.* **78**, 247–306.

Le Gac, F., Blaise, O., Fostier, A., Le Bail, P.-Y., Loir, M., Mourot, B., and Weil, C. (1993). Growth hormone (GH) and reproduction: a review. *Fish Physiol. Biochem.* **11**, 219–232.

Lee, Y., Nadal-Ginard, B., Mahdavi, V., and Izumo, S. (1997). Myocyte-specific enhancer factor 2 and thyroid hormone receptor associate and synergistically activate the α-cardiac myosin heavy-chain gene. *Mol. Cell. Biol.* **17**, 2745–2755.

Li, S. L., Termini, J., Hayward, A., Siddle, K., Zick, Y., Koval, A., LeRoith, D., and Fujita-Yamaguchi, Y. (1998). The carboxyl-terminal domain of insulin-like growth factor-I receptor interacts with the insulin receptor and activates its protein tyrosine kinase. *FEBS Lett.* **421**, 45–49.

Lied, E., and Rosenlund, G. (1984). The influence of the ratio of protein energy to total energy in the feed on the activity of protein synthesis *in vitro*, the level of ribosomal RNA and the RNA-DNA ratio in white trunk muscle of Atlantic cod (*Gadus morhua*). *Comp. Biochem. Physiol.* **77A**, 489–494.

Liu, Y. M., Lehar, S., Corvi, C., Payne, G., and O'Connor, R. (1998). Expression of the insulin-like growth factor I receptor C terminus as a myristylated protein leads to induction of apoptosis in tumor cells. *Cancer Res.* **58**, 570–576.

Londraville, R. L., and Sidell, B. D. (1996). Cold acclimation increases fatty acid-binding protein concentration in aerobic muscles of striped bass, *Morone saxatilis*. *J. Exp. Zool.* **275**, 36–44.

Lone, K. P., and Ince, B. (1983). Cellular growth response of rainbow trout (*Salmo gairdneri*) fed different levels of dietary protein and anabolic steroid hormone ethylestrenol. *Gen. Comp. Endocrinol.* **49**, 32–49.

Lone, K. P., and Matty, A. J. (1981). The effect of feeding androgenic hormones on the proteolytic activity of the alimentary canal of carp *Cyprinus carpio* L. *J. Fish Biol.* **18**, 353–358.

Louvi, A., Accili, D., and Efstratiadis, A. (1997). Growth promoring interaction of IGF-II with the insulin receptor during mouse embryonic development. *Dev. Biol.* **189**, 33–48.

Low, P. S., Rennie, M. J., and Taylor, P. M. (1997). Involvement of integrins and the cytoskeleton in

modulation of skeletal muscle glycogen synthesis by changes in cell volume. *FEBS Lett.* **417**, 101–103.
Low, S. Y., Rennie, M. J., and Taylor, P. M. (1997). Signaling elements involved in amino acid transport responses to altered muscle cell volume. *FASEB J.* **11**, 1111–1117.
Lowery, M. S., and Somero, G. N. (1990). Starvation effects on protein synthesis in red and white muscle of the barred sand bass, *Paralabrax nebulifer. Physiol. Zool.* **63**, 630–648.
Luo, D., and McKeown, B. A. (1991). The effect of thyroid hormone and glucocorticoids on carp growth hormone-releasing factor (GRF)-induced growth hormone (GH) release in rainbow trout (*Oncorhynchus mykiss*). *Comp. Biochem. Physiol.* **99A**, 621–626.
MacLatchy, D. L., Kawauchi, H., and Eales, J. G. (1992). Stimulation of hepatic thyroxine 5'-deiodinase activity in rainbow trout (*Oncorhynchus mykiss*) by Pacific salmon growth hormone. *Comp. Biochem. Physiol.* **101A**, 689–691.
Mader, S., and Sonenberg, N. (1995). Cap binding complexes and cellular growth control. *Biochimie* **77**, 40–44.
Maeda, T., Takekawa, M., and Saito, H. (1995). Activation of yeast PBS2 MAPKK by MAPKKKs or by binding of an SH3-containing osmosensor. *Science* **269**, 554–558.
Mainoya, J. R. (1982). PRL and digestive tract in teleost fishes. *J. Comp. Physiol. B* **146**, 1–7.
Markert, J. R., Higgs, D. A., Dye, H. M., and MacQuarrie, D. W. (1977). Influence of bovine growth hormone on growth rate, appetite, and food conversion of yearling coho salmon (*Oncorhynchus kisutch*) fed two diets of different composition. *Can. J. Zool.* **55**, 74–83.
Martínez, R., Estrada, M. P., Berlanga, J. J., Guillen, I., Hernandez, O., Cabrera, E., Pimentel, R., Morales, R., Herrera, F., Morales, A. E., Pina, J. C., Abad, Z., Sanchez, V., Melamed, P., Lleonart, R., and de la Fuente, J. (1996). Growth enhancement in transgenic tilapia by ectopic expression of tilapia growth hormone. *Mol. Mar. Biol. Biotechnol.* **5**, 62–70.
Martínez, M., Winger, P., Guderley, H., and Dutil, J.-D. (1999). Longitudinal variation of metabolic capacities of cod muscle (*Gadus morhua*): response to starvation. *Comp. Biochem. Physiol.* **124A**, S144.
Matschak, T. W., and Stickland, N. C. (1996). The influence of temperature on mRNA levels for muscle contractile protein and a proto-oncogene associated with cell division in Atlantic salmon (*Salmo salar*). *Can. J. Fish. Aquat. Sci.* **53**, 408–413.
Matty, A. J., Chaudhury, M. A., and Lone, K. P. (1982). Effect of thyroid hormones on the protein metabolism of *Tilapia mossambica. Gen. Comp. Endocrinol.* **46**, 387.
Matty, A. J., and Lone, K. P. (1985). The hormonal control of metabolism and feeding. *In* "Fish Energetics: New Perspectives" (Tytler, P., and Calow, P., eds.), pp. 185–209. Johns Hopkins University Press, Baltimore.
Mayer, I., McLean, E., Kieffer, T. J., Souza, L. M., and Donaldson, E. M. (1994). Antisomatostatin-induced growth acceleration in chinook salmon (*Oncorhynchus tshawytscha*). *Fish Physiol. Biochem.* **13**, 295–300.
McBride, J. R., and van Overbeeke, A. (1971). Effects of androgens, estrogens and cortisol on the skin, stomach, liver, pancreas and kidney in gonadectomized adult sockeye salmon. *J. Fish. Res. Bd. Canada* **28**, 485–490.
McCarthy, T. L., Centrella, M., and Canalis, E. (1989). Regulatory effects of insulin-like growth factors I and II on bone collagen synthesis in rat calvarial cultures. *Endocrinology* **124**, 301–309.
McCormick, S. D. (1996). Effects of growth hormone and insulin-like growth factor I on salinity tolerance and gill Na^+, K^+-ATPase in Atlantic salmon (*Salmo salar*): Interaction with cortisol. *Gen. Comp. Endocrinol.* **101**, 3–11.
McGarry, J. D. (1992). What if Minkowski had been ageusic? An alternative angle on diabetes. *Science* **258**, 766–770.
McGeer, J. C., Wright, P. A., Wood, C. M., Wilkie, M. P., Mazur, C. F., and Iwama, G. K. (1994).

Nitrogen excretion in four species of fish from an alkaline lake. *Trans. Am. Fish. Soc.* **123**, 824–829.

McLean, E., and Donaldson, E. M. (1993). The role of growth hormone in the growth of poikilotherms. *In* "The Endocrinology of Growth, Development, and Metabolism in Vertebrates" (Schreibman, M. P., Scanes, C. G., and Pang, P. K. T., eds.), pp. 43–71. Academic Press, San Diego.

McPherron, A. C., and Lee, S.-J. (1997). Double muscling in cattle due to mutations in the myostatin gene. *Proc. Natl. Acad. Sci. U. S. A.* **94**, 12457–12461.

Melamed, P., Rosenfeld, H., Elizur, A., and Yaron, Z. (1998). Endocrine regulation of gonadotropin and growth hormone gene transcription in fish. *Comp. Biochem. Physiol.* **119C**, 325–338.

Meyer-Burgdorff, K. H., and Rosenow, H. (1995). Protein turnover and energy metabolism in growing carp. 3. Energy cost of protein deposition. *J. Anim. Physiol. Anim. Nutr.* **73**, 134–139.

Migliorini, R. H., Lima-Verde, J. S., Machado, C. R., Cardona, G. M. P., Garofalo, M. A. R., and Kettelhut, I. C. (1992). Control of adipose tissue lipolysis in ectotherm vertebrates. *Am. J. Physiol.* **263**, R857–R862.

Milligan, C. L. (1997). The role of cortisol in ammonia mobilization and metabolism following exhaustive exercise in rainbow trout (*Oncorhynchus mykiss* Walbaum). *Fish Physiol. Biochem.* **16**, 119–128.

Mommsen, T. P. (1998). Growth and metabolism. *In* "The Physiology of Fishes" (Evans, D. H., ed.), pp. 65–97. CRC Press, Boca Raton, FL.

Mommsen, T. P., French, C. J., and Hochachka, P. W. (1980). Sites and pattern of protein and amino acid utilization during the spawning migration of salmon. *Can. J. Zool.* **58**, 1785–1799.

Mommsen, T. P., and Plisetskaya, E. M. (1991). Insulin in fishes and agnathans: history, structure, and metabolic regulation. *Rev. Aquat. Sci.* **4**, 225–259.

Mommsen, T. P., Vijayan, M. M., and Moon, T. W. (1999). Cortisol in teleosts: dynamics, mechanisms of action and metabolic regulation. *Rev. Fish Biol. Fish.* **9**, 211–268.

Mori, T., and Devlin, R. H. (1999). Transgene and host growth hormone gene expression in pituitary and nonpituitary tissues of normal and growth hormone transgenic salmon. *Mol. Cell. Endocrinol.* **149**, 129–139.

Murphy, L. J. (1998). Insulin-like growth factor-binding proteins: functional diversity or redundancy? *J. Mol. Endocrinol.* **21**, 97–107.

Muscat, G. E., Mynett-Johnson, L., Dowhan, D., Downes, M., and Griggs, R. (1994). Activation of myoD gene transcription by 3,5,3'-triiodo-L-thyronine: a direct role for the thyroid hormone and retinoid X receptors. *Nucleic Acids Res.* **22**, 583–591.

Mustin, W. T., and Lovell, R. T. (1993). Feeding the repartitioning agent, ractopamine, to channel catfish (*Ictalurus punctatus*) increases weight gain and reduces fat deposition. *Aquaculture* **109**, 145–152.

Nag, A. C., and Nursall, J. R. (1972). Histiogenesis of white and red muscle fibers of trunk muscle of a fish *Salmo gairdneri*. *Cytobios* **6**, 227–246.

Navarro, I., Leibush, B., Moon, T. W., Plisetskaya, E. M., Baños, N., Méndez, E., Planas, J. V., and Gutiérrez, J. (1999). Insulin, insulin-like growth factor-I (IGF-I) and glucagon: the evolution of their receptors. *Comp. Biochem. Physiol.* **122B**, 137–153.

Nazar, D. S., Persson, G., Olin, T., Waters, S., and von der Decken, A. (1991). Sarcoplasmic and myofibrillar proteins in white trunk muscle of salmon (*Salmo salar*) after estradiol treatment. *Comp. Biochem. Physiol.* **98B**, 109–114.

Ng, T. B., Leung, T. C., Cheng, C. H. K., and Woo, N. Y. S. (1992). Growth hormone binding sites in tilapia (*Oreochromis mossambicus*) liver. *Gen. Comp. Endocrinol.* **86**, 111–118.

Noel, O., and Le Bail, P.-Y. (1997). Does cyclicity of growth rate in rainbow trout exist. *J. Fish Biol.* **51**, 634–642.

O'Connor, R., Kauffmann-Zeh, A., Liu, Y. M., Lehar, S., Evan, G. I., Baserga, R., and Blättler, W. A.

(1997). Identification of domains of the insulin-like growth factor I receptor that are required for protection from apoptosis. *Mol. Cell. Biol.* **17,** 427–435.
Ojima, T., Kawashima, N., Inoue, A., Amauchi, A., Togashi, M., Watabe, S., and Nishita, K. (1998). Determination of primary structure of heavy meromyosin region of walleye pollack myosin heavy chain by cDNA cloning. *Fish. Sci.,* **64,** 812–819.
Okuda, N., Tayasu, I., and Yanagisawa, Y. (1998). Determinate growth in a paternal mouthbrooding fish whose reproductive success is limited by buccal capacity. *Evol. Ecol.* **12,** 681–699.
Olin, T., and von der Decken, A. (1987). Estrogen treatment and its implication on vitellogenin and myosin synthesis in salmon (*Salmo salar*). *Physiol. Zool.* **60,** 346–351.
Parry-Billings, M., Bevan, S. J., Opara, E. C., and Newsholme, E. A. (1991). Effects of changes in cell volume on the rates of glutamine and alanine release from rat skeletal muscle *in vitro*. *Biochem. J.* **276,** 559–561.
Párrizas, M., Baños, N., Baró, J., Planas, J., and Gutiérrez, J. (1994a). Up-regulation of insulin binding in fish skeletal muscle by increased insulin levels. *Regulat. Pept.* **53,** 211–222.
Párrizas, M., Planas, J., Plisetskaya, E. M., and Gutiérrez, J. (1994b). Insulin binding and receptor tyrosine kinase activity in skeletal muscle of carnivorous and omnivorous fish. *Am. J. Physiol.* **266,** R1944–R1950.
Párrizas, M., Plisetskaya, E. M., Planas, J., and Gutiérrez, J. (1995a). Abundant insulin-like growth factor-1 (IGF-1) receptor binding in fish skeletal muscle. *Gen. Comp. Endocrinol.* **98,** 16–25.
Párrizas, M., Maestro, M. A., Baños, N., Navarro, I., Planas, J., and Gutiérrez, J. (1995b). Insulin/ IGF-I binding ratio in skeletal and cardiac muscles of vertebrates: A phylogenetic approach. *Am. J. Physiol.* **269,** R1370–R1377.
Párrizas, M., Saltiel, A. R., and LeRoith, D. (1997). Insulin-like growth factor 1 inhibits apoptosis using the phosphatidylinositol 3′-kinase and mitogen-activated protein kinase pathways. *J. Biol. Chem.* **272,** 154–161.
Phillips, M. E., and Platt, J. E. (1994). The use of ubiquitin as a marker of thyroxine-induced apoptosis in cultured *Rana catesbeiana* tail tips. *Gen. Comp. Endocrinol.* **95,** 409–415.
Plisetskaya, E. M. (1998). Some of my not so favourite things about insulin and insulin-like growth factors in fish. *Comp. Biochem. Physiol.* **121B,** 3–11.
Plisetskaya, E. M., and Duguay, S. J. (1993). Pancreatic hormones and metabolism in ectotherm vertebrates: current views. *In* "The Endocrinology of Growth, Development, and Metabolism in Vertebrates" (Schreibman, M. P., Scanes, C. G., and Pang, P. K. T., eds.), pp. 265–287. Academic Press, San Diego.
Plisetskaya, E. M., Buchelli-Narvaez, L. I., Hardy, R. W., and Dickhoff, W. W. (1991). Effects of injected and dietary arginine on plasma insulin levels and growth of Pacific salmon and rainbow trout. *Comp. Biochem. Physiol.* **98A,** 165–170.
Plourde, G., Rousseaumigneron, S., and Nadeau, A. (1993). Effect of endurance training on beta-adrenergic system in 3 different skeletal muscles. *J. Appl. Physiol.* **74,** 1641–1646.
Pottinger, T. G., Carrick, T. R., Hughes, S. E., and Balm, P. H. M. (1996). Testosterone, 11-ketotestosterone, and estradiol-17β modify baseline and stress-induced interrenal and corticotropic activity in trout. *Gen. Comp. Endocrinol.* **104,** 284–295.
Powell, R. L., Dodson, M. V., and Cloud, J. G. (1989). Cultivation and differentiation of satellite cells from skeletal muscle of the rainbow trout *Salmo gairdneri*. *J. Exp. Zool.* **250,** 333–338.
Prack, M., Antoine, M., Caiati, M., Roskowski, M., Treacy, T., Vodicnik, M. J., and de Vlaming, V. L. (1980). The effects of mammalian prolactin and growth hormone on goldfish (*Carassius auratus*) growth, plasma amino acid levels and liver amino acid uptake. *Comp. Biochem. Physiol.* **67A,** 307–310.
Quinn, L. S., Haugk, K. L., and Grabstein, K. H. (1995). Interleukin-15: a novel anabolic cytokine for skeletal muscle. *Endocrinology* **136,** 3669–3672.
Rahman, M. A., Mak, R., Ayad, H., Smith, A., and Maclean, N. (1998). Expression of a novel piscine

growth hormone gene results in growth enhancement in transgenic tilapia (*Oreochromis niloticus*). *Transgenic Res.* **7,** 357–369.

Randall, D. J., Wood, C. M., Perry, S. F., Bergman, H., Maloiy, G. M. O., Mommsen, T. P., and Wright, P. A. (1989). Urea excretion as a stategy for survival in a fish living in a very alkaline environment. *Nature* **337,** 165–166.

Ray, A. K., and Medda, A. K. (1976). Effect of thyroid hormones and analogues on ammonia and urea excretion in Lata fish (*Ophiocephalus punctatus*). *Gen. Comp. Endocrinol.* **29,** 190–197.

Rescan, P. Y. (1997). Identification in a fish species of two Id (inhibitor of DNA binding/differentiation)-related helix-loop-helix factors expressed in the slow oxidative muscle fibers. *Eur. J. Biochem.* **247,** 870–876.

Resnicoff, M., and Baserga, R. (1998). The role of the insulin-like growth factor I receptor in transformation and apoptosis. *Ann. N. Y. Acad. Sci.* **842,** 76–81.

Roberts, S. J., and Summers, R. J. (1998). Cyclic AMP accumulation in rat soleus muscle: stimulation by β_2- but not β_3-adrenoceptors. *Eur. J. Pharmacol.* **348,** 53–60.

Ron, B., Shimoda, S. K., Iwama, G. K., and Grau, E. G. (1995). Relationships among ration, salinity, 17α-methyltestosterone and growth in the euryhaline tilapia, *Oreochromis mossambicus*. *Aquaculture* **135,** 185–193.

Ronner, P., and Scarpa, A. (1987). Secretagogues for pancreatic hormone release in the channel catfish (*Ictalurus punctatus*). *Gen. Comp. Endocrinol.* **65,** 354–362.

Sadoshima, J., Qiu, Z. H., Morgan, J. P., and Izumo, S. (1996). Tyrosine kinase activation is an immediate and essential step in hypotonic cell swelling-induced ERK activation and c-*fos* gene expression in cardiac myocytes. *EMBO J.* **15,** 5535–5546.

Saltiel, A. R. (1996). Diverse signaling pathways in the cellular actions of insulin. *Am. J. Physiol.* **270,** E375–E385.

Sandra, O., Sohm, F., De Luze, A., Prunet, P., Edery, M., and Kelly, P. A. (1995). Expression cloning of a cDNA encoding a fish prolactin receptor. *Proc. Natl. Acad. Sci. U. S. A.* **92,** 6037–6041.

Santandreu, I. A., and Díaz, N. F. (1994). Effect of 17α-methyltestosterone on growth and nitrogen excretion in masu salmon (*Oncorhynchus masou* Brevoort). *Aquaculture* **124,** 321–333.

Sharma, A., Lee, Y. B., Murray, J. D., and Oberbauer, A. M. (1996). Skeletal muscle growth of oMTIa-oGH transgenic mice. *Growth Dev. Aging* **60,** 31–41.

Shearer, K. D., Silverstein, J. T., and Plisetskaya, E. M. (1997). Role of adiposity in food intake control of juvenile chinook salmon (*Oncorhynchus tshawytscha*). *Comp. Biochem. Physiol.* **118A,** 1209–1215.

Shepherd, B. S., Sakamoto, T., Nishioka, R. S., Richman, N. H., Mori, I., Madsen, S. S., Chen, T. T., Hirano, T., Bern, H. A., and Grau, E. G. (1997a). Somatotropic actions of the homologous growth hormone and prolactins in the euryhaline teleost, the tilapia, *Oreochromis mossambicus*. *Proc. Natl. Acad. Sci. U. S. A.* **94,** 2068–2072.

Shepherd, B. S., Ron, B., Burch, A., Sparks, R., Richman, N. H., Shimoda, S. K., Stetson, M. H., Lim, C., and Grau, E. G. (1997b). Effects of salinity, dietary level of protein and 17α-methyltestosterone on growth hormone (GH) and prolactin (tPRL177 and tPRL188) levels in the tilapia, *Oreochromis mossambicus*. *Fish Physiol. Biochem.* **17,** 279–288.

Sheridan, M. A. (1986). Effects of thyroxin, cortisol, growth hormone and prolactin on lipid metabolism of coho salmon, *Oncorhynchus kisutch*, during smoltification. *Gen. Comp. Endocrinol.* **64,** 220–238.

Sillence, M. N., Reich, M. M., and Thomson, B. C. (1995). Sexual dimorphism in the growth response of entire and gonadectomized rats to clenbuterol. *Am. J. Physiol.* **268,** E1077–E1082.

Singh, A. K. (1979). Effect of thyroxine and thiourea on liver and muscle energy stores in the freshwater catfish, *Heteropneustes fossilis*. *Biol. Anim. Biochem. Biophys.* **19,** 1669–1676.

Singh, A. K., and Prasad, M. (1987). The effects of bovine prolactin on synthesis, release and mobi-

lisation of biochemical constituents responsible for somatic growth and vitellogenesis in hypophysectomized freshwater female catfish, *Heteropneustes fossilis,* during different phases of annual reproductive cycle. *Zool. Jb. Physiol.* **91,** 315–327.
Smith, M. A. K. (1981). Estimation of growth potential by measurement of tissue protein synthetic rates in feeding and fasting rainbow trout, *Salmo gairdnerii* Richardson. *J. Fish Biol.* **19,** 213–220.
Smith, M. A. K., and Thorpe, A. (1977). Endocrine effects on nitrogen excretion in the euryhaline teleost, *Salmo gairdneri. Gen. Comp. Endocrinol.* **32,** 400–406.
Smith, R. W., Houlihan, D. F., Nilsson, G. E., and Brechin, J. G. (1996). Tissue-specific changes in protein synthesis rates *in vivo* during anoxia in crucian carp. *Am. J. Physiol.* **271,** R897–R904.
Solomon, V., and Goldberg, A. L. (1996). Importance of the ATP-ubiquitin-proteasome pathway in the degradation of soluble and myofibrillar proteins in rabbit muscle extracts. *J. Biol. Chem.* **271, 26,** 690–26,697.
Solomon, V., Lecker, S. H., and Goldberg, A. L. (1998). The N-end rule pathway catalyzes a major fraction of the protein degradation in skeletal muscle. *J. Biol. Chem.* **273,** 25216–25222.
Somero, G. N., and Childress, J. J. (1990). Scaling of ATP-supplying enzymes, myofibrillar proteins and buffering capacity in fish muscle: relationship to locomotory habit. *J. Exp. Biol.* **149,** 319–333.
Sonenberg, N., and Gingras, A. C. (1998). The mRNA 5' cap-binding protein eIF4E and control of cell growth. *Curr. Opin. Cell Biol.* **10,** 268–275.
Sparks, R. T., Shepherd, B. S., Ron, B., Richman, N. H., III, Shimoda, S. K., Iwama, G. K., Ball, C., and Grau, E. G. (1999). Effects of environmental salinity and 17α-methyltestosterone on oxygen consumption and growth in the euryhaline tilapia, *Oreochromis mossambicus. Comp. Biochem. Physiol.,* submitted.
Specker, J. L., King, D. S., Nishioka, R. S., Shirahata, K., Yamaguchi, T., and Bern, H. A. (1985). Isolation and partial characterization of a pair of prolactins released *in vitro* by the pituitary of a cichlid fish, *Oreochromis mossambicus. Proc. Natl. Acad. Sci. U. S. A.* **82,** 7490–7494.
Specker, J. L., and Schreck, C. B. (1982). Changes in plasma corticosteroids during smoltification of coho salmon, *Oncorhynchus kisutch. Gen. Comp. Endocrinol.* **46,** 53–58.
Specker, J. L., and Sullivan, C. V. (1995). Vitellogenesis in fishes: status and perspectives. *In* "Perspectives in Comparative Endocrinology" (Davey, K. G., Peter, R. E., and Tobe, S. S., eds.), pp. 304–315. National Research Council of Canada Press, Ottawa.
Spierts, I. L. Y., Akster, H. A., and Granzier, H. L. (1997). Expression of titin isoforms in red and white muscle fibres of carp (*Cyprinus carpio* L) exposed to different sarcomere strains during swimming. *J. Comp. Physiol. B* **167,** 543–551.
Stickland, N. C. (1983). Growth and development of muscle fibres in the rainbow trout (*Salmo gairdneri). J. Anat.* **137,** 323–333.
Strack, A. M., Sebastian, R. J., Schwartz, M. W., and Dallman, M. F. (1995). Glucocorticoids and insulin: Reciprocal signals for energy balance. *Am. J. Physiol.* **269,** R142–R149.
Sun, L.-Z., and Farmanfarmaian, A. (1992). Biphasic action of growth hormone on intestinal amino acid absorption in striped bass hybrids. *Comp. Biochem. Physiol.* **103A,** 381–390.
Sundby, A., Eliassen, K. A., Refstie, T., and Plisetskaya, E. M. (1991). Plasma levels of insulin, glucagon and glucagon-like peptide in salmonids of different weights. *Fish Physiol. Biochem.* **9,** 223–230.
Svensson, S. P. S., Bailey, T. J., Pepperl, D. J., Grundström, N., Ala-Uotila, S., Scheinin, M., Karlsson, J. O. G., and Regan, J. W. (1993). Cloning and expression of a fish α_2-adrenoceptor. *Br. J. Pharmacol.* **110,** 54–60.
Takagi, Y., Hirano, J., Tanabe, H., and Yamada, J. (1994). Stimulation of skeletal growth by thyroid hormone administrations in the rainbow trout, *Oncorhynchus mykiss. J. Exp. Zool.* **268,** 229–238.

Takagi, Y., Moriyama, S., Hirano, T., and Yamada, J. (1992). Effects of growth hormones on bone formation and resorption in rainbow trout (*Oncorhynchus mykiss*), as examined by histomorphometry of the pharyngeal bone. *Gen. Comp. Endocrinol.* **86,** 90–95.

Tanishita, T., Shimizu, Y., Minokoshi, Y., and Shimazu, T. (1997). The β_3-adrenergic agonist BRL37344 increases glucose transport into L6 myocytes through a mechanism different from that of insulin. *J. Biochem.* **122**, 90–95.

Tashima, L., and Cahill, G. F. (1968). Effects of insulin in the toadfish, *Opsanus tau*. *Gen. Comp. Endocrinol.* **11,** 262–271.

Theill, L. E., and Karin, M. (1993). Transcriptional control of GH expression and anterior pituitary development. *Endocr. Rev.* **14,** 670–689.

Thorarensen, H., Young, G., and Davie, P. S. (1996). 11-Ketotestosterone stimulates growth of heart and red muscle in rainbow trout. *Can. J. Zool.* **74,** 912–917.

Tipton, K. D., and Wolfe, R. R. (1998). Exercise-induced changes in protein metabolism. *Acta Physiol. Scand.* **162,** 377–387.

Toyohara, H., Ito, K., Hori, N., Touhata, K., Kinoshita, M., Kubota, S., Sato, K., Ohtsuki, K., and Sakaguchi, M. (1998). Effect of steroid hormone administration on the breakdown of muscle proteins in ayu. *Fish. Sci.* **64,** 419–422.

Tujague, M., Valotaire, Y., Pakdel, F., and Ducouret, B. (1998). An extra peptide sequence within the DNA binding domain of a fish glucocorticoid receptor arising from a special exon-intron organization—analysis of its transactivating role. *Ann. N. Y. Acad. Sci.* **839,** 612–614.

Ueda, H., Hara, A., Yamauchi, K., and Nagahama, Y. (1984). Changes in serum concentrations of steroid hormones, thyroxine and vitellogenin during spawning migration of the chum salmon (*Oncorhynchus keta*). *Gen. Comp. Endocrinol.* **53,** 203–211.

Ulbright, C., and Snodgrass, P. J. (1993). Coordinate induction of the urea cycle enzymes by glucagon and dexamethasone is accomplished by three different mechanisms. *Arch. Biochem. Biophys.* **301,** 237–243.

Urbinati, E. C., Willis, M. D., and Plisetskaya, E. M. (1994). Growth-promoting activity of proinsulin in fish. *Am. Zool.* **34,** 42A.

Usher, M. L., Stickland, N. C., and Thorpe, J. E. (1994). Muscle development in Atlantic salmon (*Salmo salar*) embryos and the effect of temperature on muscle cellularity. *J. Fish Biol.* **44,** 953–964.

Valente, L. M. P., Gomes, E. F. S., and Fauconneau, B. (1998). Biochemical growth characterization of fast and slow-growing rainbow trout strains: effect of cell proliferation and size. *Fish Physiol. Biochem.* **18,** 213–224.

Van Beneden, R. J., Watson, D. K., Chen, T. T., Lautenberger, J. A., and Pappas, T. S. (1986). Cellular myc (c-myc) in fish (rainbow trout): its relationship to other vertebrate myc genes and to the transforming genes of the MC29 family of viruses. *Proc. Natl. Acad. Sci. U. S. A.* **83,** 3698–3702.

Vandenbergh, G. W., Leatherland, A., and Moccia, R. D. (1998). The effects of the beta-agonist ractopamine on growth hormone and intermediary metabolite concentrations in rainbow trout, *Oncorhynchus mykiss* (Walbaum). *Aquaculture Res.* **29,** 79–87.

Vandenbergh, G. W., and Moccia, R. D. (1998). Growth performance and carcass composition of rainbow trout, *Oncorhynchus mykiss* (Walbaum), fed the beta-agonist ractopamine. *Aquaculture Res.* **29,** 469–479.

Van Leeuwen, J. L. (1995). The action of muscles in swimming fish. *Exp. Physiol.* **80,** 177–191.

Velick, S. F. (1956). The metabolism of myosin, the meromyosins, actin and tropomyosin in the rabbit. *Biochim. Biophys. Acta* **20,** 228–236.

Villafuerte, B. C., Koop, B. L., Pao, C. I., and Phillips, L. S. (1995). Glucocorticoid regulation of insulin-like growth factor-binding protein-3. *Endocrinology* **136,** 1928–1933.

Waldegger, S., Busch, G. L., Kaba, N. K., Zempel, G., Ling, H., Heidland, A., Häussinger, D., and Lang, F. (1997). Effect of cellular hydration on protein metabolism. *Miner. Electrolyte Metab.* **23,** 201–205.

Walsh, P. J., Bedolla, C., and Mommsen, T. P. (1989). Scaling and sex-related differences in toadfish (*Opsanus beta*) sonic muscle enzyme activities. *Bull. Mar. Sci.* **45,** 143–163.
Waters, M. J., Rowlinson, S. W., Clarkson, R. W., Chen, C.-M., Lobie, P. E., Norstedt, G., Mertani, H., Morel, G., Brinkworth, R., Wells, C. A., Bastiras, S., Robins, A. R., Muscat, G. E., and Barnard, R. T. (1994). Signal transduction by the growth hormone receptor. *Proc. Soc. Exp. Biol. Med.* **206,** 216–220.
Watson, C. L., Morrow, H. A., and Brill, R. W. (1992). Proteolysis of skeletal muscle in yellowfin tuna (*Thunnus albacares*): Evidence of calpain activation. *Comp. Biochem. Physiol.* **103B,** 881–887.
Weatherley, A. H., Gill, H. S., and Rogers, S. C. (1980). The relationship between mosaic muscle fibers and size in rainbow trout (*Salmo gairdneri*). *J. Fish Biol.* **17,** 603–610.
Webster, C. D., Tiu, L. G., Tidwell, J. H., and Reed, E. B., Jr. (1995). Effects of feeding the repartioning agent L644,969 on growth and body composition of blue catfish, *Ictalurus furcatus,* fed diets containing two protein levels reared in cages. *Aquaculture* **134,** 247–256.
Wendelaar Bonga, S. E. (1997). The stress response in fish. *Physiol. Rev.* **77,** 591–625.
Whitelaw, P. F., and Hesketh, J. E. (1992). Expression of c-myc and c-fos in rat skeletal muscle. Evidence for increased levels of c-myc mRNA during hypertrophy. *Biochem. J.* **281,** 143–147.
Willis, P. E., Chadan, S. G., Baracos, V., and Parkhouse, W. S. (1998). Restoration of insulin-like growth factor I action in skeletal muscle of old mice. *Am. J. Physiol.* **275,** E525–E530.
Wilson, R. P., Poe, W. E., Nemetz, T. G., and MacMillan, J. R. (1988). Effect of recombinant bovine growth hormone administration on growth and body composition of channel catfish. *Aquaculture* **73,** 229–236.
Wong, A. O. L., Le Drean, Y., Liu, D., Hu, Z. Z., Du, S. J., and Hew, C. L. (1996). Induction of chinook salmon growth hormone promoter activity by the adenosine $3',5'$-monophosphate (cAMP)-dependent pathway involves two cAMP-response elements with the CGTCA motif and the pituitary-specific transcription factor Pit-1. *Endocrinology* **137,** 1775–1784.
Woo, N. Y. S., Chung, A. S. B., and Ng, T. B. (1991). Influence of oral administration of 3,5,3-triiodo-L-thyronine on growth, digestion, food conversion and metabolism in the underyearling red sea bream, *Chrysophrys major* (Temminck & Schlegel). *J. Fish Biol.* **39,** 459–468.
Wright, J. R., Jr., O'Hali, W., Yang, H., Han, X. X., and Bonen, A. (1998). GLUT-4 deficiency and severe peripheral resistance to insulin in the teleost fish tilapia. *Gen. Comp. Endocrinol.* **111,** 20–27.
Yamano, K., Araki, K., Sekikawa, K., and Inui, Y. (1994a). Cloning of thyroid hormone receptor genes expressed in metamorphosing flounder. *Dev. Genet.* **15,** 378–382.
Yamano, K., and Inui, Y. (1995). cDNA cloning of thyroid hormone receptor β for the Japanese flounder. *Gen. Comp. Endocrinol.* **99,** 197–203.
Yamano, K., and Miwa, S. (1998). Differential gene expression of thyroid hormone receptor α and β in fish development. *Gen. Comp. Endocrinol.* **109,** 75–85.
Yamano, K., Miwa, S., Obinata, T., and Inui, Y. (1991). Thyroid hormone regulates developmental changes in muscle during flounder metamorphosis. *Gen. Comp. Endocrinol.* **81,** 464–472.
Yamano, K., Takano-ohmuro, H., Obinata, T., and Inui, Y. (1994b). Effect of thyroid hormone on developmental transition of myosin light chains during flounder metamorphosis. *Gen. Comp. Endocrinol.* **93,** 321–326.
Yamashita, M., and Konagaya, S. (1992). Differentiation and localization of catheptic proteinases responsible for extensive autolysis of mature chum salmon muscle (*Oncorhynchus keta*). *Comp. Biochem. Physiol.* **103B,** 999-1003.
Yang, B.-Y., Chan, K.-M., Lin, C.-M., and Chen, T. T. (1997). Characterization of rainbow trout (*Oncorhynchus mykiss*) growth hormone 1 gene and the promoter region of growth hormone 2 gene. *Arch. Biochem. Biophys.* **340,** 359–368.
Yang, S. Y., Alnaqeeb, M., Simpson, H., and Goldspink, G. (1996). Cloning and characterization of an IGF-1 isoform expressed in skeletal muscle subjected to stretch. *J. Muscle Res. Cell Motil.* **17,** 487–495.

Yao, K., Niu, P.-D., Le Gac, F., and Le Bail, P.-Y. (1991). Presence of specific growth hormone binding sites in rainbow trout (*Oncorhynchus mykiss*) tissues: Characterization of hepatic receptor. *Gen. Comp. Endocrinol.* **81,** 72–82.

Yasuoka, A., Abe, K., Arai, S., and Emori, Y. (1996). Molecular cloning and functional expression of the α1A-adrenoceptor of Medaka fish, *Oryzias latipes. Eur. J. Biochem.* **235,** 501–507.

Young, G., Thorarensen, H., and Davie, P. S. (1996). 11-Ketotestosterone suppresses interrenal activity in rainbow trout (*Oncorhynchus mykiss*). *Gen. Comp. Endocrinol.* **103,** 301–307.

Yowe, D. L., and Epping, R. J. (1995). Cloning of the barramundi growth hormone-encoding gene: a comparative analysis of higher and lower vertebrate GH genes. *Gene* **162,** 255–259.

Zhou, S., Ackman, R. G., and Morrison, C. (1996). Adipocytes and lipid distribution in the muscle tissue of Atlantic salmon (*Salmo salar*). *Can. J. Fish. Aquat. Sci.* **53,** 326–332.

Zohar, Y. (1989). Endocrinology and fish farming: aspects in reproduction, growth and smoltification. *Fish Physiol. Biochem.* **7,** 395–405.

Zorzano, A., James, D. E., Ruderman, N. B., and Pilch, P. F. (1988). Insulin-like growth factor a binding and receptor kinase in red and white muscles. *FEBS Lett.* **234,** 257–262.

Zou, J. J., Trudeau, V. L., Cui, Z., Brechin, J., Mackenzie, K., Zhu, Z., Houlihan, D. F., and Peter, R. E. (1997). Estradiol stimulates growth hormone production in female goldfish. *Gen. Comp. Endocrinol.* **106,** 102–112.

INDEX

A

Acetylcholinesterase (AchE), embryo staining, 146
AchE, *see* Acetylcholinesterase
β_2-Adrenergic agonists
  aquaculture applications, 290, 293
  muscle effects, 289–290
  receptor overview, 290–291
Androgens
  concentration dependence of anabolic actions, 288
  growth activation in juveniles, 286–287
  growth hormone interactions, 288
  protein metabolism effects, 287–289
  receptors, 288
Aquaculture
  β_2-adrenergic agonist applications, 290, 293
  color visualization of filets, 177–178
  flesh
    characteristics compared with wild fish, 176
    texture optimization, 176–179
  growth hormone applications, 271
  muscle cellularity and texture, 176–177
  optimization of muscle growth, 127–128
  stress and growth, 294
  temperature effects in egg incubation, 173
ATPase, *see* Myosin

B

BMP4, head muscle development role in zebrafish, 13
buf, *see* buzz-off
buzz-off (*buf*), muscle differentiation role in zebrafish, 9

C

Cell volume
  adaptation, 261–264
  osmotic stress effects on muscle growth, 264
  signal transduction pathway alterations, 264
chinless (*chn*), head muscle development role in zebrafish, 13
chn, *see* chinless
cho, *see* choker
choker (*cho*), myogenesis role in zebrafish, 8
Condition factor, growth description, 252
Connectin, *see* Titin
Cortisol
  exercise response, 262
  functions, 283–286
  interaction with other hormones, 286
  receptors, 285
CRP, *see* Cysteine-rich protein
Cysteine-rich protein (CRP)
  gene cloning, 62
  structure, 62

D

Diameter, *see* Fiber diameter
Diploidy, *see* Ploidy

E

Echidna Hedgehog (EHH), myogenesis role in zebrafish, 6–7
Egg incubation temperature, *see* Temperature
EHH, *see* Echidna Hedgehog

EN, head muscle development role in zebrafish, 11–12
engrailed, myogenesis role in zebrafish, 4
E protein
   expression patterns in development, 20, 31–32
   structures and sequence conservation, 25–26
   transcription factor activity, 19, 21–22
Estradiol
   concentration dependence of anabolic actions, 288
   exogenous effects on muscle, 289
   growth hormone interactions, 289
Exercise
   capillarization of muscle, 233–234
   cortisol response, 262
   high-speed training effects in muscle, 235–236
   laboratory studies in fish, 232–233
   metabolic adaptation of muscle, 232, 234, 236
   mitochondrial and lipid volume effects in muscle, 235
   muscle
      growth effects, 126–127, 173–174
      morphology and phenotype effects, 233, 236
   protein synthesis effects, 262

F

Farming, *see* Aquaculture
Fast fiber, *see* White muscle
FG2
   expression in carp muscle hyperplasia, 52–53
   promoter characterization, 63–64
FGF, *see* Fibroblast growth factor
Fiber diameter
   egg incubation temperature effects, 164, 168–169, 171
   salmonid development, 148–149
   statistical analysis, 154–155
Fiber number
   egg incubation temperature effects, 161–164, 169
   relationship to fish length, 27–28
   salmonid development, 148–149
fibrils unbundled (*fub*), muscle differentiation role in zebrafish, 8–9

Fibroblast growth factor (FGF), satellite cell regulation, 174
Fish farming, *see* Aquaculture
flh, see floating head
floating head (*flh*), mesoderm allocation of cells in myogenesis, 3–4
fro, see frozen
frozen (*fro*), muscle differentiation role in zebrafish, 9
fub, see fibrils unbundled

G

GH, *see* Growth hormone
Glucose transport, fish, 278–279, 291
Growth, muscle, *see also* Fiber diameter; Fiber number; Mosaic hyperplasia; Stratified hyperplasia
   cell volume effects, 263–264
   diet effects
      juvenile growth, 120–121
      larval growth, 121
   exercise effects, *see* Exercise
   genetic variation in salmonid growth patterns
      ploidy manipulation, 159–160
      population variance, 155–156, 159
   growth history effects, 127–128
   hormonal manipulation, *see specific hormones*
   hybrid fish growth, 125
   hypertrophy
      developmental stages, 109–110
      myofibril formation, 109
      nuclei acquisition of myofibers, 110, 149
      regulatory factors, 175
   integration of factors affecting growth, 292
   investigation techniques
      enzyme activity, 105
      mitotic activity analysis, 107–109
      morphometry, 104–105, 153–154
      myosin isoform analysis, 107
      RNA levels, 107, 256–257
   optimization in aquaculture species, 127–128
   oxygen tension, *see* Oxygen tension
   phases of hyperplasia, 110–112
   relative contributions of hyperplasia and hypertrophy to post-embryonic growth
      aging effects, 253
      juvenile growth, 119–120

larval growth, 118–119
trout, 255–258
seasonal effects, 125–126
sex effects, 124
strain effects on hyperplasia, 91, 123–124
temperature effects in egg incubation
  cod, 164–165
  fiber diameter, 164, 168–169, 171
  fiber number, 161–164, 169
  hatchery implications, 173
  herring, 171–172
  mitochondrial density, 165
  nuclear density, 168
  overview of species differences, 256
  salmon, 161–169, 171–172
  satellite cell number, 168–169, 171, 174
  trout, 165
  wild fish myogenesis studies, 165–169
triploidy effects, 124–125, 159, 161
Growth hormone (GH)
  aquaculture applications, 271
  binding protein, 267
  direct actions on muscle, 267–268, 293
  functions, 264–267, 269
  gene structure and regulation, 266
  immunoneutralization effects, 268
  injection effects on fish, 268, 271
  insulin-like growth factor I response, 56, 265–267, 275, 293
  lipid regulation, 269–270
  muscle gene expression regulation in development, 55–56
  muscle growth effects, 121–123, 268–269
  receptor and signal transductiion, 267
  reproduction role, 270–271
  sex hormone interactions, 289
  structure, 266
  transgenic fish, 122

H

Head muscles, development in zebrafish, 11–14
Hepatocyte growth factor/scatter factor (HGF/SF), satellite cell regulation, 151, 174–175
HGF/SF, *see* Hepatocyte growth factor/scatter factor
Hoxa1, head muscle development role in zebrafish, 13

Hybrid fish, muscle growth, 125
Hyperplasia, *see* Growth, muscle; Mosaic hyperplasia; Stratified hyperplasia
Hypertrophy, *see* Growth, muscle
Hypoxia, *see* Oxygen tension

I

Id, structures and sequence conservation, 26–27
IGF-I, *see* Insulin-like growth factor I
IL-15, *see* Interleukin-15
Insulin
  functions, 277, 280–281
  lipid regulation, 279–288
  receptors
    muscle versus liver, 254–255
    regulation of levels, 278–279
    signal transduction, 275–276, 278
    tissue distribution, 278
  satellite cell proliferation effects, 280
  structure, 277
Insulin-like growth factor I (IGF-I)
  binding proteins, 273
  cortisol interactions, 286
  functions, 273–275
  gene cloning and sequence homology between fish species, 55–56
  growth hormone response, 56, 265–267, 277, 293
  hypertrophy effects, 177
  mechano growth factor processing, 276
  muscle gene expression regulation in development, 56–57 satellite cell proliferation effects, 93, 95, 174–175
  muscle growth effects, 122–123
  prospects for study, 277
  receptors in muscle, 255, 273–276
  regulation of expression, 265–267, 273–275
  sequence conservation between species, 272–273
  sites of synthesis, 55
  transgenic fish, 122
Interleukin-15 (IL-15), muscle growth effects, 276
Intermediate fiber, *see* Pink muscle

M

Mechano growth factor (MGF), origin and functions, 276

MEF2, see Myocyte-specific enhancer factor 2
MGF, see Mechano growth factor
Mitochondria
  egg incubation temperature effects on density, 165
  oxygen tension and response, 230–232
  red muscle density, 197–199
  temperature acclimation and volume changes, 226–227
  white muscle density, 198, 201–206
Mosaic hyperplasia, see also Growth, muscle
  age of onset, 116–117
  morphology, 116
  overview, 104, 111, 116
  satellite cell role, 74–75, 89, 97, 117–118
  source of new fibers, 117–118
MRF4, see Myf-6
Muscle growth, see Growth, muscle
Muscle satellite cell, see Satellite cell
Myf-5
  expression in development, 27–28, 30
  knockout mouse phenotype, 21
  satellite cell marker, 152
  structure and sequence conservation, 24, 62
Myf-6
  expression in development, 27, 35–36
  satellite cell marker, 152–153
Myocyte-specific enhancer factor 2 (MEF2)
  expression patterns
    development, 28, 30–31, 35
    temperature acclimation, 33
  gene cloning, 23
  isoforms, 22
  structure and sequence conservation, 25
  synergism with myogenic regulatory factors, 23
  transcription factor activity, 19–20, 22
MyoD
  expression patterns
    development, 27–28, 30–31, 35
    temperature acclimation, 33
  gene cloning, 23
  knockout mouse phenotype, 21
  myogenesis role
    salmonids, 145–146
    zebrafish, 4, 10–12
  regulation of expression, 281–282
  satellite cell marker, 151–152
  structure and sequence conservation, 24–25, 62
Myogenic regulatory factors, see also specific factors

E protein binding, 19, 21
mosaic hyperplasia role, 118
overlapping functions, 21
salmonid myogenesis role, 145–146
structures, 19–20
types, 19–21
Myogenin
  expression during during temperature acclimation, 33
  knockout mouse phenotype, 21
  muscle-specific gene upregulation, 22–23
  regulation of expression, 281–282
  satellite cell marker, 152
  structure and sequence conservation, 23, 62
Myoseptum, formation, 4–5
Myosin
  ATPase
    activity during temperature acclimation, 32–33, 45–46, 57–58, 222
    calcium regulation, 58
    differences by temperature regions, 45
    fiber type staining, 189–191, 195–196
  expression of isoforms
    development
      carp gene studies, 50, 52
      gel electrophoresis analysis in fish, 49–50
      mammals, 49
      salmonids, 149
    muscle growth analysis, 107
    muscle regeneration, 52
    temperature acclimation, 57–60, 65, 225–226
  functions, 44
  heavy chain
    domains, 47–48
    expression in development, 28, 30, 256
    FG2 expression in carp muscle hyperplasia, 52–53
    genes
      carp gene features, 33
      diversity in fish, 46, 65
      homology between species, 48
      regulatory elements, 61–64
    messenger RNA stability, 64
    structures
      ATPase loop, 60–61
      functional diversity, 60–61
      overview, 44, 46–47
    mechanism study prospects, 65
    regulation of gene expression in development

INDEX 313

growth hormone, 55–56
insulin-like growth factor I, 55–57
thyroid hormone, 55, 63
types and muscle classification, 5, 44–45, 189–190, 215
Myostatin
mutation in double muscling, 128
negative regulation of muscle mass, 291
Myotome
anatomy, 187–188
formation and muscle differentiation, 4–6, 14, 48–49, 141–142

N

Nitrogen, detoxification, 253

O

Oxygen tension
environmental factors affecting, 230
hypoxia adaptation
metabolic responses, 230, 232
mitochondria volume response, 230–232
overview of mechanisms, 230–231
muscle fiber growth pattern effects, 172–173

P

Paired fin muscles, development, 9–11
Parvalbumin
fiber type expression, 190–191
isotypes, 190–191
PAX-3, myogenesis role, 62
PCNA, see Proliferating cell nuclear antigen
Pink muscle
development and growth
larva, 209
origin, 209–211
energy metabolism, 213
function, 142, 213
histochemical analysis, 189, 211–213
myosin expression, 190–191
position and occurrence of fibers, 208–209
stratified hyperplasia, 115
ultrastructure, 212

Plasticity, muscle phenotype
environmental effects, see Oxygen tension; Temperature acclimation
exercise training effects, see Exercise
Ploidy, effects on muscle growth, 124–125, 159, 161
Prolactin, muscle growth effects, 272
Proliferating cell nuclear antigen (PCNA), satellite cell marker, 151–152
Protein catabolism
regulation, 259–260
starvation mobilization, 261
Protein synthesis
exercise effects, 262
factors affecting rates, 259
hormonal control, see specific hormones
measurement and species differences in turnover, 256–259
muscle type differences, 261–262
oxygen utilization and muscle accretion, 253–254, 258
specific dynamic action, 262–263
white muscle compared with other tissues, 263

R

Red muscle
accretion markers, 258
arrangement in Atlantic salmon, 142
development and growth
differentiation, 192–195
myofibril formation, 193–195
precursor migration, 192
energy metabolism, 195–199
function, 142, 200–201
histochemical analysis, 189, 195–196
lipid volume density, 198
mitochondrial density, 196–197
myosin expression, 189–191
numerical capillary density, 199
position and occurrence of fibers, 192
proliferation at cold temperature, 224–225, 229
rim fibers, 219–220
stratified hyperplasia, 115
ultrastructure, 195–196
vasculature, 189
RNA, levels as measure of muscle growth, 107, 254–257, 293

S

Salmonid myogenesis
  adult growth, 147–149
  embryonic phase, 145–147
  germinal zone phase, 147
  juvenile growth, 147–149
  satellite cells in muscle fiber recruitment and hypertrophy, 149–151
Satellite cell
  abundance in fish skeletal muscle, 75, 77
  aging and body weight effects on numbers, 89–91
  criteria for identification, 75, 151
  culture, 80, 82, 84, 256
  differentiation
    markers, 80, 85, 151–153
    morphology, 85
    muscle type effects, 91
    myotube formation and phenotype, 85, 87, 150–151
  distribution along fibers, 75
  egg incubation temperature effects on number, 168–169, 171, 174
  enrichment, 81–82
  feeding effects on proliferation, 95–96
  fiber recruitment role, 151, 174
  hypertrophy role, 151
  isolation, 78–79, 81
  muscle growth and regeneration role, 74–75, 89, 97, 117–118
  nuclear morphology, 149–150
  origin, 87–88
  pollutant effects, 96–97
  proliferation analysis
    in vitro, 84–85, 97–98
    in vivo, 75, 78, 97
  regulatory factors
    fibroblast growth factors, 174
    hepatocyte growth factor/scatter factor, 151, 174–175
    insulin, 278
    insulin-like growth factor effects on proliferation, 93, 95, 175
    transforming growth factor-β, 175
  strain effects
    diploid versus triploid rainbow trout, 91–92
    hyperplasia, 91
    proliferation rate, 91–93
    temperature effects, 95
Scattered dorsal and ventral muscle fibers, features, 222
Scatter factor, see Hepatocyte growth factor/scatter factor
SDA, see Specific dynamic action
SDH, see Succinic dehydrogenase
Season
  effects on muscle growth, 125–126
  temperature change effects, see Temperature acclimation
Sex, effects on muscle growth, 124
SHH, see Sonic Hedgehog
slo, see sloth
sloth (slo), muscle differentiation role in zebrafish, 9
Slow fiber, see Red muscle
Somatostatin-14 (SRIF), immunoneutralization effects, 269
Sonic Hedgehog (SHH), myogenesis role in zebrafish, 6–8, 13
spadetail (spt), mesoderm allocation of cells in myogenesis, 2–3
Specific dynamic action (SDA)
  cycloheximide inhibition, 262
  hormonal influences, 263
Specific growth rate, calculation, 252
spt, see spadetail
SRIF, see Somatostatin-14
Strain
  hyperplasia effects, 91, 123–124
  muscle fiber type distribution, 260–261
  satellite cell effects
    diploid versus triploid rainbow trout, 91–92
    hyperplasia, 91
    proliferation rate, 91–93
Stratified hyperplasia, see also Growth, muscle
  fast-white muscle, 112
  main growth zone, 112
  overview, 103
  pink muscle, 115
  slow-red muscle, 115
  superficial monolayer and external cells, 112, 115
Succinic dehydrogenase (SDH), fiber type expression, 195, 203, 209, 218

INDEX   **315**

T

Temperature
  acclimation, *see* Temperature acclimation
  cruising speed dependence, 223
  egg incubation temperature effects on muscle growth patterns
    cod, 164–165
    fiber diameter, 164, 168–169, 171
    fiber number, 161–164, 169
    hatchery implications, 173
    herring, 171–172
    mitochondrial density, 165
    nuclear density, 168
    overview of species differences, 256
    salmon, 161–169, 171–172
    satellite cell number, 168–169, 171, 174
    trout, 165
    wild fish myogenesis studies, 165–169
  myofibrillar ATPase activity effects, 45
Temperature acclimation
  acclimatization comparison with acclimation, 224
  fast-start behavior, 227–228
  lipid content of muscle response, 198, 227
  metabolic responses, 225–226, 260
  microvasculature response, 227
  mitochondrial volume changes, 226–227
  myofibrillar ATPase activity during acclimation, 32–33, 45–46, 57–58, 225
  myosin isoform expressiion, 57–60, 65, 225–226
  recruitment of fast fibers, 222–224
  red muscle proliferation at cold temperature, 223–224, 229–230
  seasonal fluctuation in freshwater temperature, 222
  twitch contraction kinetics, 224–225
TGF-β, *see* Transforming growth factor-β
Thyroid hormones
  functions, 281
  metamorphosis role, 284
  muscle gene expression regulation in development, 55, 63, 281
  muscle growth effects, 123, 282–283
  myogenic regulatory factors, hormonal control of expression, 281–282
  myosin expression regulation, 282
  receptors
    expression in development, 284
    flatfish, 283–284
    genes, 283
    skeletal growth effects, 284–285
Titin, fiber type identification, 195, 261
Tonic fiber
  development and growth, 214–215
  diameter, 214
  function, 219–220
  histochemical analysis, 215–218, 220
  myosin expression, 218
  position and occurrence, 214
  ultrastructure, 215–218
Transforming growth factor-β (TGF-β), satellite cell regulation, 175
Transitional zone muscle fiber
  function, 220–221
  histochemical analysis, 221–222
Triploidy, *see* Ploidy

W

White muscle
  arrangement in Atlantic salmon, 142
  development and growth
    embryo, 201
    larva, 201
  energy metabolism, 203–206
  fiber diameter, 148–149, 255
  function, 142, 188–189, 206–208
  histochemical analysis, 189, 202–203
  innervation, 206
  mitochondrial density, 197
  myosin expression, 189–191, 201
  numerical capillary density, 199
  position and occurrence of fibers, 201
  protein synthesis, 254, 261–263
  recruitment of fibers, 142, 144, 148, 207–208, 222–224
  stratified hyperplasia, 112
  ultrastructure, 202–203

Y

yot, *see you-too*
you-too (*yot*), myogenesis role in zebrafish, 7

Z

Zebrafish
  advantages in muscle ontogeny studies, 1
  muscle fiber types, 142
  myogenesis
    cell labeling, 1–2
    head muscles, 11–14
    mesoderm allocation of cells
      floating head, 3–4
      spadetail, 2–3
    muscle differentiation mutants, 8–9
    myotomal patterning mutants, 6–8
    myotome formation and muscle differentiation, 4–6, 14, 48–49
    paired fin muscles, 9–11
    slow muscle cell specification, 6–8

OTHER VOLUMES IN THE FISH PHYSIOLOGY SERIES

VOLUME 1 Excretion, Ionic Regulation, and Metabolism
Edited by W. S. Hoar and D. J. Randall

VOLUME 2 The Endocrine System
Edited by W. S. Hoar and D. J. Randall

VOLUME 3 Reproduction and Growth: Bioluminescence, Pigments, and Poisons
Edited by W. S. Hoar and D. J. Randall

VOLUME 4 The Nervous System, Circulation, and Respiration
Edited by W. S. Hoar and D. J. Randall

VOLUME 5 Sensory Systems and Electric Organs
Edited by W. S. Hoar and D. J. Randall

VOLUME 6 Environmental Relations and Behavior
Edited by W. S. Hoar and D. J. Randall

VOLUME 7 Locomotion
Edited by W. S. Hoar and D. J. Randall

VOLUME 8 Bioenergetics and Growth
Edited by W. S. Hoar, D. J. Randall, and J. R. Brett

VOLUME 9A Reproduction: Endocrine Tissues and Hormones
Edited by W. S. Hoar, D. J. Randall, and E. M. Donaldson

VOLUME 9B Reproduction: Behavior and Fertility Control
Edited by W. S. Hoar, D. J. Randall, and E. M. Donaldson

VOLUME 10A Gills: Anatomy, Gas Transfer, and Acid-Base Regulation
Edited by W. S. Hoar and D. J. Randall

VOLUME 10B Gills: Ion and Water Transfer
Edited by W. S. Hoar and D. J. Randall

VOLUME 11A The Physiology of Developing Fish: Eggs and Larvae
Edited by W. S. Hoar and D. J. Randall

VOLUME 11B	The Physiology of Developing Fish: Viviparity and Posthatching Juveniles *Edited by W. S. Hoar and D. J. Randall*
VOLUME 12A	The Cardiovascular System *Edited by W. S. Hoar, D. J. Randall, and A. P. Farrell*
VOLUME 12B	The Cardiovascular System *Edited by W. S. Hoar, D. J. Randall, and A. P. Farrell*
VOLUME 13	Molecular Endocrinology of Fish *Edited by N. M. Sherwood and C. L. Hew*
VOLUME 14	Cellular and Molecular Approaches to Fish Ionic Regulation *Edited by Chris M. Wood and Trevor J. Shuttleworth*
VOLUME 15	The Fish Immune System: Organism, Pathogen, and Environment *Edited by George Iwama and Teruyuki Nakanishi*
VOLUME 16	Deep Sea Fishes *Edited by D. J. Randall and A. P. Farrell*
VOLUME 17	Fish Respiration *Edited by Steve F. Perry and Bruce Tufts*